Reclaiming mine soils and overburden in the western United States

Analytic parameters and procedures

Reclaiming mine soils and overburden in the western United States
Analytic parameters and procedures

R. Dean Williams and Gerald E. Schuman, Editors

SOIL
CONSERVATION
SOCIETY
OF AMERICA

7515 N.E. Ankeny Road, Ankeny, Iowa 50021-9764

Manufactured in the United States of America

Library of Congress Catalog Number 87-9797

ISBN 0-935734-14-7

$25.00

Library of Congress Cataloging in Publication Data

Main entry under title:

Reclaiming mine soils and overburden in the western
United States.

Includes index.
1. Reclamation of land—West (U.S.) 2. Strip mining—Environmental aspects—West (U.S.) 3. Mine soils—West (U.S.) 4. Spoil banks—West (U.S.) I. Williams, R. Dean. II. Schuman, Gerald E.
S621.5.S8R428 1987 631.6'4 87-9797
ISBN 0-935734-14-7

Cover design by Brian Wuerker
Cover: High-altitude, color-infrared photograph (1977) of surface mining near Coalstrip, Montana. Courtesy of EROS Data Center, U.S. Geological Survey, Sioux Falls, South Dakota.

Contributors

FRANCIS A. AMENDOLA
Corporate Soil Scientist, NERCO Coal Corporation, Decker, Montana

R. C. BARTH
Senior Research Ecologist, Soil-Plant Systems, Golden, Colorado

WILLIAM A. BERG
Soil Scientist, Agricultural Research Service, USDA, Woodward, Oklahoma

JOHN BOKICH
Manager, Environmental Services, Santa Fe Pacific Mining, Inc., Albuquerque, New Mexico

JOE BOWDEN
Director, CDS Laboratories, Durango, Colorado

MARK G. CARLSTROM
Reclamation Soil Scientist, Montana Department of State Lands, Billings

JAMES R. DEUTSCH
Senior Environmental Scientist, Reclamation Division, North Dakota Public Service Commission, Bismarck

DOUGLAS J. DOLLHOPF
Research Soil Scientist, Montana State University, Bozeman

SCOTT E. FISHER, JR.
Biological Scientist, Office of Surface Mining Reclamation and Enforcement, Western Technical Center, Denver, Colorado

DENNIS F. FRANSWAY
Private Soils Consultant, La Mesa, California

ROGER S. FUNSTON
Supervisory Soil Scientist, McLaren Environmental Engineering, Rancho Cordova, California

DUANE H. GENTZ
Senior Project Manager, Office of Surface Mining, U.S. Department of the Interior, Denver, Colorado

LARRY P. GOUGH
Botanist, U.S. Geological Survey, Denver Federal Center, Denver, Colorado

GARY A. HALVORSON
Associate Soil Scientist, Land Reclamation Research Center, North Dakota State University, Mandan

NEIL HARRINGTON
Reclamation Supervisor/Soil Scientist, Montana Department of State Lands, Helena

JIM HERRON
Reclamation Specialist, Colorado Mined Land Reclamation Division, Denver

EVERETT HOOPER
Environmental Protection Specialist, Department of Defense, Ogden, Utah

LLOYD R. HOSSNER
Professor of Soil Chemistry, Texas A&M University, College Station

J. J. JURINAK
Professor of Soil Chemistry, Utah State University, Logan

DENNIS R. McGIRR
Reclamation Coordinator, Carter Mining Company, Gillette, Wyoming

JOAN E. McLEAN
Research Scientist, Utah Water Research Laboratory, Logan

STEPHEN D. MERRILL
Research Soil Scientist, Agricultural Research Service, U.S. Department of Agriculture, Mandan, North Dakota

LARRY MUNN
Associate Professor, Department of Plant, Soil, and Insect Sciences, University of Wyoming, Laramie

FRANK F. MUNSHOWER
Director, Reclamation Research Unit, Montana State University, Bozeman

DENNIS R. NEUMAN
Associate Research Chemist, Reclamation Research Unit, Montana State University, Bozeman

FRED PARADY
Reclamation Superintendent, Bridger Coal Company, Rock Springs, Wyoming

MICHAEL W. POLE
Manager of Environmental Affairs, North American Coal Corporation, Bismarck, North Dakota

TOM PORTAL
Soils Specialist, Wisconsin Department of Natural Resources, Madison, Wisconsin

EDWARD F. REDENTE
Associate Professor, Range Science Department, Colorado State University, Fort Collins

J. D. REEDER
Soil Scientist, Agricultural Research Service, U.S. Department of Agriculture, Fort Collins, Colorado

BURNS SABEY
Professor, Agronomy Department, Colorado State University, Fort Collins

NIRANDER M. SAFAYA
Senior Environmental Scientist, North Dakota Public Service Commission, Reclamation Division, Bismarck

FRED SAMSON
Environmental Coordinator, San Juan Coal Company, La Plata Mine, New Mexico

DAVID G. SCHOLL
Soil Scientist, Rocky Mountain Experiment Station, U.S. Forest Service, Albuquerque, New Mexico

JAMES L. SCHRACK
Senior Environmental Coordinator, Thunder Basin Coal Company, Wright, Wyoming

R. C. SEVERSON
Soil Scientist, U.S. Geological Survey, Denver, Colorado

DALE A. SHAY
Consulting Soil Scientist, Big Horn Soil and Irrigation, Sheridan, Wyoming

PETER L. SMITH
Vice-president/Soil Scientist, Stoneman-Landers, Inc., Denver, Colorado

ANDREW A. SOBEK
Technical Manager, ProMac Systems, Fabricated Polymers Division, The BFGoodrich Company, Brecksville, Ohio

DARWIN L. SORENSEN
Research Assistant Professor, Utah State University, Logan

PATRICK J. SULLIVAN
Manager, Environmental Chemistry, Western Research Institute, University of Wyoming Research Corp., Laramie

GARY B. TUCKER
Soil Scientist, Office of Surface Mining, U.S. Department of the Interior, Oklahoma City, Oklahoma

G. WEILER
Albion Laboratories, Clearfield, Utah

STEPHEN E. WILLIAMS
Associate Professor, Department of Plant, Soil and Insect Sciences, University of Wyoming, Laramie

Editorial review team

DAVID Y. BOON
Wyoming Department of Environmental Quality, Cheyenne

G. FRED GIFFORD
Department of Range, Wildlife and Forestry, University of Nevada, Reno

J. J. JURINAK
Professor, Utah State University, Logan

ED KELLEY
New Mexico Energy and Minerals Department, Santa Fe

W. W. SCHAFER
Montana State University, Bozeman

G. E. SCHUMAN
Agricultural Research Service, U.S. Department of Agriculture, Cheyenne, Wyoming

R. DEAN WILLIAMS
Senior Environmental Specialist, Utah International Inc., San Francisco, California

L. F. WOODS
Agricultural Research Service, U.S. Department of Agriculture, Cheyenne, Wyoming

Reclaiming mine soils and overburden in the western United States
Analytic parameters and procedures

Contents

Preface

Concern for reclamation of western surface mine disturbance is comparatively new, originating with the heightened environmental awareness of the late 1960s and the growth of surface mining during the 1970s. Because of public concern for the environment during and following mining, soil and overburden baseline surveys quickly became standard requirements of mine permit applications. This resulted in considerable data collection. However, limited ability for interpretation of the data restricted the use of those data in planning.

To assist permit applicants and to provide guidance for reclamation, regulatory agencies in the mid 1970s began to develop guidelines for sample collection, analytical procedures, and the suitability of overburden as a plant growth medium. Reclamation specialists found that guidelines, while beneficial in general, lacked the flexibility necessary to accommodate environmental and operational characteristics as well as advances in reclamation technology.

In response to the need for information concerning the applicability of soil and overburden data in developing reclamation plans, a group of scientists from industry, research agencies, universities, and regulatory agencies met in 1982 and formed the Western Soils and Overburden Task Group. The objectives of the task group were (1) to improve the quality and reliability of analytical data for soils and overburden and (2) to improve the interpretation and application of those data.

Following a symposium, "Soil and Overburden Requirements for Successful Revegetation," held in 1983, the task group further refined these objectives. The need for improved quality control in laboratory analytical methods had become obvious. Results of such analytical procedures varied substantially from laboratory to laboratory, and in some cases, the results were even misleading. The task group conducted a quality control program in which laboratory methodologies were refined and standardized to improve data reliability.

The second objective, and the purpose of this volume, was to conduct a thorough and impartial review of the available scientific knowledge concerning the analytical parameters commonly required for soil and overburden characterization. Interpretation of data and subsequent recommendations depend upon a number of complicating factors, such as sampling methods, analytical methods, chemical and physical characteristics of the material, climate, geology, hydrology, land use, vegetation, and mining technology. Decisions that account for these factors and their complex interaction involve interpretation, extrapolation, and judgement, tasks that we hope are aided by this compilation of manuscripts.

For the reclamation scientist, this volume provides a reference for designing and interpreting soil and overburden sampling programs. Each chapter describes the utility of specific soil and overburden analytical data as these data apply to reclamation. In addition, each chapter examines the factors that affect data interpretation and reclamation decisions and alternatives. While published suitability criteria provide a guide to the significance of analytical data, site-specific environmental factors and mining practices must be considered before final recommendations and plans are developed for any particular mining operation. This volume hopefully will improve the quality of reclamation planning by providing a guide for the interpretation of analytical data with respect to soils and overburden on a site-specific basis.

For the research scientist, we hope this volume provides some guidance for future research. To date, reclamation research has focused on specific problems that require immediate answers. This "fire-fighting" approach to research has resulted in significant technology development, but it has also resulted in significant gaps in data collection. By virtue of our efforts to characterize the current status of our scientific understanding of these analytical parameters, some areas requiring further research have been identified.

Finally, we hope this volume will aid regulatory authorities in making environmental impact assessments. Dealing with a complex situation and with incomplete knowledge, regulatory authorities have been in the position of making environmental assessments and recommendations with limited solid evidence and little precedence for judging the likelihood of success with a given reclamation strategy. While there continues to be a lack of solid evidence and precedence for making environmental impact decisions, it is hoped that this volume will be a significant source of information, providing insight, understanding, and support for environmental decision-making.

R. Dean Williams
and Gerald E. Schuman
Editors

The volume demonstrates what is possible when individuals within a profession and representing varied interests work together. Professionals representing industry, regulatory authorities, and academia wrote and critiqued each of the chapters within this volume. Objectivity was insured through this model of cooperation. Perhaps more importantly, the exercise of working together gave each of us a common basis for building better working relationships. In the end, I can honestly say the process was at least as valuable as the product.

Although more than 60 people contributed to this volume, I want to acknowledge the few whose names do not appear on any chapter but whose imprint shows throughout. Any project that involves many people requires a few "sparkplugs" to keep things moving. Jerry Schuman provided that spark by his critical review of the chapters and efforts to secure funding for publication. Dave Boon, although he joined the effort after its initiation, provided new energy and perspective to an effort that at times seemed endless. Larry Woods provided objective reviews and demanded scientific accuracy. Jerry Jurinak, Bill Schafer, Fred Gifford, and Ed Kelley, among others, also made substantial contributions. Finally, my thanks to Max Schnepf of the Soil Conservation Society for his support and gentle but firm pressure at critical moments.

R. Dean Williams

Reclaiming mine soils and overburden in the western United States
Analytic parameters and procedures

Chapter 1

pH

Gary B. Tucker, William A. Berg, and Duane H. Gentz

Perhaps the most commonly measured characteristic of soil, overburden, and water is pH. The reason: pH significantly influences other important characteristics. As the negative logarithm of the hydrogen ion concentration, pH indicates the intensity of the acidic or basic condition of the sample material (*71*). The pH scale ranges from extremely acidic (pH = 0) to extremely basic or alkaline (pH = 14). A pH of 7 is neutral. Most environmental measurements of pH range from 5 to 10.

The characteristic pH is important to the successful reclamation of mined land in the western United States. Particularly important considerations include analytical procedures, the relation of pH to revegetation, and water quality.

Different sampling and measurement techniques can yield wide variations in pH values. Some standardization of procedures is desirable. But understanding the effect of alternative procedures on reported pH values can help reclamation specialists properly interpret pH data.

Regulatory agencies and reclamation specialists are dealing with the relation of pH levels to revegetation by establishing guidelines, criteria, or standards. Because of the complex interrelationship of pH with many other soil, plant, and water factors, such general criteria often must be modified from site to site to determine the expected behavior of revegetation efforts or to select the best mitigation practice.

For revegetation efforts to be successful, desirable plant species must be selected that, in most cases, provide satisfactory forage for animals. Establishing and maintaining the desired plants in a given environmental setting depends on many factors, including the availability of both macronutrients and micronutrients and the soil physical condition. Soil pH serves as a general guide to conditions favoring the availability of nutrients, possible toxicity problems, and soil dispersion by Na. The chemical and physical characteristics of a given spoil or soil affect these general pH interpretations and may require site-specific interpretations.

Water pH likewise is addressed in general terms by standards and criteria. But specific reclamation actions must consider site-specific characteristics, including premining conditions and desired uses for the water. Factors that control the pH of surface water and groundwater are important in assessing potential future water uses. Water pH may also affect soil pH, which in turn relates to revegetation success.

Analytical procedures for pH determinations

The pH meter is the accepted instrument for making pH measurements. Colorimetric techniques are also used widely. Measuring the pH of water samples is a fairly direct procedure. But field and laboratory results often differ because of the loss of dissolved gas and/or chemical changes in the water samples. Care must be taken to recognize these factors when evaluating test results.

Determining the pH of a soil or overburden sample is a fast, relatively easy procedure. However, a number of factors, including dilution ratio, soluble salt concentration, CO_2 levels, and sulfide oxidation status, can significantly affect the pH values obtained as well as the later interpretation of the results.

Soil analysis using a pH meter. Laboratory guides are available for measuring pH in soils (*35, 47, 60, 78*) and overburden (*65, 71*). Meaningful interpretations of pH require consistent analytic procedures. Procedures should always be included or referenced when reporting pH values.

Among the important factors to be considered in selecting a specific laboratory procedure are the following:

Soil-to-water ratio. A major consideration in selecting a laboratory procedure to determine the pH of soil or overburden is the soil-to-water ratio. Soil-to-water ratios typically used in laboratory analyses include saturated paste, 1:1, 1:2, 1:5, and 1:10 by weight. As soil is diluted with water, pH usually increases, regardless of whether the soil is acidic or alkaline. The pH increase, which depends upon the soil, can range from 0.3 unit to as much as 1.5 units as the soil is diluted from a saturated paste to a ratio of 1:5 (*35*). In samples from southeastern Montana (*68*), for example, pH increased 0.3 and 0.6 unit for soil and spoil samples, respectively, with dilution from a saturated paste to 1:1 (Table 1). In another study of overburden, the pH of 1:1 dilutions averaged 0.15 pH unit higher than the pH of saturated pastes

Table 1. Mean and range in pH of soil and spoil samples from southeastern Montana, as affected by pH determination method (*68*).

	pH Determination Method					
	Saturated Paste		1:1		1:2*	
Material	Mean	Range	Mean	Range	Mean	Range
Soil (N = 16)†	7.9	6.4-8.6	8.2	6.9-8.9	7.7	6.3-8.2
Spoil (N = 31)†	7.4	6.4-8.3	8.0	7.2-8.9	7.5	6.9-7.9

*With 0.01M $CaC1_2$.
†N = number of samples.

(*48*). Overburden samples from New Mexico had saturated paste pH values ranging from 6.2 to 9.2; the pH values for 1:5 ratios ranged from 6.9 to 10.0 (*62*). The mean pH of the saturated pastes was 7.8. This compared to 9.1 for the 1:5 ratios.

Saturated paste pH has been suggested as the standard method for saline and sodic soils (*78*) and for overburden in the western United States (*65*). Use of saturated pastes for pH measurement better represents the soil water content under field conditions than do higher soil dilutions (*35, 78*). The saturation percentage represents a soil water content at which all soils are equally wet, at least in an approximate sense (*21*).

In routine laboratory analysis pH can be determined conveniently on saturated pastes prior to extraction of a sample for soluble salt and specific ion determinations. One disadvantage of a saturated paste is that fairly large samples—200 grams—are commonly used. Researchers have suggested, however, that a saturated paste can be made from a 10-gram sample (*71*).

If expediency or small sample size rules out use of saturated pastes, the 1:1 extract appears to be the next best choice because soil or overburden will be diluted less than with 1:2, 1:5, or 1:10 ratios.

Soluble salt effect. Soil pH normally declines with the addition of salt (*47, 60*). Lower pH levels have been observed in samples collected during the dry season when soluble salts tend to accumulate in and near the soil surface (*19*). Soil and spoil samples from southeastern Montana had mean pH levels 0.5 unit lower when determined in 1:2 0.01 M $CaCl_2$ than in 1:1 soil and water mixtures (Table 1). Comparison of mean saturated paste pH with pH determined by the 1:2 0.01 M $CaCl_2$ method offers advantages in interpreting soil pH data, particularly in acid soils. However, such a comparison has not been used extensively on neutral and calcareous soils in the western United States because the saturated paste method better represents actual soil water content (*78*).

Effect of CO_2. More CO_2 can substantially reduce the pH of water and calcareous soils (*47, 60*). Pure water, distilled or deionized, in equilibrium with CO_2 in the air (about 0.03%) has a pH of 5.7. Carbon dioxide dissolves in water, forming weakly ionized carbonic acid. In systems containing

$CaCO_3$ and water in equilibrium with CO_2 in the air, pH averages about 8.4. If the CO_2 content in the soil atmosphere is increased by microbial action to 1 percent, pH should be about 7.2 (*43*). Adding gypsum ($CaSO_4 \cdot 2H_2O$) to a $CaCO_3$ and water system in equilibrium with CO_2 in the atmosphere increases pH to about 7.8 (*43*).

Similar relationships should exist in soils and overburden containing $CaCO_3$ and gypsum. When excess Na is present, however, pH can exceed 8.4. Because the soil atmosphere usually contains more CO_2 than the air, the pH of soils containing $CaCO_3$ often ranges from 7.2 to 8.4. Precise measurements of pH in calcareous soils can be made only if the CO_2 pressure is standardized (*83*). For convenience, pH usually is measured at the CO_2 pressure present in the laboratory. Thus, reading pH to a precision greater than 0.1 unit is superfluous. The effect of CO_2 on soil pH is much less for acid soils than for soils containing $CaCO_3$ (*47*).

Weathering. Oxidation of sulfides to sulfuric acid in overburden can create extremely acid conditions. Such is the case when there are insufficient basic materials present to buffer the system. Acid strata in overburden are common in eastern Texas and in the midwestern and eastern United States.

Groenewold and associates suggested that even with limited sulfides in overburden in the northern Great Plains the sulfides can oxidize and hydrolyze to produce acid (*29*). The researchers concluded "...that the natural buffering capacity of the overburden in this region will eliminate, except in occasional highly localized settings, acid drainage to water courses and groundwater."

The acid-base potential is a general measure of the acid production potential from sulfides in overburden (*71*). This procedure estimates the increased acidity that might occur in fresh overburden exposed to weathering elements. The procedure assumes that all pyrite (sulfide) sulfur oxidizes to form sulfuric acid. Sulfate and organic sulfur are eliminated as sources for significant acid formation. The acidity is buffered by lime or other materials in the overburden. Measurement of acid-base potential thus indicates whether there is sufficient buffering capacity to offset potential acid production.

A second procedure involves the evaluations of cumulative acid production and neutralization over time in humidified column leach chambers (*14, 26*).

Acid overburden is much less common in the western United States than in other areas of the country. The Tertiary and Cretaceous formations being mined often contain lime that neutralizes the limited acid produced by sulfide oxidation. One study of western overburden, with an initial pH in the range of 7.5 to 9.1 and containing 2 to 23 percent $CaCO_3$ and detectable quantities of sulfides, recorded only minor (−0.4 to +0.5) or no pH changes (*30*).

The effects on pH of sample grinding and/or screening and discarding of coarse fragments has received limited attention. In one test grinding an acid soil (pH 4.7, less than 2 mm) to pass through a 0.16 mm sieve resulted in a pH increase of about 1 pH unit (*35*). Such increases may be due to the adsorption of H^+ ions by the increased surface charges on the newly created clay surface area. However, pH reductions with grinding have also been observed. Such cases may reflect a dominant role of additional dissolution or chemical reaction of the solid materials. Screening of overburden to remove coarse fragments results in a sample with greater particle surface area. A pH behavior similar to that for grinding can thus be expected.

Drying can cause CO_2 losses from calcareous soils and a subsequent increase in pH (*35*). However, samples of air-dried calcareous overburden, exposed in the laboratory for 224 days, exhibited only minor change or no change in pH (*30*).

Other effects. When determining pH on some soils with a pH meter, there may be a lower pH reading when the electrodes are inserted into the soil than when pH is measured in water (supernatant) over the soil. This difference, termed the suspension effect, has received considerable study (*47, 60*). Because most western coal overburden contains some soluble salt, however, the suspension effect should be minor if not negligible.

A pH meter may drift slowly to lower or higher values when measuring pH of calcareous materials. Electrode maintenance can minimize drift, however (*47*).

For field purposes, then, the pH of western soils and overburden probably is adequately characterized in saturated pastes without special consideration of soluble salt and CO_2 effects. If expediency or small sample size rule out use of saturated pastes, a 1:1 extract is the next choice because soil or overburden will be diluted less than if a 1:2, 1:5, or 1:10 ratio is used.

Potential acidifying effects of sulfides in noncalcareous and slightly calcareous overburden may need to be determined using appropriate acid-base potential measurements. Standardized grinding and drying techniques should be used also.

Color indicators for soil pH measurement. Colorimetric pH indicators are a convenient means of determining pH in the field. Commercially available colorimetric soil pH kits have been tested on both soils (*44*) and coal mine spoils (*5*). Most pH readings were within 0.5 pH unit of pH-meter readings using the LaMotte-Morgan kit, which uses five short-range color indicators. Colorimetric test kits using a single mixed indicator, the Hellige-Truog, Soiltex, for example, usually gives pH values within 1 pH unit of pH meter readings. Colorimetric pH determinations may be slower, less precise, and more difficult to read than determinations with a pH meter (*47*). When quantitative data rather than qualitative data are needed, therefore, a pH meter should be used.

Analysis of pH in water. Standardized procedures for measurements of pH in water require the use of a pH meter (*1*). Colorimetric indicators can be used for such measurements, but water samples must be colorless and free of turbidity (*67*). The pH meter is the method of choice.

Loss of dissolved gases, notably CO_2, during sampling is a common source of variation in pH readings. This is a particular problem for pumped groundwater samples when the pumping pressure is released (*45*). Temperature changes affect gas solubilities and have a direct effect upon pH. Therefore, pH should be measured in the field at field conditions, allowing electrodes to reach equilibrium temperature. Both pH and temperature should be recorded. Subsequent measurements in the laboratory can be used to support the laboratory test protocols. Comparisons of the two pH values, coupled with other field observations, usually are sufficient to determine the reliability of the measurements and the meaning of any differences.

The role of pH in revegetation

Alkaline pH. Surface coal mining and reclamation efforts, including topsoiling, generally result in an increase in average pH within the plant root zone in many areas of the western United States. Topsoil, which may range from slightly acid to alkaline, is diluted with subsoil and overburden that usually contains $CaCO_3$. Even in small amounts $CaCO_3$ acts to buffer soils or spoils in the alkaline pH range.

Nutritional effects. The two most limiting plant growth nutrients on mine soils in the western United States are N and P (*3*). Generally, N availability is highest in soils with a pH in the 6 to 8 range; it then decreases in the pH range of 8 to 9 (Figure 1). The availability of P declines as pH increases from 7.5 to 8.5 and then increases (*20*). These relationships are the result of biological and chemical reactions in the soil or exposed overburden. Many individual reactions are sensitive to pH.

A major N source for plants on undisturbed semiarid sites is mineralized organic N in easily decomposable plant materials and microorganisms (*16*). Such is probably the case for reclaimed areas as well (*84*). Limited research has been done on the effect of high pH levels—greater than 8.5—on N mineralization. Optimum hydrolyzation of certain proteins occurs at pH levels of 7 to 8.5 (*42*). Nitrification is independent of pH in the range 6 to 8 (*69*). Ammonium fertilizers or urea applied to alkaline soils can inhibit nitrification because NH_3 is toxic to bacteria that convert NO_2^- to NO_3^- (*69*). Ammonia loss by volatilization from surface-applied ammonium or urea fertilizers is greater from calcareous soils than from acid soils (*55*).

Effects of high pH levels—greater than 8.5—on symbiotic N fixation are not well defined. Jensen suggested that nonsymbiotic fixers (*Azotobacter, Beijerinckia, Clostridium* spp.) function over a pH range of 3 to 9 (*36*). Optimum growth of *Azotabacter* and *Clostridium* occurs near a neutral pH.

In alkaline soils dissolution of slightly soluble Ca phosphate probably is a major source of plant-available P. The P in solution from Ca phosphate decreases from a pH of 6 to about 7.5, then increases in the presence of $CaCO_3$ over the pH range from 7.5 to 9 (*43, 57*). However, plant-available P as a function of pH is complicated by reduced P fertilizer effectiveness in the presence of $CaCO_3$ (*56*). Also, plants take up less P as the dominant form of P in solution shifts from $H_2PO_4^-$ to $HPO_4^=$ as the pH increases (*64*). Mineralization of P from organic forms also contributes to plant-available P under alkaline conditions (*56*).

Mycorrhizal fungi can increase nutrient absorption by plants, protect the

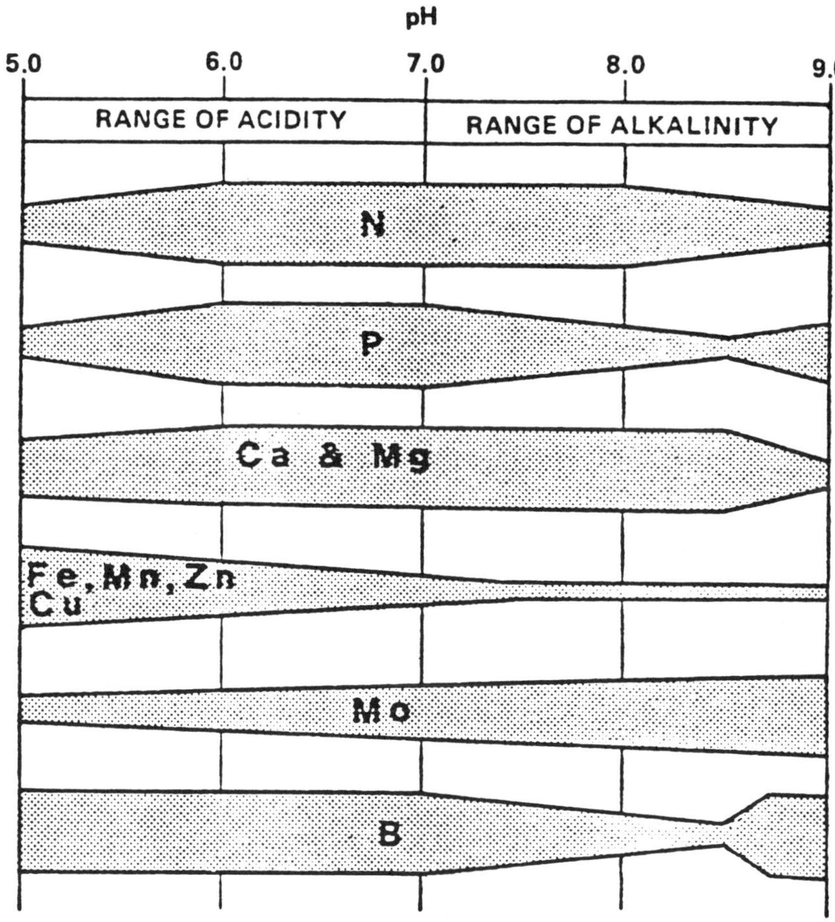

Figure 1. Relationship between mineral soil pH and nutrient availability. Band width represents relative availability (*10, 20*).

host root against pathogens, and possibly increase tolerance to drouth. Usually, mycorrhizal fungi are regarded as acidophilic (*18*). Optimum growth of most mycorrhizal fungi occurs in the pH range of 5 to 6.5 (*73*). In one test spore germination from two of three different mycorrhizae was inhibited at a pH of 9 (*28*).

The solubility of Fe and Zn declines as pH increases (*43*). Deficiencies of Fe and to a lesser extent Zn can occur on sensitive species growing in calcareous soils. Affected species are often introduced species (*17*). Excessive P applications can induce Zn deficiencies.

Toxic effects. The effects of high Na levels on ion uptake and soil physical conditions confound the physiological effects of high pH levels on plant growth in soils. In nutrient solution studies, plant growth was inhibited more at a pH of 9 than at lower pH levels (*34, 74*). The reduced growth was attributed to OH^- ions limiting water uptake (*74*).

Small amounts of B are essential to normal plant growth, but the element becomes toxic at higher levels. B is most available in the pH range of 5.0 to 7.0 and at pH 8.5 and above. Plant species vary both in their B requirements and in their tolerance to excessive B. Concentrations necessary for the growth of plants having high B requirements may be toxic to B-sensitive plants (*78*).

Mo and Cu also are essential trace elements for animals and plants. High Mo concentrations and low Cu concentrations in forage can cause molybdenosis in ruminants. Such is the case on certain alkaline soils. Cu availability in soil declines when pH exceeds 6.0 (*49*). Mo solubility, on the other hand, can increase 10-fold to 100-fold per pH unit from pH 6.0 to 8.5 (*81*).

Another trace element, Se, is often more abundant in Cretaceous shales, which are marine in origin and quite saline (personal communication with Frank Munshower, 1984). Se is not toxic to plants, but in concentrations above 5 ppm in forages, Se may be toxic to animals (*41*). Se deficiencies in animals also have been noted, for example, "white muscle disease" in sheep (*40*). Some plants accumulate Se; others do not. Se is most available in the pH range of 7.5 to 8.0 (personal communication with Frank Munshower, 1984). Low-solubility forms of selenite and elemental Se are dominant in soils with low pH. Alkaline, well-aerated seleniferous soils usually contain highly soluble Ca selenate (*49*). Where Se-concentrating plants are present, soil pH below 7.5 limits Se solubility and reduces Se concentrations in forage.

Physical effects. The physical effects of Na in dispersing soils (*78*) and overburden (*66*) are well documented. In one test only 44 percent of the variance in the exchangeable Na percentage could be associated with pH in soils ranging in pH from 6 to 10 (*21*). However, a pH of 8.5 or greater (saturated paste) almost always indicated an exchangeable Na percentage of 15 or more in soils (*78*). In a North Dakota study, 133 of 135 soil and overburden samples with pH values (saturated paste) in the range of 8.6 to 9.4 had Na adsorption ratios of 15 or greater (*70*).

Acid pH. Plant growth problems associated with toxicities in acid soils and overburden are well documented in the eastern United States. Because acid overburden is rare in the western United States, however, tolerance to acid mine soils has not been determined for many plant species used in the West.

While the effects of acid conditions on nutrient availability and biological processes in soils has been widely studied, information on overburden is more limited.

Nutritional effects. Generally, N and P availability decline at pH levels below 6 (Figure 1). Mineralization, the breakdown of organic matter, including fresh residue, supplies much of the N and P needed by plants. Kamprath and Foy (*37*) stated: "By regulating microbial activity, soil pH affects the mineralization of organic matter and the subsequent availabilities of nitrogen, phosphorus, sulfur and micronutrients...organic matter decomposes more rapidly in neutral soils than in acid soils. At soil pH values below 5.5, the fungi are most active in this decomposition, but at pH values above 6.0 the actinomycetes and bacteria are more prominently involved."

Ammonification, the production of ammonia by the breakdown of organic matter, is not as sensitive to pH as is nitrification. Nitrification, the oxidation of ammonia to nitrate by microorganisms, is sensitive to pH and drops off rapidly at pH levels below 6 (*37*). But obvious nitrification occurs in the pH range of 4 to 6 (*69*).

Nodulation on alfalfa declined when the alfalfa was grown in soil with pH levels below 6 (*63*). Legume species seeded in acid spoils nodulated well in the pH range of 5 to 5.4, but there usually were few nodules or no nodules in the pH range of 4 to 4.4 (*6*).

One review of symbiotic nitrogen fixation, concluded: "Soil pH is probably one of the most important external factors affecting the symbiosis. Some legumes, such as lucerne (*Medicago sativa* L. ssp. *sative*) and black medic (*Medicago lupulina* L.), grow best at neutral pH or slightly alkaline, whereas serradella (*Ornithopus sativus* Brot.) nodulates at pH 5 or even lower (*27*). It has been shown that low soil pH is inhibitory to root hair infection and early nodulation in peas and lucerne" (*31*).

Most plant-available P originates from mineralization of organic residues and from solubilization of slightly soluble P compounds in soils. Normally, P is most readily available to plants in the pH range of 6.5 to 7 (*57*). Under acid conditions much of the P is precipitated as Fe and Al phosphates, and the solubility of P in these compounds decreases as the soil pH decreases (*43*). Plant-available P increases as soil pH increases from about 4 to 6 (*37*).

In summary, plant availability of N and P is less under acid conditions because of a lower rate of organic matter mineralization. As acidity increases, solubility of P in Fe and Al phosphates decreases; the activity of N-fixing bacteria also decreases. A critical pH level is difficult to define because some species tolerate lower fertility conditions better than other species.

Toxic effects. Soluble Al and Mn are perhaps the major causes of plant toxicity in acid soils (*59*) and acid coal mine overburden (*7, 8, 15, 22*). Toxicity usually is expressed in terms of pH because the solubility of potentially toxic metallic ions increases rapidly as acidity increases (*43*).

It appears that pH is a better indicator of plant toxicity in acid mine soils than are concentrations of soluble or extractable Al (*46*) or Mn (*6*). The soil pH at which Al becomes soluble in toxic concentrations varies with different soils. Defining critical soil pH levels at which plants respond to liming depends upon the clay mineralogy and organic matter content of the soils. Kaolinite soils have higher critical pH levels than do clay soils dominated by vermiculite and montmorillonite. Soils high in organic matter have lower critical pH levels than soils low in organic matter (*23*). Vogel (*82*) concluded, "Although pH is an indicator for assessing acidity and predicting plantability of mine soils,...different mine soils, even with the same pH may not have the same effect on plant growth."

The lower pH limit for optimum growth of most species used in reclamation in the western United States has not been defined. Alfalfa is among the species most sensitive to acid conditions (*9*). One study showed alfalfa required a pH level of 6 or above for optimum forage production (*58*). Average alfalfa yields in a greenhouse study increased 37 percent on a soil limed to pH 6.1 than on the same soil limed to a pH of 5.3 (*13*).

Cool-season grasses tolerate acidic conditions better than alfalfa (*58*), but tolerance varies widely among species (*51*). A pH of 4.5 to 5 was reported as the lower limit on growth of a number of cool-season grasses in mine soils (*82*). Similar minimum pH levels for cool-season grasses have been reported for use on mine soils in humid regions (*4*).

Warm-season grasses have a lower pH limit of 4 to 4.5, about 0.5 pH unit less than cool-season grasses. These reports include minimum pH levels, but they do not define the lowest pH level where optimum growth occurs. In rating the production potential of soils, one team of researchers applied production indices to soil pH (0.01 M $CaCl_2$) values of 5.5 and lower (*61*). These indices might be most applicable to such crops as corn. Corn production studies have indicated a productivity index of 92 percent at pH 5.0 and 47 percent at pH 4.0.

Such information can serve as a starting point in defining critical acid pH levels for overburden in the western United States. However, more precise criteria are needed because of the different plant species and varieties used in the West. Also, varieties and strains of small grains (*25*) and grasses (*24*) selected for production on near-neutral and calcareous soils are less acid-tolerant than selections made for acid soils.

In summary, where plant species are adapted to near-neutral and alkaline soils in the western United States, the critical lower pH level appears to be about 6 for mine soils where alfalfa or other acid-sensitive species are to be grown. For cool-season grasses the lower critical pH level seems to be about 5.0 to 5.5, the level above which Al toxicity is no longer a problem for many agronomic species (*23*). Lower critical pH levels may be appropriate on acid soils where warm-season species are used for revegetation. These recommendations must be tempered by the fact that pH may vary as much as 0.5 to 1.0 unit, depending upon the pH measurement method. Further research and field observations are needed to define the critical pH level for many species used in western reclamation. Because acid coal mine overburden and mine soils are rare and are site specific in the western United States, the priority does not appear to be high. Where acid overburden is more common—in eastern portions of Texas, Oklahoma, and Kansas—research results from humid areas should be useful.

Summary. Near neutral pH (6.0 to 8.0) favors N and P availability and microbiological activities for N fixation. Soil pH levels lower than 5.0 to 5.5 for agronomic species and lower than 6.0 for alfalfa present potential toxicity risks related to soluble A1 and Mn. A pH above 8.6 indicates a Na hazard. On mine soils from certain geological formations, alkaline pH conditions favor greater solubility of B, Mo, and Se. Under specific conditions these elements can pose toxicity problems. Other generalizations can be developed for given areas to screen out problem areas through reconnaissance studies. Such screening then could be followed by additional tests to confirm or refute the suspected problems.

Soil pH is a valuable indicator parameter to monitor reclamation efforts at mine sites. For example, periodic pH measurements that yield values about the same as baseline values suggest that many other soil characteristics are also about the same. On the contrary, if an unexpected pH shift occurs, then other changes may be expected. Additional parameters may then need to be measured to determine the suitability of the soil material. Where such a pH shift occurs, the quantity and areal extent of soil material involved can be determined in the field using pH measurements. Such data then can be used to select appropriate special handling. For example, a small decline in pH in a small quantity of material might be resolved by soil dilution without the need for special handling. On the other hand, a low pH in a sizeable quantity of material might require segregation and burial.

Use of pH as a primary indicator parameter for reclamation offers the advantage of increased frequency of monitoring because of the ease and rapidity of measurement. This, in turn, offers the possibility for better, ongoing control of reclamation activities. Increased use of pH as a primary indicator parameter, both for soil screening and reclamation monitoring, likely will become common in the western mining industry. This is consistent with the recommendation for a realistic, comprehensive soil/overburden testing program with greater emphasis on selected parameters, such as pH (*50*).

Water quality and pH

One of the prime environmental concerns facing the coal mining industry is the potential impact of mining on water quality, both surface water and groundwater supplies. Direct contributions of suspended solids from disturbed lands have rather straightforward solutions. But chemical changes resulting from water's interaction with newly exposed geologic materials and with weathered spoil materials are more complex. Consequently, chemical water quality changes are more difficult to predict and to control.

Adverse water quality effects from coal mining in the West are not common. A National Research Council report noted that mining situations in the East are so complex that it is difficult to generalize; however, "in arid basins in the West, local variabilities result in some complexity, but the existence of persistent flow regimes permits analyses of effects to be extrapolated spatially" (*53*). That report also noted the general uniformity of soil and overburden materials over relatively large geographic areas. This expected uniformity can be routinely verified using an indicator parameter, such as pH. That is, for relatively uniform sample characteristics, pH should remain uniform. Abrupt shifts in pH indicate changes in other characteristics that necessitate additional testing (*38, 76, 77, 79, 80*).

Because pH strongly influences the solubility of materials, the parameter plays a key role in mine-related water quality changes. To consider pH, it is also necessary to consider the terms acidity and alkalinity. Acidity is caused by the presence of weak acids or strong (mineral) acids. Carbon dioxide from the atmosphere and from bacterial decomposition of organic matter forms weakly ionized carbonic acid, which increases acidity. Groundwaters containing decomposing organic material may develop high levels of dissolved CO_2 and elevated acidities. Other naturally occurring, weak acids include organic acids from decomposing vegetation. Strong mineral acids in the environment are often related to industrial waste discharges. But in the case of mining such acids are normally the result of oxidation and hydrolysis of sulfides. The hydrolysis of other heavy-metal salts, such as Al^{+3}, may also add to the mineral acidity (*67*).

Alkalinity of natural waters is due principally to bicarbonates. Carbonates and hydroxides also contribute to alkalinity. As salts of weak acids, these substances act as buffers to pH changes (*67*). Consequently, alkalinity is an indicator of water's buffering capacity.

The general aridity in many western areas strongly influences acidity and alkalinity relationships. High levels of dissolved solids—reflecting a chemical condition near saturation of lime and gypsum (related to the lack of water)—also influence water quality relationships in the West.

The pH of natural waters is controlled primarily by the equilibrium between CO_2 and the form of alkalinity: hydroxyl, bicarbonate, or carbonate. A shift in concentration of these forms will affect pH and vice versa (*67*). The following general equations show the reaction and equilibrium between CO_2 and the alkalinity forms, where M represents metal ions:

$$CO_2 + H_2O \leftrightarrow H_2CO_3 \leftrightarrow HCO_3^- + H^+$$
$$M(HCO_3)_2 \leftrightarrow M^{+2} + 2HCO^-$$
$$HCO_3^- \leftrightarrow CO_3^{-2} + H^+$$
$$CO_3^{-2} + H_2O \leftrightarrow HCO_3^- + OH^-$$

Figure 2 graphically presents these relationships.

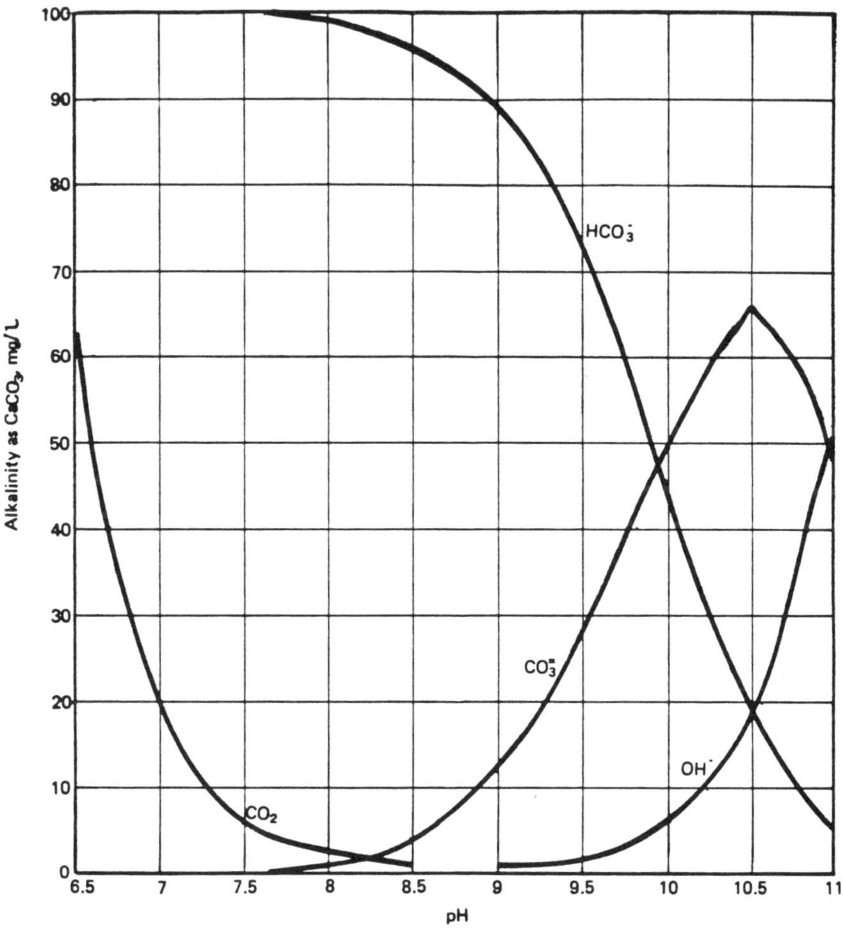

Figure 2. Relationship between CO_2 and the three forms of alkalinity at various pH levels; values calculated for a water with a total alkalinity of 100 mg/liter at 25 °C (*67*).

Acid mine drainage is a water quality problem commonly associated with coal mining. Such drainage and land disturbance were the major factors that brought about reclamation legislation. Acid mine drainage results from the oxidation and hydrolysis of sulfides in coal and spoil materials. It is neutralized by alkaline materials in spoils to the extent that such materials are present and by alkalinity in water. In cases where the pH of water is low, any heavy metals present are more soluble; elevated concentrations of metals are often associated with acid mine drainage (*11, 12, 52*).

Acid drainage problems are common in eastern mining areas, where the amounts of sulfide materials are high and alkaline materials low. Where soil alkalinity is low, surface water and groundwater commonly have only limited buffering capacity. In the West where spoil materials often are calcareous and sulfide levels are low, acid mine drainage problems are rare except in isolated, localized areas.

Acid mine drainage has been defined as water with a pH below 2.3, anion (sufates) concentrations exceeding 10,000 mg/liter, and acidities averaging 5,000 mg/liter as $CaCO_3$ (*14*).

In the presence of water and oxygen, Fe and other metallic sulfides oxidize to form soluble hydrous Fe sulfates. In contact with free aerobic water ferrous Fe oxidizes to the ferric state. The ferric ion further complexes with ferrous and ferric oxyhydroxides to form a precipitate, commonly called yellow boy, in surface streams.

The chemical reactions that produce H+ ions are represented by the oxidation of pyrite, FeS_2:

$$2FeS_2 + 70_2 + 2H_2O \rightarrow 2Fe^{+2} + 4SO_4^{-2} + 4H^+$$
$$Fe^{+2} + \tfrac{1}{4}0_2 + H^+ \rightarrow Fe^{+3} + \tfrac{1}{2} H_20$$
$$Fe^{+3} + 3H_20 \rightarrow Fe\,(OH)_3 + 3H^+$$

Fe bacteria can bring about the second reaction above and cause more H^+ ions to form, thus further depressing the pH of the system (*14*). Three bacteria species actively catalyze pyrites: *Thiobacillus thiooxidans, Thiobacillus ferrooxidans,* and *Ferrobaccillus ferroxidans.* These bacteria occur in groundwater environments with a pH range of 2.8 to 3.2.

Once these reactions begin, they are self-propagating and synergistic. Another factor affecting the relative reactivity of oxidizable sulfide is morphology. Framboidal pyrite generates orders of magnitude more acid than coarse-grained pyrite (*14, 26*).

Fortunately, most coal overburdens in the West are low in pyritic S, and the naturally calcareous overburdens tend to neutralize acid formation by the above process before drainage leaves the immediate vicinity of its formation. The alkalinity of the system, produced by the dissolution of calcite and dolomite plus some oxidation of clay minerals, buffers the system. With the dissolution of the calcite or dolomite, H^+ ions are consumed, which increases pH. Slow-moving groundwater aids in the dissolution of these minerals when the concentrations remain at levels below saturation (*29*).

Dissolution of calcite is also influenced by CO_2 and whether the system is open or closed. An open system allows a continuing supply of ambient CO_2 to promote dissolution of calcite/dolomite. A closed system limits the supply of CO_2. Thus, H^+ ions are consumed in the solution reaction, and dissolution of calcite/dolomite declines (*29*).

In some cases, Al hydrolysis may also produce H^+ ions (*75*):

$$Al^{+3} + H_2O \leftrightarrow Al\,(OH)^{+2} + H^+$$
$$Al(OH)^{+2} + H_2O \leftrightarrow Al(OH)_2^+ + H^+$$
$$Al(OH)_2^+ + H_2O \leftrightarrow Al(OH)_3 + H^+$$

Water with a pH below 4.0, which is common in acid mine drainage, may contain several thousand mg/liter of Al (*32*).

In the arid West most rainfall or snowmelt events do not cause infiltration and percolation below the root zone. Only during exceptional events does percolation below the root zone occur. In such cases, CO_2 produced from oxidation of surficial organic materials may reduce the pH of groundwater. Also, if unbuffered pyrites are available for oxidation, groundwater pH may decline. Lime and, to a greater extent, gypsum in association with lime usually will neutralize acidified water. As rainfall or snowmelt water enters the organic-rich horizons of the solum, the water acquires H^+ ions (*29*). Decomposition of organic matter and root respiration enrich the CO_2 environment, increase the partial pressure, and increase the H^+ ion concentration. This phenomenon plays an important role in the geochemical evolution of groundwater. Pyrite oxidation is rarely an important source of H^+ ions in Fort Union formation spoils and overburden in the northern Great Plains (*29*).

Natural water usually contains dissolved CO_2 and bicarbonate ions, which form a buffer system with the CO_2. Thus, there are variations in pH, generally between 5.0 and 8.0 (*45*). High pH may occur with high carbonate concentrations.

Controlling acid mine drainage is very much a site-specific undertaking. Acid formation requires the simultaneous presence of reasonable amounts of Fe sulfides, water, oxygen, sometimes S bacteria, and low amounts of neutralizing materials (*26*). Control efforts may involve preventing access of water (and oxygen) to the iron pyrite by sealing the spoil materials and sloping the surface to promote rapid runoff. Conversely, problems can be corrected by flooding a mined area with water. Neutralization of acid after its formation is occasionally feasible with lime (*33*).

Fortunately, Fe sulfides are limited and neutralizing materials abound in most western settings. Water availability is often low. Surface water containing oxygen rarely percolates into zones where significant acid formation can occur (*39*). Groundwater recharge of backfilled spoils at large western surface mines rarely occurs by vertical recharge with oxygen-bearing water. Such recharge commonly occurs laterally and at very slow rates with water containing little or no oxygen from coal seams and/or overburden that abuts the backfilled area (*39*). In this environmental setting, potential acid-forming materials situated below the reclaimed water level are effectively controlled by the lack of available oxygen and the usually excessive capability for neutralization of the recharge water and surrounding spoils. Consequently, groundwater pH problems are rare in western coal mining situations.

State governments have established surface water quality standards in response to the Federal Water Pollution Control Act Amendments of 1972 [section 101(a)(2)]. The pH standards for surface water vary according to water use. Table 2 displays pH ranges, by water use, for coal-producing areas in seven western states.

These water quality standards must be interpreted in conjunction with natural or baseline water quality data. If a natural stream's pH levels fall outside the range of the standards, further evaluation is needed to determine the advisable use. Impacts from mining or any other activity are determined according to a change in water use. If no change in water use has occurred, the surface-disturbing activity should not have an impact on water quality.

For domestic water use, the pH standard is commonly 6.0 to 8.5 because the pH of most unpolluted water falls within this range. A pH of 6.0 to 8.5 is common for farm water use because corrosion of plumbing becomes a problem when the pH drops below 6.0. Above a pH of 8.5, Na concentrations become a problem (*54*).

The pH range of water for fish and wildlife is also in the slightly acid to alkaline range because the major buffering system in nature is the carbonate system, as discussed above. This is important to the fertility of a stream for aquatic organisms (*54*). The commonly suggested pH range, therefore, is 6.0 to 9.0, but that range varies by state or region depending upon specific indigenous organisms.

Groundwater quality standards do not exist in several western states. In

Table 2. Surface water quality standards for pH.

State	pH Standard by Water Use		
	Domestic	Agricultural/Livestock	Fish and Wildlife
Montana	6.5-8.5	6.5-9.5	6.5-8.5
North Dakota	7.0-8.5	6.0-9.0	6.0-9.0
Wyoming	6.5-9.0	6.5-9.0	6.5-9.0
Utah	6.5-9.0	6.5-9.0	6.5-9.0
New Mexico	6.6-8.8	6.6-8.0	6.6-8.8
Oklahoma	6.5-9.0	6.5-9.0	6.5-9.0
Texas	6.0-9.0	6.0-9.0	6.5-9.0

Table 3. Groundwater pH standards.

State	pH Standard by Water Use		
	Domestic	Agricultural/Livestock	Fish and Wildlife
Montana	No change*	No change*	No change*
North Dakota	None	None	None
Wyoming	6.5-9.0	4.5-9.0† 6.5-8.5	6.5-9.0
Utah	6.5-9.0	6.5-9.0	7.0-9.0
New Mexico	6.0-9.0	None	None
Oklahoma	None‡	None‡	None‡
Texas	No change*	None	None

*No change from baseline.
†Agricultural and livestock uses, respectively.
‡Pending standard.

other western states the standards are similar to the standards for surface water. Wyoming and Utah have established groundwater pH standards for domestic, agricultural/livestock, and fish and wildlife uses of water. Montana and Texas require no change from baseline water quality characterization. Table 3 outlines groundwater quality standards in the principal western coal-producing states. The states that have developed groundwater quality criteria used a rationale similar to that used for surface water quality standards.

Management implications and recommendations

Analytical procedures. Use of a pH meter with a saturated paste is the preferred method for determining pH in soils and overburden in the western United States. If expediency or limited available sample size rules out use of saturated paste, use of a 1:1 overburden-to-water ratio is the next logical choice because the overburden will be diluted less than if other ratios are used. Special consideration of soluble salts and CO_2 are unnecessary. Because a number of factors can influence the pH of soil and overburden, analytical procedures, including grinding and drying, must be consistent. Moreover, these procedures should be reported or referenced when the pH data are presented. Appropriate acid/base potential tests may be useful in evaluating potential pH changes due to pyrite weathering. However, nonacid-forming S forms must be eliminated to allow proper interpretation of the results.

pH relationships and soil management. The pH ranges for plant nutritional requirements, pH levels at which trace elements may become toxic, and pH levels at which Na may cause soil physical problems are summarized in table 4. These recommendations must be tempered by the knowledge that pH values may vary by as much as 0.5 to 1.0 unit and, as an extreme, by a much as 1.5 units, depending upon the method used to measure pH. In addition, other site-specific characteristics could modify these generalized conclusions. However, these generalizations are useful for assessing potential nutrient availability, toxicity, or soil physical problems. In addition, pH is a prime indicator of the overall condition of plant growth media.

The pH of most natural waters ranges from 6.0 to 8.0. Western surface waters and groundwater are usually rather well buffered by the CO_2-HCO_3 equilibrium system and are reasonably resistant to pH changes. Fortunately, acid mine drainage problems are rare on coal mines in the West because of the typically low amounts of sulfides, high amounts of lime, and limited access to sufficient amounts of oxygen-bearing water. Where significant acid formation does occur, the ensuing neutralization will increase the concentration of dissolved solids in groundwater (*29*). These increases usually are not significant, however, because high concentrations of dissolved solids are common in the background water quality. For cases where significant pH reductions do occur, increased solubilities of heavy metals may lead to increases in metal ion concentrations. Such occurrences are strongly site-specific, however. This is supported by numerous studies of the potential impacts of western coal mining on water quality.

Recommendations. Unquestionably, pH plays a major role in determining the potential impacts of western coal mining on revegetation success and post-mining water quality. In summary, the following recommendations seem appropriate:

Table 4. Summary of potential pH effects on revegetation.

Parameter	pH
Nutritional	*Desirable Range*
Nitrogen	6-8
Phosphorus	6.5-7.0
Iron	Less than 7.4
Zinc	Less than 7.4
Toxicity	*Desirable pH to avoid growth inhibition*
Boron	See text
Molybdenum	See text
Aluminum	
Cool-season grasses	Greater than 5.0 to 5.5
Warm-season grasses	Greater than 4.5 to 5.0
Alfalfa	Greater than 5.5 to 6.0
Manganese	
Alfalfa and sweetclover	Greater than 5.0 to 5.5
Se	None for plants*
Physical	*Desirable pH to avoid clay dispersion*
Sodium	Less than 8.5

*pH less than 7.5 for seleniferous soils limits Se solubility to reduce forage concentrations. See text.

► Saturated pastes for soil and overburden should be used with standard sample collection techniques.

► Both field and laboratory measurements should be reported for water quality pH values.

► Proper overburden acid/base evaluations should also be reported where significant pH anomolies occur.

► Use of pH as the primary indicator parameter for screening soil characteristics and for monitoring reclamation operations should be encouraged.

► Expanded regional data bases for western plant species should be developed. Additional information on species requirements for pH and other parameters in specific western regions would facilitate improvements in reclamation practices. Experience gained at current western mining operations could be compiled for this purpose by an appropriate organization.

► A consensus needs to be developed among scientists regarding the proper test techniques and use of test results for acid/base potential. Continued uncertainty about the test on the part of some groups increases the extent of analyses required for western mine permitting and may increase the cost of special overburden handling.

► Expanded regional data bases for potential mining impacts on postmining water quality need to be developed by appropriate organizations. These data bases should be based on current mining experience. The problems identified should be used to define the scope of environmental analysis required for mine permitting in those regions. Many benefits can be derived from the extensive data base analyses and postmining observations developed for key mining districts. At some point, additional studies that produce the same results as previous studies in a given mining district should not be required. In such cases, only limited studies to confirm that a site-specific anomoly does not exist should be adequate.

REFERENCES

1. American Public Health Association. 1975. *Standard method for examination of water and wastewater.* Washington, D.C.
2. Bartholomew, W. V., and F. E. Clark. 1965. *Soil nitrogen.* American Society of Agronomy, Madison, Wisconsin. 615 pp.
3. Bauer, A., W. A. Berg, and W. L. Gould. 1978. *Correction of nutrient deficiencies and toxicities in strip-mined lands in semiarid and arid regions.* In F. W. Schaller and P. Sutton [editors] *Reclamation of Drastically Disturbed Lands.* American Society of Agronomy, Madison, Wisconsin. pp. 451-466.
4. Bennet, O. L., E. L. Mathias, W. H. Armiger, and J. N. Jones, Jr. 1978. *Plant materials and their requirements for growth in humid regions.* In F. W. Schaller and P. Sulton [editors] *Reclamation of Drastically Disturbed Lands.* American Society of Agronomy, Madison, Wisconsin. pp. 285-306.
5. Berg, W. A. 1969. *Determining pH of strip-mine spoils.* Research Note NE-98. Forest Service, U.S. Department of Agriculture, Broomall, Pennsylvania. 7 pp.
6. Berg, W. A., and W. G. Vogel. 1968. *Manganese toxicity of legumes seeded in*

Kentucky strip-mine spoils. Research Paper NE-119. Forest Service, U.S. Department of Agriculture, Broomall, Pennsylvania. 12 pp.
7. Berg, W. A., and W. G. Vogel. 1973. *Toxicity of acid coal-mine spoils to plants.* In R. J. Hutnik and G. Davis [editors] *Ecology and Reclamation of Devastated Land.* Gordon and Breach, London, England. pp. 57-68.
8. Beyer, L. E., and R. J. Hutnik. 1969. *Acid and aluminum toxicity as related to strip-mine spoil banks in western Pennsylvania.* Research Report SR-72. Office Coal Research Administration, Harrisburg, Pennsylvania.
9. Brown, J. C., and J. H. Graham. 1978. *Requirements and tolerance to elements by alfalfa.* Agronomy Journal 70: 367-373.
10. Buckman, H. O., and N. C. Brady 1969. *The nature and properties of soils.* MacMillan Company, New York, New York. 623 pp.
11. Burdick, J. C., and F. L. Parker 1971. *Estimation of water quality in a new reservoir.* Report No. 8. National Center for Research and Training in the Hydrologic and Hydraulic Aspects of Water Pollution Control, Vanderbilt University, Nashville, Tennessee.
12. Burdick, J. C., J. R. Burris, A. J. Meyer, and K. L. Roberts. 1973. *A study of the sources, fate, and effects of heavy metals in the environment.* Environmental Engineering Project Report 73-8. University of Tennessee, Knoxville.
13. Buss, G. R., J. A. Lutz, Jr., and G. W. Hawkins. 1975. *Yield response of alfalfa cultivars and clones to several pH levels in tatum subsoil.* Agronomy Journal 67: 331-334.
14. Caruccio, F. T., G. Geidel, and M. Pelletier. 1981. *Occurrence and prediction of acid drainages.* Journal of the Energy Division, American Society of Civil Engineers 107(EYI): 167-168.
15. Chadwick, M. J., and J. K. Salt. 1969. *Population differentiation within* Agrostis tenuis *in response to colliery spoil substrate factors.* Nature 224: 186.
16. Clark, F. E. 1977. *Internal cycling of* ^{15}*nitrogen in shortgrass prairie.* Ecology 58: 1,322-1,333.
17. Clark, R. B. 1981. *Iron deficiency in plants grown in the Great Plains of the United States.* Journal of Plant Nutrition 5: 251-268.
18. Cline, M. L. 1980. *Genetic and mycorrhizal influence of pine seedling growth under stress.* Ph.D. dissertation, Colorado State University, Fort Collins. 224 pp.
19. Collins, J. B., E. P. Whiteside, and C. E. Cress. 1970. *Seasonal variability of pH and lime requirements in several southern Michigan soils when measured in different ways.* Soil Science Society of America Proceedings 34: 56-61.
20. Donahue, R. L., R. H. Follet, and R. W. Tullock. 1976. *Our soils and their management.* Interstate Printers and Publishers, Inc., Danville, Illinois. 791 pp.
21. Fireman, M., and C. H. Wadleigh. 1951. *A statistical study of the relation between pH and the exchangeable-sodium-percentage of western soils.* Soil Science 71: 273-285.
22. Fleming, A. L., J. W. Schwartz, and C. D. Foy. 1974. *Chemical factors controlling the adaptation of weeping lovegrass and tall fescue to acid mine spoils.* Agronomy Journal 66: 715-719.
23. Foy, C. D. 1974. *Effects of aluminum on plant growth.* In E. W. Carson [editor] *The Plant Root and its Environment.* University Press of Virginia, Charlottesville. pp. 601-642.
24. Foy, C. D., P. W. Voigt, and J. W. Schwartz. 1980. *Differential tolerance of weeping lovegrass genotypes to acid coal mine spoils.* Agronomy Journal 72: 859-862.
25. Foy, C. D., W. H. Armiger, L. W. Briggle, and D. A. Reid. 1965. *Differential aluminum tolerance of wheat and barley varieties in acid soils.* Agronomy Journal 57: 413-417.

26. Geidel, G. 1979. *Alkaline and acid production potentials of overburden material: The rate of release.* Reclamation Review 2: 101-107.
27. Gibson, A. H. 1977. *The influence of the environment and managerial practices on the legume-rhizobium symbiosis.* In R.W.F. Hardy and A. H. Gibson [editors] *A Treatise on Dinitrogen Fixation. Section IV. Agronomy and Ecology.* John Wiley and Sons, Inc., New York, New York. pp. 393-450.
28. Green, N. E., and S. O. Graham. 1976. *The influence of pH on the germination of vesicular-arbuscular mycorrhizal spores.* Mycologia 68: 929-934.
29. Groenewold, G. H., R. D. Koob, G. J. McCarthy, B. W. Rehm, and W. M. Peterson. 1983. *Geological and geochemical controls on the chemical evolution of subsurface water in the undisturbed and surface-mined landscape in western North Dakota.* U.S. Bureau of Mines, Contract No. J0275010. Minerals Environmental Technology, Washington, D.C. 167 pp.
30. Hall, Alan. 1979. *Weathering of sodic coal overburden.* MS thesis. Colorado State University, Fort Collins. 130 pp.
31. Havelka, U. D., M. G. Boyle, and R.W.F. Hardy. 1982. *Biological nitrogen fixation.* In F. J. Stevenson [editor] *Nitrogen in Agricultural Soils.* American Society of Agronomy, Madison, Wisconsin. pp. 365-422.
32. Hem, J. D. 1970. *Study and interpretation of chemical characteristics of natural water.* Water Supply Paper 1473. U.S. Geological Survey, Washington, D.C pp. 88-170.
33. Hill, R. D. 1973. *Control and prevention of mine drainage cycling and control of metals.* In Proceedings, Environmental Resource Conference. National Environmental Research Center, U.S. Environmental Protection Agency, Cincinnati, Ohio. 187 pp.
34. Islam, A.K.M.S., D. G. Edwards, and C. J. Asher. 1980. *pH optima for crop growth.* Plant and Soil 54: 339-357.
35. Jackson, M. L. 1958. *Soil chemical analysis.* Prentice-Hall Inc., Englewood Cliffs, New Jersey. 498 pp.
36. Jensen, H. L. 1965. *Nonsymbiotic nitrogen fixation.* In W. V. Bartholomew and F. E. Clark [editors] *Soil Nitrogen.* American Society of Agronomy, Madison, Wisconsin. pp. 436-480.
37. Kamprath, E. J., and C. D. Foy. 1971. *Lime-fertilizer plant interactions in acid soils.* In *Fertilizer Technology and Use.* Soil Science Society of America, Madison, Wisconsin. pp. 105-151.
38. Keefer W. R., and R. F. Hadley. 1976. *Land and natural resource information and some potential environmental effects of surface mining of coal in the Gillette Area, Wyoming.* Geological Survey Circular 743. U.S. Geological Survey, Washington, D.C.
39. Kerr-McGee Coal Corporation. 1984. *Jacobs ranch mine permit application—hydrology.* Oklahoma City, Oklahoma.
40. Krehl, W. A. 1970. *Selenium the maddening mineral.* Nutrition Today X(X): 26-32.
41. Kubata, J., and W. H. Allaway. 1972. *Geographic distribution of trace element problems.* Soil Science Society of America, Madison, Wisconsin. pp. 525-551.
42. Ladd, J. N., and R. B. Jackson. 1982. *Biochemistry of ammonification.* In F. J. Stevenson [editor] *Nitrogen in Agricultural Soils.* American Society of Agronomy, Madison, Wisconsin. pp. 173-228.
43. Lindsay W. L. 1979. *Chemical equilibria in soils.* John Wiley and Son, New York, New York. 449 pp.
44. Mason, D. D., and S. S. Obenshain. 1939. *A comparison of methods for the determination of soil reaction.* Soil Science Society of America Proceedings 3: 129-137.

45. Matthess, George. 1982. *The properties of groundwater.* John Wiley and Sons, New York, New York.
46. McCormick, L. H., and F. A. Amendola. 1983. *Soil pH, extractable aluminum and tree growth on acid minesoils.* Communications in Soil Science and Plant Analyses 14: 249-262.
47. McLean, E. O. 1982. *Soil pH and lime requirements.* In A. L. Page [editor] *Methods of Soil Analysis. Part 2.* American Society of Agronomy, Madison, Wisconsin. pp. 199-224.
48. Montana Agricultural Experiment Station. 1981. *Selective placement of strip mine overburden in Montana. V. Summary Report.* Bureau of Mines Contract H0262032. Bozeman.
49. Munshower, F. F. 1983. *Microelements and their role in surface mine plannings.* U.S. Bureau of Land Management, Denver, Colorado. 41 pp.
50. Munshower, F. F. 1983. *Soil and overburden trace element concentrations.* In M. G. Berg [editor] *Conference on Soil and Overburden Requirements for Successful Revegetation.* Denver, Colorado. pp. 115-147.
51. Murray, J. J., and C. D. Foy. 1978. *Differential tolerances of turf-grass cultivars to an acid soil high in exchangeable aluminum.* Agronomy Journal 70: 769-774.
52. National Environmental Research Center. 1973. *Cycling and control of metals.* Proceedings of an Environmental Resources Conference. National Environmental Research Center, U.S. Environmental Protection Agency, Cincinnati, Ohio. 187 pp.
53. National Research Council. 1981. *Coal mining and groundwater resources in the United States.* Committee on Groundwater Resources in Relation to Coal Mining, Board on Mineral and Energy Resources, Commission on Natural Resources, National Academy Press, Washington, D.C.
54. National Technical Advisory Committee. 1968. *Report of the committee on water quality criteria.* Federal Water Polluton Control Administration, U.S Department of the Interior, Washington, D.C. 234 pp.
55. Nelson, D. W. 1982. *Gaseous losses of nitrogen other than through denitrification.* In F. J. Stevenson [editor] *Nitrogen in Agricultural Soils.* American Society of Agronomy, Madison, Wisconsin. pp. 327-363.
56. Olsen, S. R., and A. D. Flowerday. 1971. *Fertilizer phosphorus interactions in alkaline soils.* In *Fertilizer Technology and Use.* Soil Science Society of America, Madison, Wisconsin. pp. 153-185.
57. Olsen, S. R., and M. Fried. 1957. *Soil phosphorus and fertility.* In *Soil. Yearbook of Agriculture.* U.S. Department of Agriculture, Washington, D.C. pp. 94-100.
58. Palazzo, A. J., and R. W. Duell. 1974. *Responses of grasses and legumes to soil pH.* Agronomy Journal 66: 678-682.
59. Pearson, R. W., and F. Adams, editors. 1967. *Soil acidity and liming.* American Society of Agronomy, Madison, Wisconsin. 275 pp.
60. Peech, M. 1965. *Hydrogen-ion activity.* In C. A. Black [editor] *Methods of Soil Analysis. Part 2.* American Society of Agronomy, Madison, Wisconsin. pp. 914-926.
61. Pierce, F. J., W. E. Larson, R. H. Dowdy, and W.A.P. Graham. 1983. *Productivity of soils: Assessing long-term changes due to erosion.* Journal of Soil and Water Conservation 38: 39-44.
62. Rai, D., P. J. Wierenga, and W. L. Gould. 1974. *Chemical and physical properties of soil samples from a coal-bearing formation in San Juan County, New Mexico.* Research Report 287. New Mexico Agricultural Experiment Station, Las Cruces. 12 pp.
63. Rice, W. A., C. D. Penney, and M. Nyborg. 1977. *Effects of acidity on rhizobia*

numbers, nodulation and nitrogen fixation by alfalfa and sweet clover. Canadian Journal of Soil Science 57: 197-203.
64. Safaya, N. M. 1978. *Delineation of mineral stresses in mine spoils and screening plants for adaptability.* In M. K. Wali [editor] *Ecology and Coal Resource Development, Volume 2.* Pergammon Press, New York, New York. pp. 830-849.
65. Sandoval, F. M., and J. F. Power. 1978. *Laboratory methods recommended for chemical analysis of mined-land spoils and overburden in western United States.* Agriculture Handbook 525. U.S. Department of Agriculture, Washington, D.C.
66. Sandoval, F. M., and W. L. Gould. 1978. *Improvement of saline—and sodium—affected disturbed lands.* In F. W. Schaller and P. Sutton [editors] *Reclamation of Drastically Disturbed Lands.* American Society of Agronomy, Madison, Wisconsin. pp. 485-504.
67. Sawyer, C. N. 1980. *Chemistry for sanitary engineers.* McGraw-Hill Book Company, Inc., New York, New York. 367 pp.
68. Schafer, W. M., G. A. Nielson, D. J. Dollhopf, and K. Temple. 1979. *Soil genesis, hydrological properties, root characteristics and microbial activity of 1-to 50-year-old stripmine spoils.* EPA-600/7-79-100. U.S. Environmental Protection Agency, Washington, D.C.
69. Schmidt, E. L. 1982. *Nitrification in soil.* In F. J. Stevenson [editor] *Nitrogen in Agricultural Soils.* American Society of Agronomy, Madison, Wisconsin. pp. 253-288.
70. Schroer, F. W. 1978. *Characterization of coal overburden and strip-mine spoils in North Dakota.* Research Report 68. Soils Department, North Dakota State University, Fargo. 17 pp.
71. Sobek, A. A., W. A. Schuller, J. R. Freeman, and R. M. Smith. 1978. *Field and laboratory methods applicable to overburdens and minesoils.* EPA-600/2-78-054. U.S. Environmental Protection Agency, Washington, D.C.
72. Stevenson, F. J., editor. 1982. *Nitrogen in agricultural soils.* American Society of Agronomy, Madison, Wisconsin. 940 pp.
73. Theodorow, C., and G. D. Bowen. 1969. *The influence of pH and nitrate on mycorrihizal associations of* Pinus radiata *D. Don.* Australian Journal of Botany 19: 13-20.
74. Thorup, J. T. 1969. *pH effect on root growth and water uptake by plants.* Agronomy Journal 61: 225-227.
75. Tisdale, S. L., and W. L. Nelson. 1966. *Soil fertility and fertilizers.* The Macmillan Company, Toronto, Ontario. 694 pp.
76. U.S. Bureau of Land Management. 1981. *Energy Transportation System, Inc. (ETSI) final environmental impact statement.* U.S. Department of the Interior, Washington, D.C.
77. U.S. Bureau of Land Management. 1981. *Powder River—final environmental impact statement—coal.* U.S. Department of the Interior, Washington, D.C.
78. U.S. Salinity Laboratory. 1954. *Diagnosis and improvement of saline and alkali soils.* Handbook 60. U.S. Department of Agriculture, Washington, D.C.
79. Van Voast, W. A., and K. S. Thompson. 1982. *Estimates of post-mining water quality for the Upper Tongue River, Montana and Wyoming.* Montana Bureau of Mines and Geology, Helena.
80. Van Voast, W. A., R. B. Hodges, and J. J. McDennott. 1976. *Hydrologic aspects of strip mining in the subbituminus coal fields of Montana.* Montana Bureau of Mines and Geology, Helena.
81. Vlek, P.L.G., and W. L. Linsay. 1977. *Thermodynamic stability and solubility of molybdenum minerals in soils.* Soil Science Society of America Journal 41: 42-46.

82. Vogel, W. G. 1981. *A guide for revegetating coal minesoils in the eastern United States.* General Technical Report NE-68. Forest Service, U.S. Department of Agriculture, Broomall, Pennsylvania.
83. Whitney, R. S., and R. Gardner. 1943. *The effect of carbon dioxide on soil reaction.* Soil Science 55: 127-141.
84. Woodmansee, R. G., J. D. Reeder, and W. A. Berg. 1978. *Nitrogen in drastically disturbed lands.* In C. T. Youngberg [editor] *Forest Soils and Land Use.* Department of Forest and Wood Sciences, Colorado State University, Fort Collins. pp. 376-392.

Chapter 2

Electrical conductivity

J. J. Jurinak, Joe Bowden, Fred Samson, and Tom Portal

Saline soils are common features of arid and semiarid landscapes. These soils are associated with climates in which annual evapotranspiration greatly exceeds annual rainfall; therefore, essentially no water percolates through the soil under normal conditions. The result is that, although the lack of water reduces the intensity of soil mineral weathering, the products of the weathering that does occur, for example, salts, tend to accumulate in the soil.

Saline soils contain cations and anions derived from the weathering of primary, secondary, alluvial, and fluvial materials. Lesser elemental contributions come from atmospheric deposition of N, C, S, and other elements. Soils developed from weathering under subhumid conditions differ in composition from their parent material because of the variable solubility of their constituents. Whereas Na and K salts of chloride and sulfate are soluble and readily mobile, Ca and Mg tend to interact with carbonate to form less soluble secondary minerals.

Because water is the vector for salt, salt accumulation in soil commonly reflects the relief and geomorphological conditions of the area. Measured salinity in soils or overburden material varies spatially and temporally as a result of this relationship between salinity and water movement. Within any sampling site, measurements can vary in both horizontal and vertical directions. Seasonal climatic changes also affect salinity levels within the profile.

In the context of disturbed-land reclamation, a salinity hazard exists when there is sufficient soluble salt in a soil to interfere with the growth of desired vegetation. This definition of salinity hazard, which necessarily differs from the agronomic definition, precludes establishing rigid threshold limits for salinity and makes management decisions a site-specific process.

Measuring soil salinity

The physical and chemical properties of a saline material reflect the amount and type of salt present. Ideally, the salt present should be determined under field conditions without need for sample collection or chemical analyses. There is no convenient method to attain this goal, although estimates of total salinity can be made in the field using various salinity sensors (*19, 21*). Portable, commercially available bulk soil salinity field sensors include the four-electrode sensor (*20*) and the electromagnetic induction sensor (*22*). The time domain reflectrometric sensor is a recent development that measures both salinity and water content (*4*). These field methods should find increasing use for survey or inventory purposes in disturbed-land reclamation.

When reestablishing vegetation, the chemical composition of the soil solution, for example, Na adsorption ratio and B concentration, is of as much interest as total salinity. Because of the need for chemical data, the chemical analyses of aqueous extracts of soil or overburden samples is still the most commonly used technique to monitor salinity and other potential hazards.

In principle, any soil-water ratio can be used to obtain an aqueous extract. However, because both the total and relative amounts of solutes are determined by the ratio used to obtain the extract, some standardization of methodology is required for universal interpretation. Most salinity data are obtained from extracts of saturated soil pastes obtained by vacuum filtration (*29*). An advantage of this technique is that the chemical analyses of the saturation extract is the basis of the majority of crop and forage salt tolerance data (*13*). Also, the percentage of water that the soil holds at saturation is correlated to other soil parameters and properties (see Chapter 3, Saturation Percentage). Thus, within limits, the electrical conductivity of the saturation extract (EC_e), a measure of soil salinity, is related to the electrical conductivity of the soil solution under field conditions. For example, in medium- and heavy-textured soils the electrical conductivity at field capacity is approximately twice the EC_e value obtained in the laboratory.

Extracts can be obtained more easily with other soil-water ratios, such as 1:1 or 1:2, but they are less related to field moisture content. Use of these wider extraction ratios, however, can be advantageous when relative field changes are of principle interest.

Use of electrical resistance or its reciprocal, electrical conductivity, to measure salinity or the salt concentration of solutions (aqueous extracts) is a simple, well-established method. Electrolytes in solution dissociate into charged ions that carry electric current. Thus, as the ionic concentration increases so does the electrical conductivity.

The units of electric conductivity are reciprocal ohms or ohms $^{-1}$, traditionally known as the mho. Under the international movement to standardize scientific dimensions—Systeme International d' Units, commonly known as Sl units—the unit for electrical conductivity is siemens (S). The working unit for the soil electrical conductivity is decisiemens per meter (dS/m), which is equivalent to the former dimension, millimhos per centimeter (mmhos/cm).

Electrical conductivity values are sensitive to the temperature of the solution. In the temperature range of 15° to 35 °C each degree increase in temperature increases electrical conductivity about 2 percent. To compare data from various sources, all electrical conductivity data are standardized to the reference temperature of 25 °C. This can be done automatically if the electrical conductivity measuring instrument has a functional thermocompensator. If not, the temperature correction in the 15° to 35 °C range can be approximated by $EC_{25} = EC_T - 0.02(T - 25)EC_T$, where EC_T is the electrical conductivity at temperature T and EC_{25} is the value corrected to 25 °C. Knowing the temperature of the extract or solution being analyzed is important for correct data interpretation.

In certain salt systems the measured electrical conductivity is only an approximate measure of the total dissolved solids due to the formation of soluble complexes called ion pairs (*9*). The presence of ion pairs lowers the current carrying capacity of the solution. This is particularly important in solutions where SO_4^{2-} is the dominant anion because of its tendency to form ion pairs. This is demonstrated in a saturated solution of gypsum ($CaSO_4 \cdot 2H_2O$) where about one-third of the total Ca^{2+} and SO^{2-} in solution exists as the neutral ion pair $CaSO_4^0$. Thus, the electrical conductivity of a saturated gypsum solution is about one-third less than expected. An abbreveated list of the ion pairs common to saline soils includes $CaSO_4^0$, $CaHCO_3^+$, $CaCl^+$, $NaSO_4^-$, $NaHCO_3^0$, $MgSO_4^0$, and KSO_4^-, among others. The net result of ion speciation is that the electrical conductivity measurement underestimates the total dissolved solids. That is, the commonly used relationship TDS (mg/l) ≈ 640 EC (dS/m) (*29*) is not valid for a mixed salt system (soil solution) when the electrical conductivity is greater than 5-6 dS/m or for sulfate-dominated systems.

Affect of salinity on plant growth

The major adverse effect of soil salinity is to reduce the availability of soil water to plants. Briefly, this is because the presence of salt in water increases the work that the plant must do to extract water from the soil solution. This work is referred to as the osmotic potential; it is additive to the work required by the plant to extract water from a nonsaline soil solution (the matric potential). The sum of the two potentials, osmotic plus matric, is called the soil-water potential, which determines the total work required by the plant to extract water (*27*). The result is that for any given soil moisture level the presence of salt increases the work required of the plant to maintain its normal turgor. This, however, is not the complete picture.

Research has suggested that a plant grown in a saline environment makes an osmotic adjustment in the direction of maintaining a constant water potential gradient between the plant and the soil solution. The ability of the plant to make the physiological changes associated with this adjustment is related to its salt tolerance (*3, 8*). The osmotic adjustment comes at the expense of plant growth and biomass production. Thus, the osmotic effect on plant growth does not result simply from a decrease in water availability as the osmotic potential of the soil solution decreases. The response of plants to salinity is a complex biological phenomenon. This response is correlated to the osmotic potential of the soil solution, which, in turn, is related to electrical conductivity by the relationship: osmotic potential (bars) = −0.36 electrical conductivity (dS/m) (*29*).

Salinity-stressed plants exhibit no distinctive symptoms. The most common effect of salinity stress is a general reduction or stunting of plant growth. Both the growth rate and size of most plants decrease with increasing salinity. Under severe conditions, plant leaves may have a purple, dark-green color and a waxy appearance. Salinity damage is most prevalent during germination and in the early stages of growth (*13*). Thus, during revegetation of overburden, transplanted stock may survive but seeding may not be successful because of poor germination or seedling survival.

Salt tolerance of an agronomic crop relates to the yield decline under a given salinity level (EC_e) relative to a normal crop. This approach is not realistic when dealing with native vegetation where salt tolerance is measured on the basis of survival. Because the response of vegetation to salinity varies by stage of plant growth, soil texture, spatial and temporal variability of soil salinity, and climatic conditions (*14*), ranking plants according to their salt tolerance is at best a qualitative guide, except at the extreme ends of the salinity spectrum. Plants that survive successfully in saline soils are those capable of internally compensating for the low (negative) osmotic potential and nutrient imbalance of the soil solution.

The impact of salinity on the nutrient uptake is usually considered a secondary negative effect on vegetative growth relative to the osmotic effect. But because passive nutrient uptake is related to water uptake, any decrease in available water would be reflected in a reduced intake of nutrient elements (*28*). Also, the imbalance in the elemental composition of a saline soil solution can affect the absorption of selected nutrients. For example, there is considerable evidence that the presence of excess chloride ion reduces the uptake of nitrates (*12, 30*). Similarily, research has shown that K uptake declines in the presence of excess Na (*6, 26*). Researchers also have reported that lack of adequate Ca can enhance substitution of Na for K in plants (*11*). That study also suggested that the ability of a cell to retain K in the presence of high Na concentration could be an important regulating mechanism in salt-tolerant species.

Response of soils to saline conditions

Electrical conductivity levels associated with the reclamation of mine spoils must be considered not only with regard to biomass production but also in light of its effect on soil physical properties.

Of considerable interest is the interaction among salinity; exchangeable Na percentage, as measured by the Na adsorption ratio; hydraulic conductivity; and infiltration rate. The rate at which water infiltrates into soil bears importantly on the amount of erosion and sedimentation encountered during reclamation, as well as the water that moves into the root zone.

The presence of electrolytes in the soil solution is essential to maintaining the permeability or hydraulic conductivity of soil in the presence of Na, that is, the exchangeable sodium percentage (*7, 18, 25*). A high hydraulic conductivity can be attained even at high exchangeable Na percentages if the solution electrical conductivity is above a threshhold value, which is a function of mineralogy, texture, and other factors. For a medium-textured soil, data indicate that with an exchangeable Na percentage or Na adsorption ratio of 20 to 25 the hydraulic conductivity of the soil can be maintained at an acceptable level if the electrical conductivity threshold of the soil solution is kept at 1 to 2 dS/m.

Correspondingly, when the electrical conductivity falls to 0.3 dS/m, decreases in the hydraulic conductivity and clay dispersion can occur in soils at exchangeable-Na-percentage or Na-adsorption-ratio values less than 10 or 12. If the electrical conductivity falls to near that of rainwater (distilled water), clay dispersion and resultant hydraulic conductivity reduction can occur at exchangeable Na percentages as low as 2 (*25*).

Two related electrical conductivity-exchangeable Na percentage phenomena are responsible for the reduction in soil permeability. During leaching, clay dispersion and migration occur in soils with a low exchangeable Na percentage or Na adsorption ratio (<10) and low electrical conductivity (<0.5 dS/m). This results in the blockage of pores and a corresponding reduction in soil permeability. This process is considered irreversible. On the other hand, the swelling of clay, a reversible process, is important in reducing the permeability of soil to water when the soil has a high exchangeable Na percentage and contains smectite clay minerals (*7, 16, 17*).

Control and maintainence of a predetermined, moderate soil salinity level thus has a definite place in surface mine land reclamation management, particularly with fine-textured materials. These data suggest that after a rainy season (winter), the decrease in electrical conductivity in the surface soil can result in lower soil infiltration rates and increased surface crusting, which promotes erosion.

Mitigation of salinity problems

Proper fertilization of saline soils may increase the salt tolerance of plants. Research has shown that at a given fertility level plant growth slows with increasing salinity. But at low to moderate salinity levels fertilization can, to a limited degree, ameliorate the adverse effects of salinity (*10*). Recently, Majerus found that yields nearly tripled in the case of beardless wild rye (*Elymus triticoldes*) grown on a saline soil with the application of 70 kg of N/acre.[1]

Applying fertilizer that is a soluble inorganic salt to topsoil, subsoil, or mine spoil can aggravate salinity problems (*5*). Such problems can be minimized by using fertilizers with a low salt index during reclamation. Normally, the higher the analysis of the fertilizer, the lower the salt index per unit of nutrient supplied (*28*). Most inorganic N fertilizers have relatively high salt indexes, as does the chloride salt of K. On the other hand, all phosphate carriers have low salt indexes (*10*). In addition to using high-analysis fertilizer, only the amount of fertilizer indicated by soil analysis should be applied to reduce the salinity input.

Mulching soil surfaces to reduce evaporation and minimize the upward convective/diffusive movement of salt is an accepted practice (*1*). Use of organic amendments, such as bark or straw, for mulch further serves to reduce soil bulk density, which improves infiltration and moisture retention. Together with practices that incorporate organic matter into the soil, such as furrowing, mulching improves soil structure and enhances the downward flux of salt within the profile, thereby decreasing the electrical conductivity (*15*). Bender and Opeka (unpublished data, Horticulture Department, Virginia Tech University) demonstrated that enhanced infiltration was due largely to the physical separation of the soil particles by organic material. However, the process of organic matter decay also produces long-chain organic compounds that enhance and stabilize soil structure (*24*).

Revegetation success in saline soils can be improved by proper plant selection. This essentially precludes the use of guidelines intended for agricultural conditions (*2, 29*). Vegetation adapted to conditions inherent to relatively harsh natural environments is quite different from agricultural cultivars. In this sense classifying a soil as saline or nonsaline is not pertinent so long as the native plants used for reclamation of a given soil can survive and reproduce. Grasses, such as beardless wild rye, can produce well in soil with electrical conductivities exceeding 18 dS/m.[2] In the saltbrush family a 50

[1]Majerus, M. 1985. "Salt-Tolerant Plant Materials-Species Selection and Establishment Techniques." Paper presented at the Society for Range Management Annual Meeting, Salt Lake City, Utah.

[2]Ibid.

percent reduction in leaf biomass was not reached until the soil electrical conductivity reached 24 and 36 dS/m for fourwing saltbush (*Atriplex canescens*) and cuneat saltbush (*Atriplex cuneata*), respectively (*23*). These data indicate that use of salt-tolerant vegetation should have high priority in reclamation projects and can significantly increase the amount of topsoil considered as suitable for reclamation.

Measurement of electrical conductivity in a soil or mine spoil is a simple, rapid determination. It suffers from the weakness that the interpretation of the data incorporates all of the errors associated with obtaining a representative field sample. Even with this limitation, if electrical conductivity is considered in the context of the ionic composition of the soil solution and the soil mineralogy and texture, it is a valuable tool in determining reclamation strategies.

REFERENCES

1. Aldon, E. F. 1979. *Cultural treatments.* In *User Guide to Vegetation: Mining and Reclamation in the West.* General Technical Report INT-64. Intermountain Forest and Range Experiment Station, U.S. Forest Service, Ogden, Utah.
2. Ayers, R. S., and D. W. Westcot. 1985. *Water quality for agriculture.* Irrigation and Drainage Paper 29. Food and Agriculture Organization, United Nations, Rome, Italy.
3. Bernstein, L. 1963. *Osmotic adjustment of plants to saline media: Dynamic phase.* American Journal of Botany 50: 360-370.
4. Dalton, F. N., W. Herelrath, D. Rawlins, and J. D. Rhoades. 1984. *Time domain reflectometry for simultaneous measurement of bulk medium water and electrical conductivity.* Science 224: 989-990.
5. Dollhopf, D. J. 1983. *Chemical amendment and irrigation effects on salt migration in sodic mine spoils in the Northern Great Plains.* In *Symposium on Soils and Overburden in Arid/Semiarid Mine Land Reclamation.* Wyoming Agricultural Experiment Station, Laramie.
6. Epstein, E. 1972. *Mineral nutrition of plants: Principles and perspectives.* John Wiley and Sons, Inc., New York, New York.
7. Frenkel, H., J. O. Goertzen, and J. D. Rhoades. 1978. *Effects of clay type and content, exchangeable sodium percentage, and electrolyte concentration on clay dispersion and soil hydraulic conductivity.* Soil Science Society of America Journal 42: 32-39.
8. Gale, J. 1975. *Water balance and gas exchange of plants under saline conditions.* In A. Poljakoff-Mayber and J. Gale [editors] *Plants in Saline Environments.* Springer-Verlag, New York, New York.
9. Jurinak, J. J. 1984. *Salt-affected soils: Thermodynamic aspects of the soil solution.* In I. Shainberg and J. Shalhevet [editors] *Soil Salinity under Irrigation, Processes and Management.* Springer-Verlag, New York, New York.
10. Jurinak, J. J., and R. J. Wagenet. 1981. *Fertilization and salinity.* In D. Yaron [editor] *Salinity in Irrigation and Water Resources.* Marcel Dekker, Inc., New York, New York.
11. Kafkafi, U. 1984. *Plant nutrition under saline conditions.* In I. Shainberg and J. Shalhevet [editors] *Soil Salinity under Irrigation, Processes and Management.* Springer-Verlag, New York, New York.
12. Kafkafi, U., N. Veloras, and J. Letey. 1982. *Chloride interaction with nitrate*

and phosphate nutrition in tomatoes. Journal of Plant Nutrition 5: 1,369-1,385.
13. Maas, E. V., and J. F. Hoffman. 1977. *Crop salt tolerance-current assessment.* Journal of the Irrigation Drainage Division, American Society of Civil Engineers 103: 115-134.
14. Meiri, A. 1985. *Plant response to salinity: Experimental methodology and application to the field.* In I. Shainberg and J. Shalhevet [editors] *Soil Salinity under Irrigation, Processes and Management.* Springer-Verlag, New York, New York.
15. Opeka, T., and R. Morse. 1979. *Use of green manure and tillage to improve minespoil productivity.* EPA-600/7-79-257. U.S. Environmental Protection Agency, Washington, D.C.
16. Park, C. S., and G. A. O'Connor. 1980. *Salinity effects on hydraulic properties of soils.* Soil Science 130-167.
17. Pupinsky, H., and I. Shainberg. 1979. *Salt effects on the hydraulic conductivity of a sandy soil.* Soil Science Society of America Journal 43: 429-433.
18. Quirk, J. P., and R. K. Schofield. 1955. *The effect of electrolyte concentration on soil permeability.* Journal of Soil Science 6: 163-178.
19. Rhoades, J. D. 1976. *Measuring, mapping and monitoring field salinity and water table depths with soil resistance measurements.* Soils Bulletin 31: 159-186. Food and Agriculture Organization, United Nations. Rome, Italy.
20. Rhoades, J. D. 1979. *Inexpensive four-electrode probe for monitoring soil salinity.* Soil Science Society of America Journal 43: 817-818.
21. Rhoades, J. D. 1984. *Principles and methods of monitoring soil salinity.* In I. Shainberg and J. Shalhevet [editors] *Soil Salinity under Irrigation, Processes and Management.* Springer-Verlag, New York, New York.
22. Rhoades, J. D., and D. L. Corwin. 1981. *Determining soil electrical conductivity-depth relations using an inductive electromagnetic soil conductivity meter.* Soil Science Society of America Journal 45: 255-260.
23. Richardson, S. G., and C. M. McKell. 1980. *Salt tolerance of two saltbush species grown in processed oil shale.* Journal of Range Management 33: 460-462.
24. Russell, E. W. 1973. *Soil conditions and plant growth.* Longsman Group Limited, London, England.
25. Shainberg, I., and J. Letey. 1984. *Response of soils to sodic and saline conditions.* Hilgardia 52(2): 57.
26. Stuiver, C.E.E., P.J.C. Kuiper, H. Marschner, and A. Kylin. 1981. *Effects of salinity and replacement of K^+ by Na^+ on lipid composition in two sugarbeet inbred lines.* Physiology of Plant 52: 77-82.
27. Taylor, S. A., and G. L. Ashcroft. 1972. *Physical edaphology.* W. H. Freeman and Co., San Francisco, California.
28. Tisdale, S. L., and W. L. Nelson. 1975. *Soil fertility and fertilizers.* Macmillan Publishing Co., Inc., New York, New York.
29. U.S. Salinity Laboratory Staff. 1954. *Diagnosis and improvement of saline and alkali soils.* Handbook 60. U.S. Department of Agriculture, Washington, D.C.
30. Weigel, Jr., R. C., J. A. Schillinger, B. A. McCaw, H. G. Gauch, and E. Hsiao. 1973. *Nutrient nitrate levels and the accumulation of chloride in leaves of snapbeans and roots of soybeans.* Crop Science 13: 411-412.

Chapter 3

Saturation percentage

Stephen D. Merrill, James R. Deutsch, and Michael W. Pole

Saturation percentage is the amount of water, by weight, necessary to form a minimally flowable mud from dry soil materials. That amount of water typically is greater than that held by in situ soil materials that contain the maximum amount of water possible. Thus, saturation percentage, as an analytic parameter, should be distinguished from various measures of water saturation of in situ soil.

Saturation percentage can and is used as an indicator of soil water-holding properties. But it developed as an adjunct to a standard methodology for making extracts for analyses of soluble soil constituents. Before the 1940s, soil scientists used a variety of water-to-soil extraction ratios to analyze soluble components. Different soils, however, respond in qualitatively dissimilar ways to alterative water-to-soil extraction ratios (*27*). Schofield then proposed using extracts from saturated pastes as a standard procedure. U.S. Department of Agriculture scientists (*37*) recommended use of saturation paste and saturation extracts, citing the fact that saturation percentage is the lowest soil water content that is easily and conveniently extractable. Other researchers also found that saturation pastes and percentages were ''fairly reproducible'' (*20*), citing Schofield (*31*) for description of the procedure. Thus, saturation percentage arose through the need for reproducibility in various soil chemical analyses.

Saturation percentage has significance both as an index of soil water-holding capacity and swelling capacity and as an indicator of the soil-to-water ratio at which the saturation extract is made. In this latter role the saturation percentage value influences chemical analyses of extracted soil solution and aids in interpreting results of such analyses.

Saturated paste and saturation percentage determinations

Current standard methods. Schofield (*31*) gave one of the earliest descriptions of saturation percentage determination:

"One hundred grams of the dry soil are weighed into a soil can having a tight cover. Distilled water is added and the soil is stirred with a spatula until a condition of saturation is reached. In its saturated condition, the soil mass should be plastic enough to flow slightly when the container is tipped. The surface should glisten as it reflects light, and the air should all be displaced. The can with its saturated soil is again weighed, and the increase in weight from the dry condition is recorded as 'water required to saturate' in terms of the percentage of the dry weight of the soil."

The saturation percentage value would then be used to prepare a saturated paste with 1 to 2 kg of soil, an amount necessary to yield sufficient saturation extract to support macroanalytic procedures for soluble constituents. The current standard procedure, which is geared to microanalytic or semimicroanalytic procedures for determination of ions, uses from 100 to 250 g of soil to prepare the saturated paste for subsequent extraction. Saturation percentage is determined from the same paste mass as used for extraction.

One of the basic references for preparation of saturated paste and determination of saturation percentage is *Agriculture Handbook 60* (*37*). Other references (*9, 28, 29, 35*) are closely aligned with or derived from this source. *Agriculture Handbook 60* specifies the following procedure: Soil materials intended for general analysis should be air dried, clods broken, and the soil passed through a 2-mm sieve. Then a 250-g sample is placed in a 250-ml or larger-sized container. Distilled water is added in increments while stirring with a spatula. The container is tapped on the workbench from time to time. At saturation, the soil paste glistens as it reflects light, flows slightly when the container is tipped, and the paste slides freely and cleanly off the spatula for all soils but those with a high clay content. One particular problem in making saturated pastes is that some fine-textured soils tend to puddle. Adding water with a minimum of stirring will abate this effect. After mixing, the paste is allowed to stand for a period of time. Then the criteria for saturation is rechecked, that is, free water should not collect nor should the paste stiffen markedly or lose its glisten.

Once the saturated paste is prepared, the saturation percentage may be determined with one of four methods (*28, 29, 35, 37*). One of these techniques requires a separate measurement of the percentage of water in air-dry soil. *Agriculture Handbook 60* recommends that air-dry water content be determined on at least a 25-g sample with heating to constant weight at

105 °C. The four methods include:

1. *Direct determination.* A subsample of the saturated paste is weighed, dried at 105 °C to constant weight, and reweighed. This method does not require a separate determination of air-dried water content.

2. *Determination from volume of water added.* The volume of water added to a known weight of air-dried soil is measured with a burette or graduated cylinder during paste preparation. A separate determination of air-dried water content is required.

3. *Determination from the weight of a known volume of saturated paste.* A sample of saturated paste is placed in a cup of known volume, leveling the surface of the paste with the cup lip. A value for soil particle density must be assumed.

4. *Determination from weight of air-dry soil, weight of paste after preparation, and separately determined air-dry water content.* This method, a variation of method 2, does not allow for addition of more dry soil in the event of oversaturation during paste mixing.

While method 4 is marginally more accurate than method 1, method 1 is more practical for high-volume analytic throughput. Methods 2 and 3 are used less frequently. Method 3 is vulnerable to error if caution is not used in removing entrapped air.

References on saturated paste procedures do not specify clearly the method of reducing dried soil clods and aggregates to screened material. Such grinding or breaking operations can affect subsequent results, as will be discussed later. Thus, these operations should be specified as a relevant part of saturated paste and saturation percentage procedure. In general, the breaking and reduction to a 2-mm size should be done so as to minimize breakage of truly consolidated rock material but while effectively maximizing breakage of moderately bonded aggregates. There are a number of grinding or soil breaking devices commercially available that are capable of reducing dried soil in an appropriate manner.

The time elapsed between the initial mixing and saturation percentage determination and between the additions of soil or water to paste and rechecking of saturation criteria can affect the saturation percentage value and subsequent results of extract analyses. Saturation criteria should be rechecked only after 1 or more hours have elapsed after adjusting the paste by water or soil addition (*9*). The paste then should stand at least 4 or more hours after the last mixing operation before extraction is made (*29*). Peat soils and fine-textured, swelling soils should sit overnight before rechecking saturation criteria (*28, 37*). Normally, water will need to be added to such soils. In general, it is best to wait overnight or for at least 14 hours before rechecking saturation criteria. As discussed later, this is particularly important when using saturation percentage as an indicator of sodic hazard. Two references (*34, 35*) recommend an overnight or equivalent waiting period before rechecking saturation criteria without exception. Thus, procedures outlined in these sources are preferred for use in mined-lands analyses.

Alternative methods. The commonly used saturation paste procedure, in which water is added to air-dried soil by mixing with a spatula, yields a fairly reproducible water content (*20*). The accuracy of the procedure depends upon operator skill and experience, particularly with regard to determination of saturation criteria. Alternative procedures have been proposed that either reduce the need for operator skill in determining saturation endpoint or reduce the time and effort needed during paste mixing, or achieve both.

Allison proposed an oversaturation technique for preparing paste by mixing (*1*). For a given amount and textural type of air-dried soil, the operator places 10 to 20 percent more distilled water in a mixing vessel than is necessary for saturation. Procedures are prescribed for adding soil until saturation is reached. The oversaturation method reportedly reduces the time and effort necessary to mix paste with repeated additions of water and prevents puddling with fine-textured soils.

Longenecker and Lyerly proposed a capillary saturation-table method (*18*). Dried, screened soil is placed in shallow filter-paper cups and allowed to absorb water as it lies upon a bed of wet quartz sand. This absorption technique thus eliminates the problem of determining when saturation occurs. In comparison with the commonly used technique, Longenecker and Lyerly reported that the capillary saturation method produced a 29 percent saturation percentage value in loamy sand soil compared to 28.5 percent with the common technique. For silt loam soil, the values were 39.0 percent for the common technique and 40.8 percent for the capillary saturation method. However, with clay-textured soil the capillary saturation method produced a significantly lower value than the common technique, 56.3 percent versus 71.4 percent, respectively.

Beatty and Loveday described a modified capillary saturation technique (*4*). Saturation percentage is determined on a subsample by absorption, while the main sample is subsequently mixed with the determined amount of water. This prevents the bulk of the paste from contamination by possible salt movement during water uptake by capillary saturation, as in the procedure advanced by Longenecker and Lyerly. Screened and dried soil is placed in a 4.5-cm diameter by 1-cm high retainer ring that rests on a blotter-paper tension table sitting in a dish with a lid.

Capillary wetting techniques cannot be recommended for use in mined-land analyses. This is because of the apparent sensitivity of saturation percentage values from such techniques to aggregation characteristics of the sample, which is discussed in the next section, and low saturation percentage values for fine-textured soils. The oversaturation technique, however, might improve the conventional mixing technique for paste preparation.

Procedural effects on quality

Effects on saturation percentage results. The capillary saturation procedure can produce saturated paste and saturation percentage results without relying on operator judgment and with less apparent effort than the common mixing procedure. Few studies, beyond Longenecker and Lyerly's (*18*), have compared the results of the two procedures.

Campbell and Richards (*7*) compared saturation percentage results for several different levels of grinding of a peat soil. Saturation percentages for coarse, medium, and fine grinding using the common mixing procedure were 152.4, 147.2, and 142.4, respectively. Using the capillary saturation procedure, saturation percentages were 90.9, 149.8, and 166.5, respectively. Thus, the active mixing action of common paste preparation yields saturation percentage values that are considerably less sensitive to aggregate size differences generated by variation of grinding procedure.

Longenecker and Lyerly (*18*) compared saturation percentage values of three soils crushed and sieved to a series of different finenesses and saturated by capillarity. Saturation percentages for soils sieved through 0.25-mm screens were 2 to 6 percentage points greater than those for soils passed through 2-mm screens.

Loveday (*19*) compared saturation percentage values for capillary wetting of six soils ground and sieved to pass 2-mm, 1-mm, and 0.5-mm screens. Grinding and sieving resulted in saturation percentages 4 percentage points higher for the finest screening of the coarser textured soils, and 2 percentage points lower for the finest screening of the finer textured soils. In that same study Loveday subjected soils to grinding with three different machines, which produced significant differences in the size distribution of dry aggregates. Saturation percentages for the coarser aggregate mixtures were 1 to 10 units higher than the finer aggregate mixtures, depending on soil texture.

These observations appear contradictory. But they do demonstrate the complex but significant dependence of capillarity-derived saturation percentages upon particle and aggregate size distribution. This limited evidence suggests that values obtained with the common mixing technique are less dependent upon sieving and grinding practices. Longenecker and Lyerly suggest that adoption of the capillary wetting technique will depend upon standardization of sieving and grinding practices.

Again, few researchers have studied the effect of waiting time on saturation percentage. One series of measurements by Longenecker and Lyerly (*18*) using capillary wetting indicated slight increases in the amount of water absorbed after 110 hours compared to 18 hours. Gains between 9 hours and 18 hours were more significant, especially for finer textured soil. In another series of measurements the increase in water absorbed between 22 hours and 40 hours was 0.7 saturation percentage units or less. These results appear to be due to swelling phenomena and slower water movement through finer textured soils because river sand showed little additional absorption beyond 1- or 2-hour initial time periods. Longenecker and Lyerly's data and undocumented laboratory experience firmly support the value of overnight or 24-hour waiting times before rechecking saturation criteria during saturated paste preparation.

Effects on chemical results. Preparation of saturated pastes almost always is connected with analysis of saturation extracts, with saturation percentage determined as a subsidiary measurement. Thus, effects of saturated paste methodologies on subsequent chemical results should be considered. Grinding and sieving methods and the total equilibration time before paste is extracted directly affect saturation percentage values, while techniques of making saturation extractions do not. Thus, the former aspects of methodology are of greatest concern. An important consideration connecting saturation percentage values with analyses of saturation extracts is the effect of variations in saturation percentage values on extracted solution data.

Effects on chemical results can be examined both in theory and by reference to studies. Some physiochemical events that determine composition of saturation extracts, such as ion exchange, occur with considerable rapidity. Other processes, such as dissolution of sparingly soluble salts, do not.

Jurinak and associates (*17*) found that a 1:1 paste of a saline mineral soil reached 98.5 percent of its final long-term electrical conductivity value after 10 hours. Within 1 to several hours, fast dissolution and other reactions produce soil solution concentrations within 5 percent or less of final values. Thereafter, chemical weathering and other processes alter solution composition in a diffusion-limited manner, that is, at rates proportional to the square root of time elapsed (*12, 17*). Some research (*27*) has shown slow increases in soluble contents of saturation extracts for incubation periods of several days or more. Microbiological generation of CO_2 and the effect of this upon lime solubilization may be the cause. Altogether, the information available justifies overnight or equivalent total elapsed times between initial mixing and extraction and 4-hour or greater times between subsequent water or soil additions to paste and extraction.

Jacober and Sandoval conducted a study of the effects of grinding and extraction methods on salt concentration of saturation extracts (*15*). Electrical conductivity of saturation extracts from more finely ground soils were about 0.5 dS/m higher than that for more coarsely ground soils. This information emphasizes the desirability of using standardized grinding and sieving devices, such as a 2-mm screen or 2-mm crushing machine settings. Operators also should use caution so as to not overgrind moderately consolidated materials.

To understand the effect of the quality of saturation percentage measurements upon saturation extract analyses, one must consider the known effects of water content variation upon soil solution composition, then the reproducibility of saturation percentage values themselves.

Theoretically, a soil composed of an inert mineral phase with no ion adsorption or exchange capacity and neutral, highly water-soluble salts should produce water extracts with salt concentrations inversely proportional to extract volumes. That is, the ratio of salt extracted to soil is constant over various soil-to-water ratios. In reality, variations in saturation percentage will produce variations in the amount of various soluble constituents extracted.

As the soil solution becomes more dilute, the sodium adsorption ratio (SAR) decreases, and Ca is adsorbed, lowering the amount of Ca in solution and raising that of Na. Increases in soil solution volumes will increase the amount of sparingly soluble salts, such as $CaCO_3$ and gypsum, that are taken into solution. The amount of conservative anions, such as chloride, appearing in extracted solution will be subject to anion exclusion phenomena that will appear to counteract the effect of dilution in saline soils (*5*).

Table 1 illustrates the various phenomena responsible for changes in soluble constituents as water content varies about the saturation point (*27*). The data show both the change of molar quantities per mass of dry soil per saturation percentage unit change and the change of concentration per saturation percentage unit change.

Soils 1, 4, and 5 have significant $CaCO_3$ contents. They show significant increases of Ca and total electrolyte with increasing saturation percentage values, reflecting dissolution of lime and possibly other sparingly soluble salts. Soil 3, with no $CaCO_3$ equivalent, shows a decrease of Ca with increased saturation percentage, illustrating ion exchange effects. Chloride decreases with increasing saturation percentage are partially the result of anion exclusion effects. This is most apparent in soil 3, which is saline and noncalcic, showing a decrease of both chloride and total salts with increased saturation percentage.

Changes in SAR on a percentage basis for the example soils range from 0.1 to 1.1 percent, with the most saline soil (soil 5) showing the lowest decrease of SAR with saturation percentage increase. Although saline soils undergo larger absolute increases of lime dissolution (*6*), the effect on SAR is greater as total salt concentration decreases (*25*). Electrical conductivity decreases for unit saturation percentage increase range from 1.3 to 3.4 percent. Some of the lower percentage decreases of electrical conductivity (soils 4 and 5) are associated with large increases in the total amount of salt appearing in soil solution on a dry weight basis.

Reproducibility of saturation percentage measurements. According to Rhoades (*28*), saturation percentage measurements generally are reproduc-

Table 1. Change of soluble ion concentration with change of water content at the saturation percentage point for six soils. Data extracted by graphical techniques from Reitemeier (27) and given both on a soil weight basis and a soil solution basis.

Soil	Saturation Percentage	EC (dS/m)	$CaCO_3$ Equivalent (%)	Total Ions (mEq/kg)	Change of Electrolyte per Percent Water Change							
					Total	Na	Ca	Mg	SO_4	Cl	NO_3	Alk*
					mEq/kg/% water x 1,000							
1	63.7	3.6	13.1	25	70	51	7.4	7.7	82	−12	−6.0	6.9
2	41.4	2.9	4.7	13	39	41	0.5	0.09	−3.6	−5.1	−1.7	46
3	44.9	10.5	0	58	−30	93	−95	−27	9.6	−54	−5.4	12
4	40.7	7.6	13.0	37	425	42	267	78	426	−9.3	−25	17
5	45.6	15.3	3.3	90	669	153	339	205	634	−13	−5.0	10
6	28.4	1.3	0.1	3.8	6.8	4.2	1.6	0.3	-	−0.3	−5.7	8.9

Soil	EC (dS/m)	SAR $[(mmol/l)^{1/2}]$	Total Ion Concentration (mEq/l)	Change of Concentration per Percent Water Change							
				Total	Na	Ca	Mg	SAR $[(mmol/l)^{1/2}/\%H_2O]$	SAR $(\%/\%H_2O)$	EC $(dS/m/\%H_2O)$	EC $(\%/\%H_2O)$
				mEq/l/% H_2O							
1	3.6	5.3	40	−0.51	−0.19	−0.21	−0.11	−0.020	−0.4	−0.046	−1.3
2	2.9	33.8	32	−0.65	−0.60	−0.03	−0.01	−0.287	−0.8	−0.063	−2.2
3	10.5	22.5	128	−2.87	−1.79	−0.78	−0.23	−0.104	−0.5	−0.235	−2.2
4	7.6	2.3	91	−1.17	−0.24	−0.22	−0.68	−0.025	−1.1	−0.097	−1.3
5	15.3	9.8	197	−2.79	−1.34	0.17	−1.67	−0.0111	−0.1	−0.217	−1.4
6	1.3	0.75	13	−0.43	−0.04	−0.28	−0.08	−0.007	−0.9	−0.073	−3.4

*Alk = HCO_3 plus CO_3.

ible to 5 percent (as percentage of whole, not as saturation percentage units). Analysis of referee samples by various laboratories appears to be a credible way to test such figures and to gain an appreciation of the nature of variance in saturation percentage and analyses dependent upon saturated paste. Table 2 shows results of two studies that compared analyses of standard samples.

In a study by Sandoval,[1] identified public agency laboratories produced saturation percentage values for the two coarser textured samples (A and B) with standard deviations of about 5 saturation percentage units; for sample C, the standard deviation was 10 saturation percentage units. When expressed as coefficients of variation, the deviations average to 12 percent, double the figure given by Rhoades (*28*). The nonidentified commercial and public laboratories produced analyses with standard deviations of about 10 percent, double the 5 percent value indicated by Rhoades (*28*).

Longenecker and Lyerly's (*18*) data from five identified university laboratories illustrate the increased variation of saturation percentage that may be expected for very coarse or fine-textured soils compared with that of medium-textured soils. Taken as a whole, these data show that saturation percentage measurements routinely may vary by 5 percent to 10 percent in saturation percentage units for materials that are not at textural extremes. Variations in saturation percentages for materials high in clay content or of very coarse texture can exceed 10 saturation percentage units.

Saturation percentage and other soil properties

General relationships. Saturation percentage is the complex aggregation of water adsorption on particle surfaces, capillary retention of water in larger interparticle spaces, and mineral and organic matter particle swelling. Saturation percentage, therefore, can have complex correlations with other soil properties. It should respond to factors dependent upon particle and pore sizes at either end of the range, as opposed to those dependent upon the middle range of sizes.

A study of the comprehensive relationships among physiochemical properties of a balanced set of 33 Israeli soils confirms these complex correlations (Table 3) (*2*). Saturation percentage was highly correlated with properties strongly dependent upon surface area. This included surface area itself, clay content, cation exchange capacity (CEC), and hygroscopic water. Interestingly, saturation percentage was nearly as well correlated with clay percentage as were the much drier water-holding parameters, 15-bar water

[1]Sandoval, F. M., Letter of April 11, 1979, entitled "Report to cooperating laboratories on results of analyses of three standard or referee samples." Agricultural Research Service, U.S. Department of Agriculture, Mandan, North Dakota. 10 pp.

Table 2. Analysis of standard samples for saturation percentage at public and commercial laboratories. Note that range, average, and standard deviation are given in saturation percentage units (percent), while coefficient of variation is given as percentage of the average.

Source, Number and Type of Participating Laboratories	Sample Designation	Sample Properties: Texture	Sample Properties: EC (dS/m)	Saturation Percentage Values: Range	Saturation Percentage Values: Average	Saturation Percentage Values: Standard Deviation	Saturation Percentage Values: Coefficient of Variation (%)
				SP units (%)			
Sandoval:* 4 government agency and university labs, identified, 1 value per lab	A	Clay loam	1	35-45	41.3	4.8	11.6
	B	Clay loam	5	52-68	59.0	6.8	11.5
	C	Clay loam	7	64-85	71.5	9.5	13.3
Sandoval:* 9 public institution and commercial labs, not identified, 1 value per lab	A	Clay loam	1	30-72	46.2	12.0	19.4 (n = 9)
				30-55	43.0	7.7	17.9 (n = 8)
	B	Clay loam	5	46-65	60.0	7.1	11.8
	C	Clay loam	7	50-83	70.1	11.3	16.1
Longenecker and Lylerly (*18*): 5 state agricultural experiment station labs, identified, 2 values per lab	D	Loamy sand	27	24.3-32.3	28.5	9.0	31.4
	E	Silt loam	8	33.8-42.5	39.0	3.6	9.2
	F	Clay	18	65.3-91.5	76.4	9.7	12.7

*Sandoval, F. M., Letter of April 11, 1979, entitled "Report to cooperating laboratories on results of analyses of three standard or referee samples." Agricultural Research Service, U.S. Department of Agriculture, Mandan, North Dakota. 10 pp.

and hygroscopic water. These latter properties are better correlated with soil surface area than saturation percentage.

Saturation percentage was not correlated (Table 3) with properties dependent upon the middle range of particles and pore sizes, as exemplified by the lack of correlation with percent silt. The lack of correlation between saturation percentage and $CaCO_3$ equivalent and organic matter percentage falls within this category. Both saturation percentage and hygroscopic water did not correlate with percent silt. But available water and 1/3-bar water were significantly correlated with percent silt. Saturation percentage, like all the other water-holding indices, was negatively and very significantly correlated with percent sand.

The usefulness of saturation percentage as a predictive and diagnostic index of soil physical condition can be assessed using regression relationships. Table 4 compares regression relationships between saturation percentage and water-holding and sodic-dispersive properties extracted from four data sets. All relationships except those indicated have been examined for significance of quadratic and cubic terms. Equations showing these are given where significant.

For water-holding indices, regressions between saturation percentage and 15-bar water, which is more surface area-dependent, are stronger than for 1/3-bar water or available water. The 1/3-bar water-holding data used here were from disturbed material. As such, the material would be more sensitive to swelling and dispersive tendencies than in situ soil. To use saturation percentage for predictive purposes, regressions with available water are of greatest interest. R-squared values for both of the nonlinear saturation

Table 3. Correlation coefficients for relationships between saturation percentage and other water-holding indexes and various physiochemical properties of 33 Israeli soils (*2*).

Property	Saturation Percentage	Available Water	1/3-Bar Water	15-Bar Water	Hygroscopic Water
Surface area	.865B*	.702B	.886B	.922B	.972B
Cation exchange capacity	.882B	.733B	.885B	.897B	.900B
Calcium carbonate equivalent	−.194	.354A	.183	.033	−.187
Organic matter percentage	.286	.130	.290	.374A	.419A
Clay percentage	.906B	.719B	.881B	.900B	.962B
Silt percentage	.224	.671B	.540B	.379A	.282
Sand percentage	−.847B	−.829B	−.929B	−.895B	−.918B
Hygroscopic water	.899B	.710B	.892B	.924B	
15-bar water percentage	.848B	.738B	.915B		
1/3-bar water percentage	.834B	.911B			
Available water percentage	.686B				

*A = significant at 0.05-level of probability and B = significant at the 0.01-level of probability.

percentage-available water regressions are sufficient for general usefulness (Table 4). Figure 1 shows a scatter diagram for the saturation percentage-available water relationship (*2*).

Saturation percentage should respond to swelling tendency more than any

Table 4. Regression relationships between various soil or mine spoil properties and saturation percentage as the independent variable. Where significant, quadratic or cubic equations have also been displayed.

Dependent Variable	Units	Regression Equation	R^2	n	Source, Material*
15-bar water	Percent by weight	$Y = -1.83 + 0.263X$	.750	191	(*3*), Mollisols†
15-bar water	Percent by weight	$Y = 9.90 + 0.094X$	.289	787	(*3*), mine spoils†
15-bar water	Percent by weight	$Y = 10.1 + 0.0759X$ $Y = -8.81 + 0.664X - 0.00509X^2 + 0.0000125X^3$	.306 .523	92	(*3*), surface mine soils
15-bar water	Percent by weight	$Y = -4.91 + 0.311X$	.824	33	(*2*)
15-bar water	Percent by weight	$Y = -4.09 + 0.305$ $Y = -22.0 + 0.935X - 0.00525X^2$	.803 .855	45	(*26*)
15-bar water	Percent by weight	$Y = 2.31 + 0.1601X$	.763	113	‡
1/3-bar water	Percent by weight	$Y = -2.43 + 0.498X$ $Y = -67.7 + 3.68X - 0.0465X^2 + 0.000208X^3$	.802 .884	33	(*2*)
1/3-bar water	Percent by weight	$Y = 5.23 + 0.404X$ $Y = -72.9 + 3.86X - 0.0470X^2 + 0.000195X^3$	.736 .879	45	(*26*)
Available water	Percent by weight	$Y = 2.47 + 0.187X$ $Y = -49.5 + 2.72X - 0.0368X^2 + 0.000164X^3$	.486 .712	33	(*2*)
Available water	Percent by weight	$Y = 9.32 + 0.0986X$ $Y = -70.0 + 3.94X - 0.0591X^2 + 0.000290X^3$	.271 .660	45	(*26*)
SAR	$(mmol/liter)^{1/2}$	$Y = -3.27 + 0.112X$	.169	191	(*3*), Mollisols†
SAR	$(mmol/liter)^{1/2}$	$Y = 9.50 + 0.259X$	.296	787	(*3*), mine spoils†
SAR	$(mmol/liter)^{1/2}$	$Y = -1.77 + 0.113X$ $Y = -2.82 + 0.00715X + 0.00276X^2 - 0.0000131X^3$	.353 .531	92	(*3*), surface mine soils
SAR	$(mmol/liter)^{1/2}$	$Y = -6.71 + 0.231X$ $Y = 14.7 - 0.754X + 0.0131X^2 - 0.0000528X^3$	.742 .781	102	‡
Log (HC) undistributed	(cm/day)	$Y = 3.15 - 0.0520X$ $Y = 4.22 - 0.0887X + 0.000260X^2$	.684 .696	406	‡§
Log (HC) distributed	(cm/day)	$Y = 1.61 - 0.0298X$ $Y = -0.276 + 0.0616X - 0.00129X^2 + 0.00000548X^3$	.733 .750	406	‡§

*Sources: Bauer et al. (*3*), 3 datasets, North Dakota sites; Banin and Amiel (*2*), balanced, comprehensive set of 33 Israeli soils; Reeve et al. (*26*), predominantly saline, irrigated Utah soils, three sites; Doering and Merrill (unpublished), North Dakota sites, disturbed soils and mine spoils.
†These relationships were not tested for quadratic or higher regressions.
‡Unpublished data, USDA-ARS, Mandan, North Dakota.
§The dependent variables for these regressions are decimal logs of undisturbed and disturbed hydraulic conductivity.

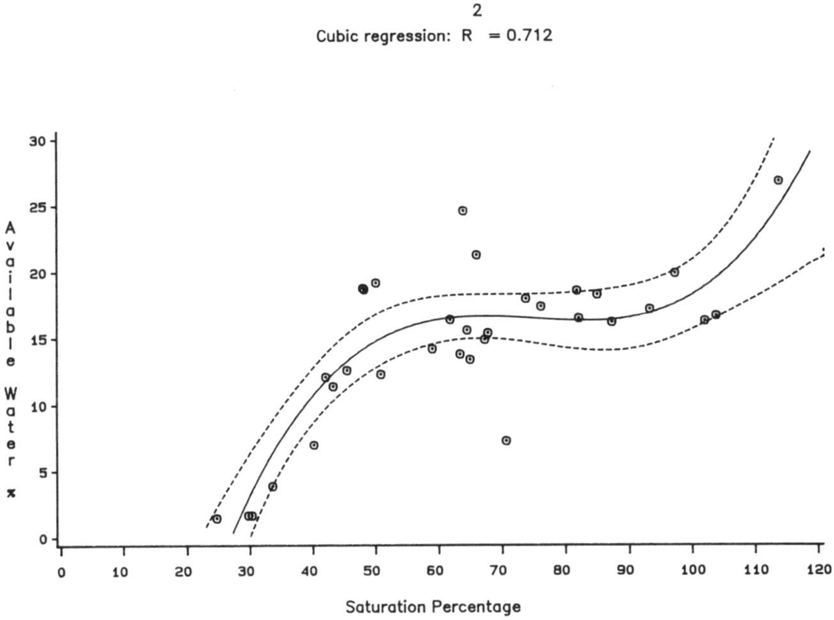

Figure 1. Regression relationship between saturation percentage and available water content for data set of Banin and Amiel (2) consisting of 33 Israeli soils. Dashed lines show 95 percent confidence limits.

other water-holding parameter. Standard methods for saturated paste preparation permit swelling, while removing structural determinants that influence water-holding indexes of in situ soil. To the extent that SAR represents swelling tendencies of sodicity, there should be a firm regression relationship between SAR and saturation percentage. Soil clays can differ greatly in their tendency to disperse (*21*) or swell (*22*) under high SAR, low electrolyte conditions. Out of seven soils examined, the extracted soil clay of three soils did not exhibit macroscopic swelling (*22*). These soils all contained 25 percent or less montmorillonite in the clay fraction. Thus, saturation percentage can be an even more direct indicator of dispersive/swelling conditions than SAR. For predictive purposes, it would seem appropriate to use ratios of saturation percentage to clay percentage because saturation percentage also is strongly correlated with clay content, as discussed above.

Regressions between saturation percentage and SAR are of particular interest in mined-land reclamation because SAR commonly is used as the prime indicator of sodic hazard. The range of R-squared values for the saturation percentage-SAR regressions is considerable (Table 4). All are highly significant as correlations. Figure 2 shows SAR-saturation percentage rela-

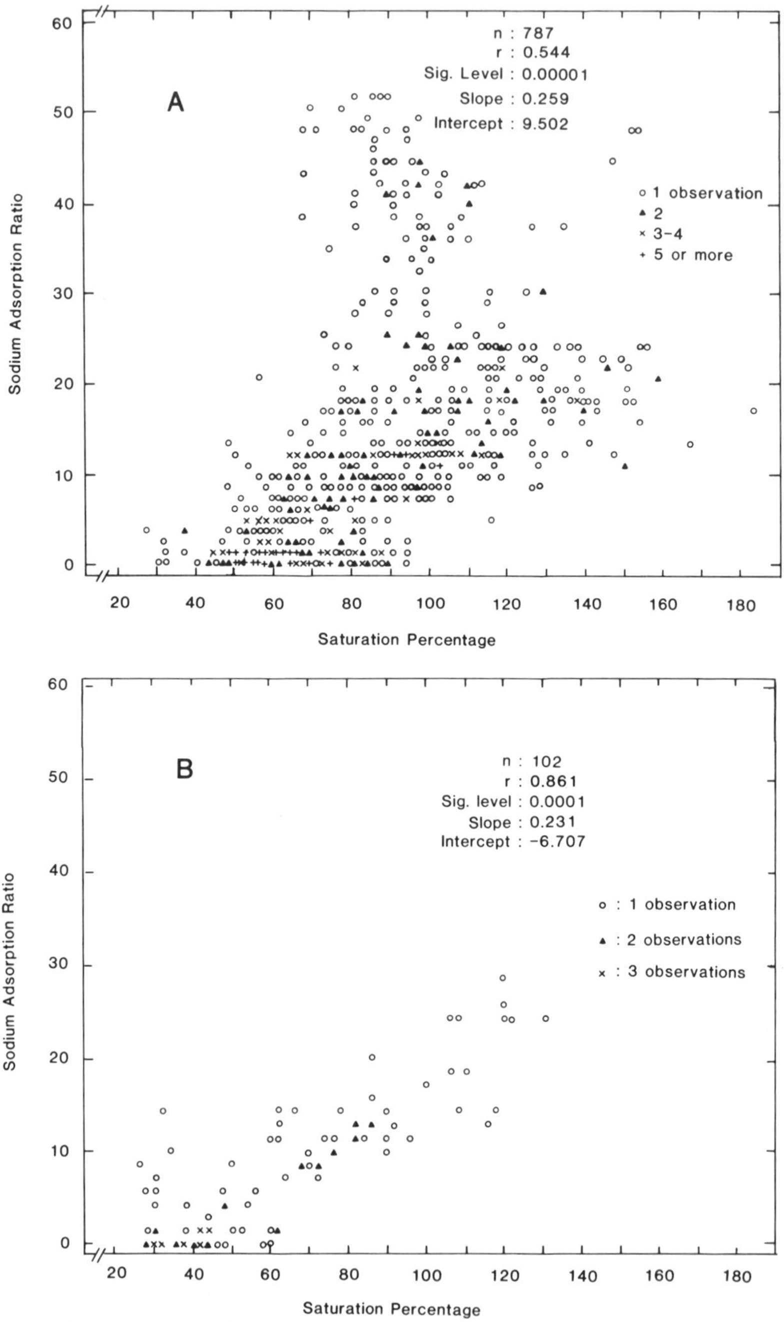

Figure 2. Scatter plots of SAR versus saturation percentage: A. Data from mine spoil samples of Bauer et al. (3); B. Data from disturbed soils and mine spoils from Doering and Merrill (unpublished, USDA-ARS, Mandan, North Dakota).

tionships for two data sets. Data from Doering and Merrill[2] (Figure 2B) show considerably higher R-squared values than the data from Bauer and associates (*3*). The data of Bauer appear to contain a wider range of clay mineralogical types than that of Doering and Merrill. The scatter diagram for the Doering and Merrill data indicate that the materials were more uniform in their swelling response to elevated SAR. The two different data sets came from the same general combination of sites, but mine spoil materials sampled by Bauer and associates included samplings that were deeper in the profiles, apparently resulting in greater physical heterogeneity of the materials. The SAR-saturation percentge regressions generally appear strong enough to recommend saturation percentage as a useful subsidiary indictor of sodicity, but without real quantitative reliability.

The Doering and Merrill study provides data not found elsewhere, in particular, relationships between saturation percentage and hydraulic conductivity. While SAR is an indicator of sodicity hazard, hydraulic conductivity is an important indicator of the physical consequences of sodicity in soils. The hydraulic conductivity-saturation percentage regressions (Table 4) have highly significant R-squared values, both of which are close to the R-squared values of the SAR-saturation percentage regression of this data set. Thus, saturation percentage can be a useful, subsidiary indicator of swelling effects in soil and the low hydraulic conductivity associated with such effects.

Ratios of saturation percentage to other water-holding indices. One of the claims made for the usefulness of saturation percentage as an index of soil water-holding capacity is that there is an approximate relationship between saturation percentage and both 1/3-bar water, which approximates field capacity, and 15-bar water, which is the wilting point. The ratios of saturation percentage to 1/3-bar water and 15-bar water are about 2 and 4, respectively (*37*). Table 5, which shows univariate statistics for ratios of saturation percentage to 1/3-bar, 15-bar, and other water-holding indices, indicates the reliability of these relationships. Also shown are statistics on saturation percentage data for each data set.

Median values for the ratios between saturation percentage and 15-bar water are all above 4, with the comprehensive but sodicity-affected data set of Doering and Merrill having a median ratio of 4.82 (Table 5). First decile values of the saturation percentage to 15-bar water ratios are considerably nearer the *Agriculture Handbook 60* value of 4 (*37*) than are last (90%) decile values. Median values of ratios between saturation percentage and 1/3-bar water for two data sets are near 2, with first decile values of about 1.7 and last decile values of 2.5 and 3.8 for the two data sets. Ratios be-

[2]Doering, E. J., and S. D. Merrill. Unpublished data. Agricultural Research Service, U.S. Department of Agriculture, Mandan, North Dakota.

tween saturation percentage and other water-holding indexes vary with both clay percentage and swelling tendency. Thus, ratios of saturation percentage to lower water-holding indices generally are higher for coarse-textured materials and lower for nonswelling, high-clay-content material. This is shown by the regression between the saturation percentage-to-15-bar water ratio (SP/W15B) and clay percentage (CLAY) for Banin and Amiel's data (*2*), less one outlier, $n = 32$: SP/W15B = 7.67 − 0.749*CLAY, with an R-squared value of 0.449. In summary, these data indicate that commonly assumed values of ratios between saturation percentage and other water-holding indexes are useful if textural and swelling-tendency extremes are not involved.

The ratios between saturation percentage and undisturbed, saturated water contents are of interest because they provide a comparison between

Table 5. Ratios between saturation percentage and other water-holding indexes displayed as univariate statistics for four data sets.

Ratio	Source	n	First Decile	Median	Last Decile	Standard Deviation	Coefficient of Variation (%)
SP/hygroscopic water	Banin and Amiel (*2*)	33	8.28	12.64	21.08	10.9	117.8
SP/15-bar water	Banin and Amiel (*2*)	33	3.36	4.28	6.69	4.97	88.3
SP/15-bar water	Bauer et al. (*3*)	92	3.35	4.31	6.81	1.65	35.1
SP/15-bar water	Reeve et al. (*26*)	45	3.51	4.45	5.73	0.89	19.4
SP/15-bar water	Doering and Merrill*	102	3.93	4.82	6.96	1.44	28.1
SP/1/3-bar water	Banin and Amiel (*2*)	33	1.67	2.19	3.78	1.84	68.7
SP/1/3-bar water	Reeve et al. (*26*)	45	1.63	2.02	2.47	0.28	13.9
SP/undisturbed, saturated water	Doering and Merrill*	102	1.05	1.22	1.56	0.22	17.4

Reference	Materials	Saturation Percentage Statistics n	10%	Median	90%
Banin and Amiel (*2*)	33 Israeli arid zone soils	33	31.7	64.5	100.1
Bauer et al. (*3*)	Surface mine spoils and soils, North Dakota	92	47.3	64.0	115.0
Reeve et al. (*26*)	Irrigable, saline soils, 3 sites, Utah	45	39.3	50.7	74.7
Doering and Merrill*	Reconstructed mine soils, North Dakota	102	30.2	48.0	106.6

*Unpublished data, USDA-ARS, Mandan, North Dakota.

the influence of unconfined swelling in saturated paste, the saturation percentage, and the largely confined swelling of in situ soil samples in core retainers. The first decile value of this ratio (Table 5) approaches a value of 1, while the last decile value, 1.56, indicates considerable unconfined swelling in saturated pastes.

Role of saturation percentage in sodic hazard assessment. Saturation percentage has an important role in sodic hazard assessment that is seldom recognized. Exchangeable sodium percentage (ESP) most directly indicates the effective level of exchangeable sodium in soil and, hence, the potential for sodic hazard. Because of analytical costs and reproducibility problems, scientists usually use SAR instead of ESP for sodic hazard evaluation. A relationship between SAR and ESP must be assumed to use SAR for sodicity assessment. The saturation percentage of the saturated paste used for the cation analyses from which SAR is calculated directly and significantly affects the SAR value. The swelling characteristics of soil materials strongly affect saturation percentage and SAR values and, thus, affect empirical ESP-SAR relationships.

The ESP-SAR relationship is better understood by using exchangeable sodium ratio (ESR) instead of ESP.[3] To understand ESR-SAR relationships, a clear distinction must be made between data sets obtained by separate determinations of ESR and SAR on samples, with the SAR determined at saturation percentage soil water content, and ion-exchange measurements, in which ESR and SAR are determined in an effectively simultaneous manner on soil suspensions. The former data sets may be labelled "field" relationships. The latter data, which give more fundamental information about ESR-SAR relationships, are best labelled "ion-exchange experiments."

The mass action principle is the basis for expecting a linear relationship between ESR and SAR. The slope of an ESR-SAR regression, the ESR/SAR ratio, is also known as the Gapon selectivity coefficient (KG). The best known and most widely used field-type regression between ESR and SAR was based on 59 selected soils (*37*). The KG value of the *Agriculture Handbook 60* regression is 0.01475 $(\text{mmol/liter})^{1/2}$.

Although the *Agriculture Handbook 60* value is used widely to assess sodic hazard, its use has come under doubt and criticism, particularly in the field of mined-land reclamation. For example, Shay and Parady reported KG values of 0.01297 and 0.0043 for ESR/SAR relationships from field-type data for two different types of mine spoil (*33*). They attributed the low value to dominance by kaolinitic clay mineralogy. In contrast, Frenkel and Alperovitch found that KG values for coarse-textured, calcareous, arid

[3]ESR is the ratio of exchangeable Na to exchangeable divalent cations and is thus related to ESP as ESR = ESP/(100-ESP).

zone soils tended to be significantly greater than the *Agriculture Handbook 60* value (*13*). For coarse-textured, nonsaline soils, many of the KG values exceeded 0.01475 and ranged up to 0.1 or more. In terms of sodic hazard assessment, KG values greater than the standard value of 0.0148 indicate that use of the *Agriculture Handbook 60* value to estimate ESP from SAR would underestimate the sodic hazard. KG values less than the standard would lead to overestimation of the hazard.

Generally, saturation percentage affects field-type ESR-SAR relationships through the dilution effect. As the soil solution becomes more dilute, that is, as the saturation percentage value increases, the SAR value decreases. This tendency is reduced in materials that are low in salts and sparingly soluble minerals, such as lime and gypsum. Most western mine spoils with sodic tendencies, however, are moderately saline and typically calcareous. Materials dominated by swelling-type clay minerals will have higher saturation percentage values at a given level of ESP. Higher saturation percentage levels will result in lower SAR values and, thus, KG values will be higher for these materials. Conversely, materials dominated by clay minerals with low swelling potential will have relatively low saturation percentage values. At given ESP levels, SAR values will be relatively higher and KG values correspondingly lower. Shay and Parady's relatively low KG value of 0.0043 for field-type ESR-SAR data for kaolinitic-dominated minespoil materials undoubtedly reflects this latter saturation percentage effect on ESR-SAR relationships (*33*).

In contrast to the general variation of KG values for different field-type ESR-SAR data sets, KG values do not vary as much for various ion-exchange measurements of soils, mine spoils, and clay mineral specimens. Saturation percentage is not a factor in ion-exchange experiments because of the functionally simultaneous determination of ESR and SAR, practically achieved through use of a wide water-to-soil ratio. Bower performed ion-exchange measurements on six soils and specimen clays, finding initial KG values ranging from 0.013 to 0.021 (*8*). His data closely followed predictions of the diffuse double-layer theory. According to that theory, surface charge density—the ratio of CEC to surface area—is the major determinant of the ion-exchange coefficient. Bower's work demonstrated that KG values in ion-exchange measurements are affected by surface charge density and not differences in clay mineralogy, which lead to differences in swelling potential and, hence, to differences in saturation percentage values.

A factor of lesser practical importance is salt concentration. Two studies have shown that increasing salt concentration decreases KG values of ESR-SAR relationships (*10, 16*). A considerable portion of this concentration effect can be quantitatively predicted. Other ion-exchange experiments have produced KG values for various materials that varied from 0.008 to 0.023. Shainberg and associates found KG values of 0.011 and 0.023 for montmorillonitic and illitic specimen clays, respectively (*32*). Jurinak and associates determined average KG values ranging from 0.0079 to 0.0097 for one soil and two mine spoils (*16*). Doering and associates measured a KG value of 0.015 for a mine spoil at low salt concentration (*10*).

In summary, assessing sodic hazard from SAR data can be improved significantly by using computational procedures that would predict the effect of saturation percentage on SAR. The approach would be to calculate an effective SAR value based on a normalized saturation percentage value, which, in turn, would be derived from the clay content of the sample in question. Recent examples of graphical and algorithmic summarization of complex soil-chemical calculations (*36*) show that reasonably simple and accessible computational procedures can be developed for these purposes. Use of an effective SAR value, calculated by an appropriately constructed procedure, should yield ESP estimates that are considerably less subject to the inaccuracies of prediction caused by variations in saturation percentage value among different soil and mine spoil materials. Development of procedures to make such improved sodic hazard estimates is an important research priority.

Saturation percentage in western mine spoil reclamation

During the past 15 years, researchers have accumulated experience in applying saturation percentage analysis to mine spoil reclamation in western North America. Correlations of saturation percentage with water-holding capacities, soil texture, and swelling tendency provide valuable information to reclamation planners. Where suitable plant growth materials are limited, saturation percentage can be used to evaluate overburden materials as soil substitutes. Saturation percentage is useful in understanding the relation of SAR to exchangeable Na, which is important in assessing sodic hazard.

Dollhopf and associates studied water and salt movement and plant growth on a sodic mine spoil site that had been reclaimed by soil respreading (*11*). While the SAR of the spoil was 25, its saturation percentage was 75, somewhat lower than would be typically associated with that SAR level. The mine spoil was largely kaolinitic. The researchers observed significant water movement from the covering topsoil into mine spoil. As a result, they recommended that material with a saturation percentage greater than 90 percent should not be placed in the root zone.

Merrill and associates studied salt movements at four sites on which non-sodic soils were spread over sodic, montmorillonite-dominated mine spoils (*23, 24*). Saturation percentage of mine spoil at two sites averaged 83 percent and 92 percent. At those sites, seasonal leaching events reduced upward Na movement. At two other sites, with mine spoil saturation percentage values of 113 percent and 134 percent, Na and salt accumulation in covering soil was more stable and potentially more deleterious. These results, therefore, supported use of a saturation percentage value of about 95 percent as a guideline for separating more deleterious from less deleterious sodic mine spoils.

Schafer, reviewing methods to delimit materials suitable for soil reconstruction, recommended use of materials only within the saturation percentage range of 25 percent to 80 percent (*30*). In the same review Schafer recommended that procedures outlined in *Agriculture Handbook 60* be used for analytical standards in reclamation usage.

The Forest Service also has recommended saturation percentage analysis as one of eight analytical tests for characterizing mined-land overburden (*14*), especially as an indicator of sodic hazard.

Summary and conclusions

Saturation percentage is the amount of water in a saturated soil paste on a dry weight basis. Because saturation percentage is the lowest convenient water content for extracting soil solution for analysis, use of saturation percentage as a parameter has evolved from its coincidental availability. It is useful as an index for gauging the dilution factor in saturation extract analyses and as an indicator of soil water-holding capacity and swelling potential.

The standard method for preparing a saturated paste is to stir water into the soil sample until a saturated condition is attained, as judged by such criteria as glistening appearance and minimum ability to flow. To assure that saturation percentage values are reproducible and fully indicate swelling tendency, the saturated paste should sit overnight before rechecking saturation criteria. Extraction for chemical analyses should not be performed until at least 4 hours after the last addition of water or soil. An oversaturation technique will prevent puddling and save time in saturated paste preparation. A Soil Conservation Service report provides the best reference for saturation percentage methods for mined-lands analysis (*35*).

Paste preparation by capillary wetting in theory should remove much of the effect of operator skill on saturation percentage results. Saturation percentage values from capillary wetting generally match values from conventional paste preparation for moderate- to coarse-textured materials, but not for fine-textured materials. Saturation percentage values from capillary wetting are more sensitive to aggregate size and degree of sample grinding than are those from conventional paste preparation. Thus, capillary wetting cannot be recommended as a general replacement of the current conventional technique for saturated paste preparation.

Methods of soil sample handling and saturated paste preparation can affect subsequent chemical analyses of the extract and the saturation percentage itself. Studies have shown that extracted soil solution is at 95 percent or greater of final salt concentration within 1 to several hours of initial paste preparation. Further increases in concentration occur much more slowly.

Saturation percentage will affect SAR and total ion concentration rather differently in soils of different chemical type. In noncalcic, less weatherable soils, SAR will decrease much less as saturation percentage values increase than simple dilution would indicate, while electrical conductivity changes will approximate those indicated by dilution. In calcic, more weatherable materials, dissolution of lime and other soil minerals will maximize the decline of SAR as saturation percentage increases.

Studies of the reproducibility of saturation percentage values indicate that variations of 5 to 10 saturation percentage units (percent) may be expected in the middle textural range of soil materials. For very fine- and coarse-textured materials and for materials that have significant swelling potential, variations of 10 saturation percentage units or more can be expected.

Saturation percentage is strongly correlated with clay percentage, hygroscopic water, CEC, 15-bar water retention, surface area, macroscopic swelling, and percent sand. It is uncorrelated or only weakly correlated with percent organic matter, percent silt, and $CaCO_3$ equivalent. Saturation percentage appears to be correlated strongly enough with available water so that regression equations applied to saturation percentage for prediction of available water are generally useful.

Because SAR is used commonly as an indicator of sodic hazard, saturation percentage-SAR relationships are of great interest to mined-land reclamationists. The strength of saturation percentage-SAR correlations varies among data sets. Such regression relationships are useful for general indications, but not quantitative predictions. Soils and overburdens of different clay-mineralogical types vary greatly in their swelling response to sodic conditions, resulting in the divergence of SAR-saturation percentage regressions. Because saturation percentage is an excellent practical indicator of one of the most important physical manifestations of sodic hazard, soil swelling, it should be used along with SAR to assess sodic hazard.

As a convenient approximation, the saturation percentage value may be assumed to be 2 times the 1/3-bar water retention (field capacity) and 4 times the 15-bar water retention (wilting point). Actual ratios will be somewhat lower for nonswelling, finer textured material, and up to 50 percent or more greater for coarse-textured or high-swelling materials. The ratio of saturation percentage to 15-bar water retention is significantly and negatively correlated with the clay content of soil materials.

While ESP is a more direct measure of sodic hazard, the analytically more economical and practical SAR is commonly used instead. Evaluations of sodic hazard frequently are made by relating SAR to ESP (or ESR, an alternative expression of ESP) by a value published in *Agriculture Handbook 60*. But many soils and overburdens have ESR/SAR ratios different from the standard value. Because the saturation percentage values of materials reflect their tendency to swell and because this significantly affects

SAR values and hence ESR/SAR relations, more accurate predictions of ESP from SAR could be made if saturation percentage and chemical data were used to calculate the effect of saturation percentage on SAR and the effect of total salt concentration on ESR/SAR ratio. This can be accomplished by developing procedures to calculate an effective SAR value from a normalized saturation percentage value, based on sample clay content.

Use of saturation percentage for evaluating soil, overburden, and mine spoils in mined-land studies has been recommended by a number of authors. Saturation percentage is a useful parameter for detecting low water-holding capacity and clarifying the degree of sodic hazard as indicated by SAR. Several authors have recommended that 25 percent saturation percentage be used as an indicator point for significantly low water-holding capacity, and that 80 to 95 percent saturation percentage be used as an indicator point for swelling tendency associated with sodic hazard.

REFERENCES

1. Allison, L. E. 1973. *Oversaturation method for preparing saturation extracts for salinity appraisal.* Soil Science 116: 65-69.
2. Banin, A., and A. Amiel. 1969. *A correlative study of the chemical and physical properties of a group of natural soils of Israel.* Geoderma 3: 185-198.
3. Bauer, Armand, Glendon W. Gee, and John E. Gilley. 1976. *Physical, chemical and biological aspects of reclamation of strip-mined lands in western North Dakota.* Old West Regional Commission, Billings, Montana. 605 pp.
4. Beatty, H. J., and J. Loveday 1974. *Soluble cations and anions.* In J. Loveday [editor] *Methods for Analysis of Irrigated Soils.* Technical Communication No. 54. Commonwealth Agricultural Bureaux, Farnham Royal, Bucks, England. pp. 108-117.
5. Bolt, G. H., and M.G.M. Bruggenwert. 1976. *Soil chemistry. A. Basic elements.* Elsevier Scientific Publishing Company, Amsterdam, The Netherlands. 281 pp.
6. Bresler, E., B. L. McNeal, and D. L. Carter. 1982. *Saline and sodic soils. Principles-dynamics-modeling.* Springer-Verlag, Berlin, West Germany. 236 pp.
7. Campbell, R. B., and L. A. Richards. 1950. *Some moisture and salinity relationships in peat soils.* Agronomy Journal 42: 582-585.
8. Bower, C. A. 1959. *Cation-exchange equilibria in soils affected by sodium salts.* Soil Science 88: 32-35.
9. Bower, C. A., and L. V. Wilcox. 1965. *Soluble salts.* In C. A. Black [editor] *Methods of Soil Analysis. Part 2. Chemical and Microbiological Properties.* American Society of Agronomy, Madison, Wisconsin. pp. 933-951.
10. Doering, E. J., W. O. Willis, and S. D. Merrill. 1982. *Effect of total cation concentration and sodium absorption ratio on exchangeable sodium ratio.* Canadian Journal of Soil Science 62: 177-185.
11. Dollhopf, D. J., E. J. DePuit, and M. G. Klages. 1980. *Chemical amendment and irrigation effects on sodium migration and vegetation characteristics in sodic mine soils in Montana.* Bulletin 736. Montana Agricultural Experiment Station, Bozeman. 109 pp.
12. Frenkel, H., C. Amrhein, and J. J. Jurinak. 1983. *The effect of exchangeable cations on soil mineral weathering.* Soil Science Society of American Journal 47: 649-653.
13. Frenkel, H., and N. Alperovitch. 1984. *The effect of mineral weathering and soil*

solution concentration on ESR-SAR relationships of arid and semi-arid zone soils from Israel. Journal of Soil Science 35: 367-372.

14. Intermountain Forest and Range Experiment Station. 1979. *User guide to soils. Mining and reclamation in the West.* General Technical Report INT-68. Forest Service, U.S. Department of Agriculture, Logan, Utah. 80 pp.
15. Jacober, F., and F. Sandoval. 1971. *Effect of soil grinding, suction, and extraction time on salt concentration of saturation extracts.* Soil Science 112: 263-266.
16. Jurinak, J. J., C. Amrhein, and R. J. Wagenet. 1984. *Sodic hazard: The effect of SAR and salinity in soils and overburden materials.* Soil Science 137: 152-159.
17. Jurinak, J. J., J. C. Whitmore, and R. J. Wagenet. 1977. *Kinetics of salt release from a saline soil.* Soil Science Society of America Journal 41: 721-724.
18. Longenecker, D. E., and P. J. Lyerly. 1964. *Making soil pastes for salinity analysis: A reproducible capillary procedure.* Soil Science 97: 268-275.
19. Loveday, J. 1972. *Moisture content of soils for making extracts and the effect of grinding.* Division of Soils Technical Paper No. 12. Commonwealth Scientific and Industrial Research Organization, Melbourne, Australia. 9 pp.
20. Magistad, O. C., R. F. Reitemeier, and L. V. Wilcox. 1945. *Determination of soluble salts in soils.* Soil Science 59: 65-75.
21. McNeal, B. L., and N. T. Coleman. 1966. *Effect of solution composition on soil hydraulic conductivity.* Soil Science Society of America Journal 30: 308-312.
22. McNeal, B. L., W. A. Norvell, and N. T. Coleman. 1966. *Effect of solution composition on the swelling of extracted soil clays.* Soil Science Society of America Proceedings 30: 313-317.
23. Merrill, S. D., E. J. Doering, and J. F. Power. 1980. *Changes of sodicity and salinity in soils reconstructed on strip-mined land.* North Dakota Agricultural Experiment Station Farm Research 37(6): 13-16.
24. Merrill, S. D., E. J. Doering, J. F. Power, and F. M. Sandoval. 1983. *Sodium movement in soil-minespoil profiles: Diffusion and convection.* Soil Science 136: 308-316.
25. Nadler, A., and M. Magaritz. 1981. *Expected deviations from the ESP-SAR empirical relationships in calcium- and sodium-carbonate-containing arid soils: Field evidence.* Soil Science 131: 220-225.
26. Reeve, R. C., L. E. Allison, and D. F. Peterson, Jr. 1948. *Reclamation of saline-alkalai soils by leaching—Delta area, Utah.* Bulletin 335. Utah Agricultural Experiment Station, Logan. 53 pp.
27. Reitemeier, R. F. 1946. *Effect of moisture content on the dissolved and exchangeable ions of soils of arid regions.* Soil Science 61: 195-214.
28. Rhoades, J. D. 1982. *Soluble salts.* In A. L. Page et al. [editors] *Methods of Soil Analysis. Part 2. Chemical and Microbiological Properties.* American Society of Agronomy, Madison, Wisconsin.
29. Sandoval, F. M., and J. F. Power. 1978. *Laboratory methods recommended for chemical analysis of mined-land spoils and overburden in western United States.* Agriculture Handbook No. 525. Science and Education Administration, U.S. Department of Agriculture, Washington, D.C. 31 pp.
30. Schafer W. M. 1979. *Guides for estimating cover-soil quality and mine soil capability for use in coal strip-mine reclamation in the western United States.* Reclamation Review 2: 67-74.
31. Schofield, C. S. 1932. *Measuring the salinity of irrigation waters and of soil solutions with the Wheatstone bridge.* Circular 232. U.S. Department of Agriculture, Washington, D.C. 9 pp.
32. Shainberg, I., J. D. Oster, and J. D. Wood. 1980. *Sodium/calcium exchange in*

montmorillonite and illite suspensions. Soil Science Society of America Journal 44: 960-964.

33. Shay, Dale A., and Fred E. Parady. 1983. *Significance of refinement in the state-of-the-art of predicting sodic hazard in soils and overburden.* In Edward J. DePuit and Gerald E. Schuman [editors] *Proceedings of Symposium, Soils and Overburden in Reclamation of Arid/Semiarid Mined Lands.* University of Wyoming, Laramie. pp. 11-14.
34. Soil Conservation Service. 1972. *Soil survey laboratory methods and procedures for collecting soil samples.* Soil Survey Investigations Report No. 1. U.S. Department of Agriculture, Washington, D.C. 63 pp.
35. Soil Conservation Service. 1984. *Procedures for collecting soil samples and methods of analysis for soil survey.* Soil Survey Investigations Report No. 1. U.S. Department of Agriculture, Washington, D.C. 68 pp.
36. Suarez, D. L. 1982. *Graphical calculation of ion concentrations in calcium carbonate and/or gypsum soil solutions.* Journal of Environmental Quality 11: 302-308.
37. U.S. Salinity Laboratory Staff. 1954. *Diagnosis and improvement of saline and alkalai soils.* Agriculture Handbook No. 60. U.S. Department of Agriculture, Washington, D.C. 160 pp.

Chapter 4

Particle size distribution

Burns Sabey, Jim Herron, David G. Scholl, and John Bokich

Particle size distribution in soils, spoils, or tailings has a profound effect on the success of reclaiming mined or other drastically disturbed land just as it does on managing any agronomic area. Many environmental factors that influence plant growth in a developing ecosystem are governed, or at least affected, by particle size distribution.

Critical factors in revegetation and management of disturbed or degraded sites that are affected by particle size are moisture, temperature, and air relations, as well as physical impedance of roots, chemical reactivity, and erosion.

In most soil, spoil, or tailing materials, particles are classified as primary and secondary. Primary particles are individual sands, silts, and clays that exist separately with no forces of attraction binding two or more of the particles together. Secondary particles form when two or more primary particles are bound or attracted together by physical or chemical forces.

The relative amounts of primary sand, silt, and clay particles usually is referred to as soil texture. The texture of mineral materials is classified by most scientific groups concerned with mineral particles, but the size limits of sands, silts, and clays vary. Figure 1 shows some of the variations among these classification systems. The U.S. Department of Agriculture classification is used most often by persons concerned with plant growth. Sands are characterized by relatively large particle size, 0.05 mm to 2 mm in diameter; a small specific surface, that is, surface area per unit weight; low chemical activity; lack of plasticity (moldability); low total pore space, consisting of mostly macro pores; adequate aeration; low water-holding capacity; a

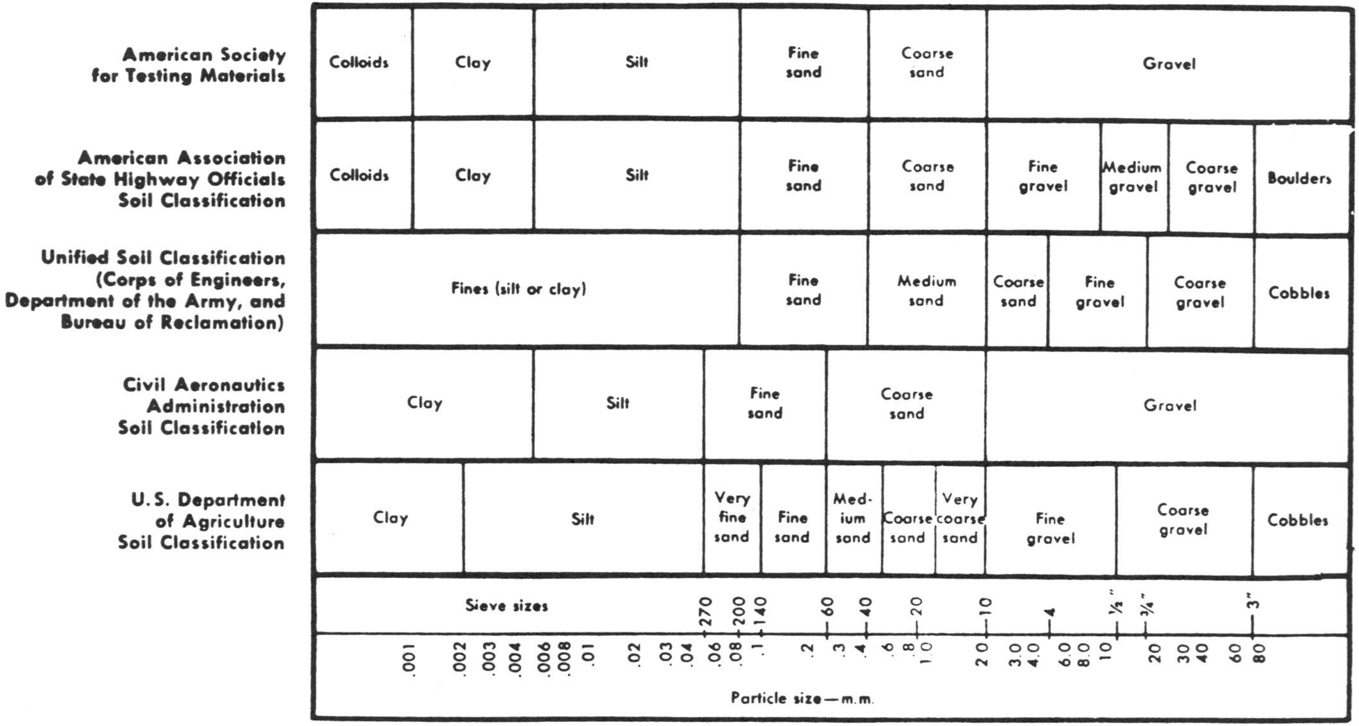

Figure 1. Comparison of some of the textural classification systems used in the United States. Modified diagram reprinted with permission of the Soil Primer, Portland Cement Association, 33 West Grand Avenue, Chicago, Illinois.

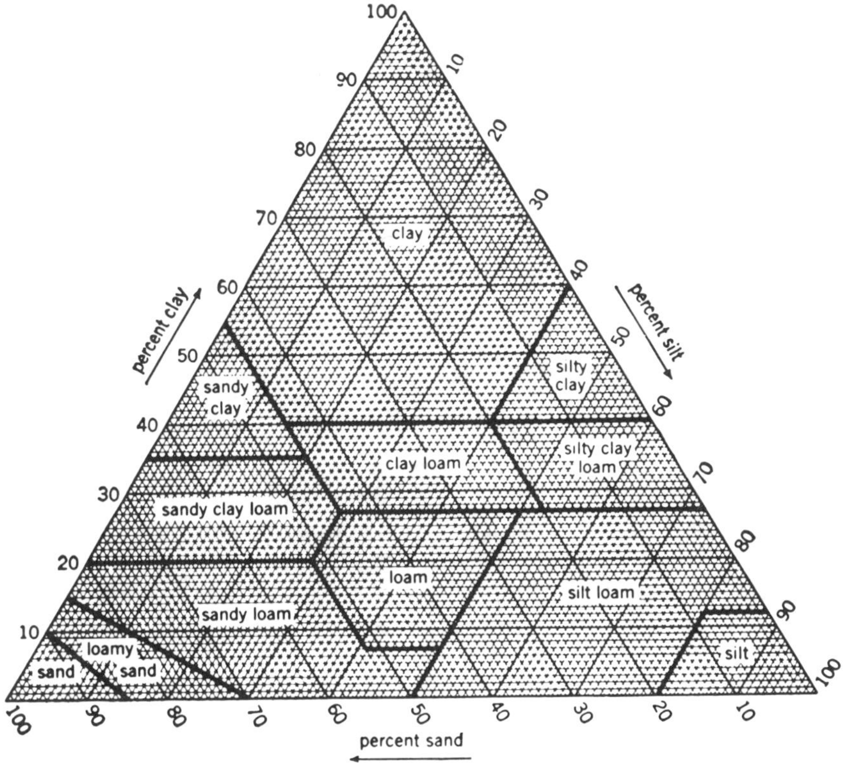

Figure 2. Textural classification guide, U.S. Department of Agriculture.

rough, coarse, gritty feel; and few plant available nutrients. Clays, on the other hand, have the following characteristics: small particle size, less than 0.002 mm in diameter; platelike shape; large specific surface; harsh and hard when dry; plastic and moldable when moderately wet and sticky when very wet; high total pore space, usually with many very small pores; high water-holding capacity and often poorly drained and aerated; chemically and physically active when moist; and many plant available nutrients if not too weathered. Silt particles are smooth, powdery, floury, and talc-like to the feel and have properties intermediate between sand and clay.

If the percentages of sand, silt, and clay are known, the soil texture can be classified into several categories using the USDA textural triangle (Figure 2).

There is not an optimum texture for all purposes and all plants because requirements vary greatly. But a plant growth medium with enough sand to allow for aeration and looseness to permit plant root growth and development and enough clay for adequate nutrient and water-holding capacity would be ideal for most situations.

Unconsolidated sand, silt, and clay particles develop as a result of weathering and physical degradation of larger consolidated rocks and minerals. This physical weathering, due to temperature changes, water freezing and the resultant forces of expansion, abrasion of wind and water, and other mechanisms, tends to split and crumble larger rocks and minerals. As smaller particles are produced, not only does physical weathering occur more rapidly, but chemical weathering also occurs. The physical and chemical forces produce the sand, silt, and clay particles that are so necessary for adequate plant growth and revegetation.

Rocks and minerals weather at differing rates, depending on internal and external conditions. Some, such as sandstones, form sands initially, while shales tend to produce silts and clays. A perusal of an elementary geology or mineralogy text will indicate some of the variations that exist in the ease of weathering and the resulting proportion of sand, silt, and clay products.

Course fragments (pebbles and rocks) influence growth of plants, but to a lesser extent than smaller particles (< 2mm size). These larger fragments contribute relatively few nutrients when considered on a short-term basis, because the specific surface is small. They do, however, influence physical properties of the soil or spoil, thus affecting plant growth. With sufficient fine particles, however, even with a relatively high proportion of coarse pebbles and rocks in the soil or spoil, vegetation can still be established.

Soil structure results when primary sand, silt, and clay particles are bound or cemented together to form secondary or compound particles. Soil scientists classify soil structure according to dominant patterns of cleavage plane formation—the shape of particle—in a naturally developed soil profile. The major categories in soil structure classification are granular or crumb (porous), plate-like, block-like, and prismatic or columnar. Generally, the granular or crumb structure is optimum for plant growth. Otherwise, these structural categories may have little significance to revegetation, except that plant roots grow in the cracks along the cleavage planes.

Although scientists have gained some knowledge about the kinds of forces that attract, bind, or cement primary particles together, there is much to be learned about the exact mechanisms of soil formation and genesis that result in variations in structure. Furthermore, the relationship between soil structural types and plant growth are vague and qualitative. Quantitative relationships are difficult to formulate and measure. Measurements of bulk density (weight per unit volume), which reflects the relative proportion of the heavier minerals to the lighter airspace (pore space); penetrability, which reflects the resistance to disruption of the secondary particles due to an outside force; and aggregate stability, measured by agitation in a water bath, indicate the influence of soil structure on plant growth. Most soil physicists would caution that these relationships are useful but are not of adequate reliability. Such measurements may give an indication of the effect that secondary particles will have on plant growth, but that is about all.

Soil structure is important in plant growth because it influences soil moisture, temperature, and air, which in turn affect other soil-degrading and/or soil-developing processes. But sound, quantified relationships are rare. On the other hand, the relationships between soil texture and plant growth are considerably more definitive.

Exchangeable cations can affect aggregate formation and soil structure. Clays with large quantities of exchangeable Ca generally will be flocculated, a stepping stone to aggregate formation. Sodium, however, promotes deflocculation or dispersion of colloidal particles and thus has a degrading influence on structure. The effect of Na on the structure of coarse-textured soil and spoil is minimal. But as Na accumulates it can decrease greatly the infiltration and permeability of fine-textured materials. Bernstein (*1*) found that when exchangeable Na exceeds about 10 percent of the cation exchange capacity (CEC) of fine-textured soils and about 20 percent of coarser textured soils, soil structure deteriorates and permeability decreases.

Moisture

Soils, spoils, and tailings store water on or below the land surface. The amount of moisture stored depends on several factors, including particle size and structure, both of which affect pore size distribution; drainage; organic matter content; soil and above-ground temperature; above-ground surface humidity and wind. Growing plants can use some of this stored water. The amount of stored water available for plants depends on the amount and rate of extraction from the matrix of the growth medium.

Moisture losses can occur before or during plant use. Vapor losses occur due to evaporation from soil or spoil surfaces. As more cracks develop in the soil surface, more soil surface is exposed to the air; thus, more evaporation occurs, other factors being constant. As the soil surface dries water will move to the exposed surfaces due to capillary action, the rate of movement being governed by soil water potential. Plant transpiration also causes vapor losses. Such losses, however, are desirable because they are a necessary function of plant use and growth. Moisture loss from transpiration depends on above ground relative humidity; air temperature; wind; available water supply to the roots; and on plant density and plant species, both of which affect plant tissue surface area.

Liquid water is lost when surface runoff occurs, causing erosion and soil particle removal. Water also may be lost due to percolation through the root zone. When water percolates below the root zone, plants do not have access to it unless there is a water table below from which capillary rise may occur, bringing the water up into the root zone again. Such a mechanism is functional in subirrigation of forages on river bottom lands or other poorly drained areas.

Capillary rise is one of the causes of surface salt accumulation. As the water rises due to capillary adhesion, the force between unlike materials, and cohesion, the force between like molecules, soluble salts move with the liquid. When the liquid reaches the soil surface and evaporates, salts are deposited on the surface particles. Saline and/or sodic soils form when these salt accumulations are large enough.

Particle size influences vapor losses because of its effect on capillary forces in soil. Capillary forces affect the water available to plant roots. Figure 3 shows how particle size affects the field capacity and the wilting point of a disturbed soil. The difference between these two parameters represents the approximate plant available water, although this may vary among plants, soils, and field situations. For many plants, the wilting point will be about -15 bars, whereas field capacity will be near $-1/3$ bar soil moisture potential (negative soil moisture tension). Some adapted plants, such as native desert species, can extract water in the soil held at potentials lower than -15 bars. Available moisture increases with decreasing particle size up to a point, then decreases with a very high clay content. This is because the water held by the soil at or below -15 bars rapidly increases in the clay range. When the soil moisture potential is ≤ -15 bars, the rate of water uptake by many plants is not sufficient to sustain growth, and the plants will die unless water is added to the soil or spoil.

Particle size also affects vapor losses due to surface evaporation. Because more water is held near the surface of a clayey soil, more water will evapo-

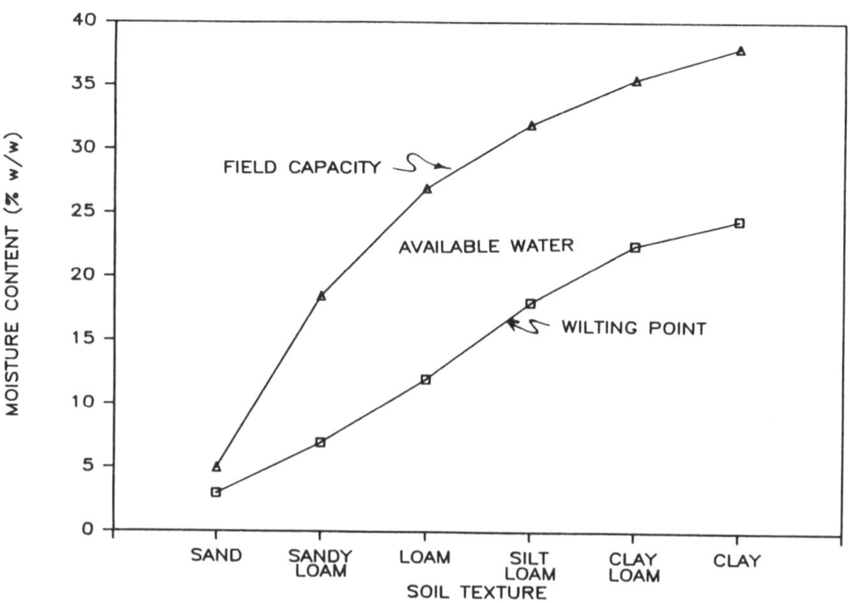

Figure 3. Relationship of particle size to field capacity and wilting point of a soil.

rate from a clay than a sand. In a dry climate water falling on the surface of a sandy soil likely will percolate into the root zone. Thus, with less water near the surface to evaporate, there may be more available water for semiarid plant species on sandy soils than on finer textured materials.

Soils with expanding and contracting smectite clays will crack extensively as the soil dries and as the clays contract or shrink after being expanded when wet. This causes greater surface area exposure to the air and subsequent evaporation. Silty and sandy soils are less prone to develop large cracks than clays, resulting in lower evaporative losses.

Because of low infiltration and percolation rates, surface runoff generally will be greater on a clayey soil than on a sandy soil. Management practices can control erosion and conserve moisture on the fine-textured soils. Mulches and stabilizers, along with rapidly germinating and growing species that provide a more permanent surface stabilization, can be effective on such soils. Sandy soils, on which greater infiltration and percolation occur, may not need such intensive management for erosion control. On the other hand, sandy soils that dry out rapidly may require wind erosion control practices.

Because coarser textured soils have larger individual pore spaces, percolation may result in water losses below the root zone. In humid regions of the United States, plant-available moisture generally is greater in moderately fine soils than in coarse-textured soils. Rainfall usually fills the water storage capacity of all soils, and the medium-fine materials have a greater field capacity and a lower wilting point. However, in the arid and semiarid West, without irrigation, soils and spoils seldom reach their water-holding capacity. Thus, plant-available water is determined not by water holding capacity but by such properties as runoff, infiltration, percolation away from the surface, and surface evaporation rates. The plant growth relationships in coarse-textured soils generally are superior to fine-textured materials under arid and semiarid conditions.

These soil texture and water relationships can be modified somewhat by secondary particles or soil structure. Soil structure can change solids-to-pore-space ratios and thus influence moisture reactions in the material. When soil structure is altered either suddenly (compaction due to heavy equipment) or over extended time periods as occurs with soil development processes, the pore size distribution is altered.

Physical impedence of roots—compaction

Topsoil stripping and subsequent replacement greatly alter soil structure. The most often encountered problem is compaction due to the heavy equipment used in these operations. This compaction, if not mitigated, can make it difficult if not impossible for some reclamation species to germinate and grow due to the resistance against root penetration. Highly compacted soils also restrict moisture and air supply to plant roots. Soils and spoils high in clays and low in sands compact more drastically due to manipulative forces than coarser materials, especially if the compaction activities occur when the materials are wet. Compaction occurs when the wet plate-like-shaped clay particles are reoriented with their flat surfaces parallel to each other, rather than the random arrangement that occurs naturally during soil formation. Compaction also expels air and decreases the volume of pore space. Upon drying, the compacted material can be extremely dense and relatively impenetrable. Although seasonal wetting and drying and freezing and thawing tends to loosen the compacted material, this can be a slow process. In the meantime, without plant growth, erosion can be severe.

Mechanical ripping or similar physical soil treatments can loosen compacted material and provide a better environment for root growth and development. Ripping increases (at least temporarily) water infiltration and percolation into the root zone and decreases root growth impedence by decreasing bulk density. While ripping can decrease bulk density greatly, ultimate density is affected by many factors and may increase quickly.

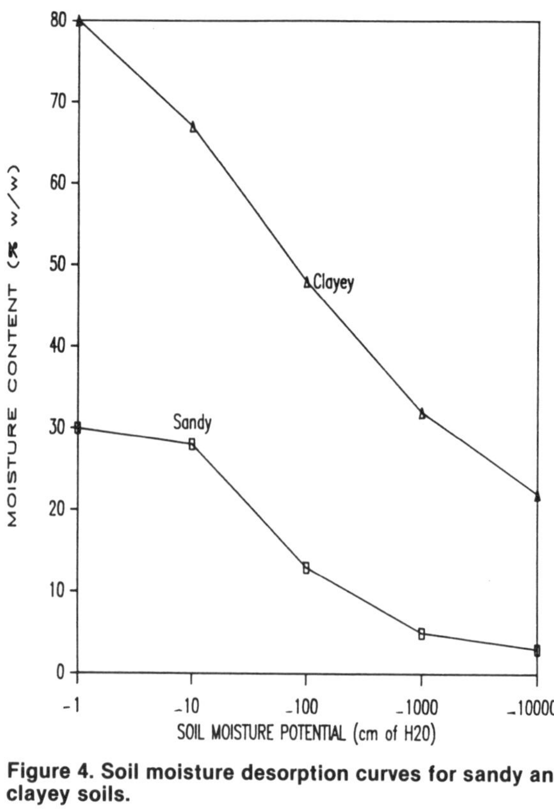

Figure 4. Soil moisture desorption curves for sandy and clayey soils.

Incorporating organic matter into the soil or spoil also can improve the plant rooting environment. Organic matter enhances optimal structure formation and when incorporated during ripping or disking results in retention of lower bulk densities for longer periods. Also, organic matter increases nutrient and water-holding capacity, both of which are beneficial to revegetation. Potential sources of organic matter include municipal sewage sludge, animal manure, wood chips, straw, and other plant and animal residues that may be available from nearby suppliers.

Pore size distribution and aeration

Pore space (nonsolids) commonly is divided into capillary (micropores) and noncapillary (macropores) space. When there is adequate moisture present, aeration—the supply of oxygen and removal of carbon dioxide—occurs mainly in the macropores because the micropores are filled with water. This classification of pore size distribution is artificial even if somewhat practical for introductory considerations. The real situation in most soil, spoil, and tailing material is that there is a continuous variation from large to very small pore sizes. The variation of pore size distribution is influenced by particle size distribution and structure. Generally, with greater amounts of clay small pore spaces predominates; the more sand, the greater the number of large pore spaces. Structure, however, can modify this generalization.

Pore size distribution affects moisture relationships. Micropores will hold water more tenaciously than macropores because of increased surface area contact. In fact, the relationship is so close that the pore-size distribution curve that reflects the pattern of pore-size variation in a porous medium can be determined from a soil moisture potential (soil moisture desorption) curve (Figure 4). The bottom curve in the figure is for a sandy soil with a predominance of larger pore spaces; the top curve is for a clay soil with a preponderance of micropores. These curves show the interdependence of particle size distribution on pore size distribution, both of which have a profound influence on plant growth and revegetation success.

Temperature

Several factors influence the temperature of the land's surface, such as the angle of solar radiation, intensity of solar radiation (thus slope and aspect), soil moisture content, and soil color. Particle size distribution influences moisture content, as pointed out previously, and color. During soil formation, soils with higher clay contents often accumulate more organic matter and thus may be darker. With a given amount of solar radiation, a moist, porous medium will have a lower temperature than a drier one due to the high heat capacity of the water and to the heat energy used to evaporate water. On the other hand, if a soil is high in organic matter, thus, having a darker color, it will absorb greater solar radiation and reflect less, giving rise to higher temperatures. The temperature of an area results from a combination of these factors.

Chemical reactivity

Small particles of a particular mineral will have greater chemical activity because of the greater specific surface area (area per unit weight, cm^2/g). For example, a 30-cm × 30-cm × 30-cm rock would have a surface area of 5,400 cm^2 (900 cm^2 × 6 sides) (about 6 square feet). However, if cut into particles that were of a medium clay size (0.1 micron on all three dimensions), the same piece of rock would have a surface area of 162×10^{14} square microns or about 167 hectares (400 acres). The amount of mineral material (weight or volume of solids) is exactly the same. Figure 5 shows the relationship between specific surface area and particle size (*5*).

Because of the large surface area, many chemical bonds would have to be broken and exposed at each surface. Thus, chemical activity is much greater for clays than sands or silts. This calculation assumes a mineral of uniform composition cut into cubes. However as larger minerals weather into clay-size particles, their configuration becomes plate-like instead of cubic, resulting in an even greater specific surface than the above calculation illustrates. The chemical activity often is measured in terms of CEC, which

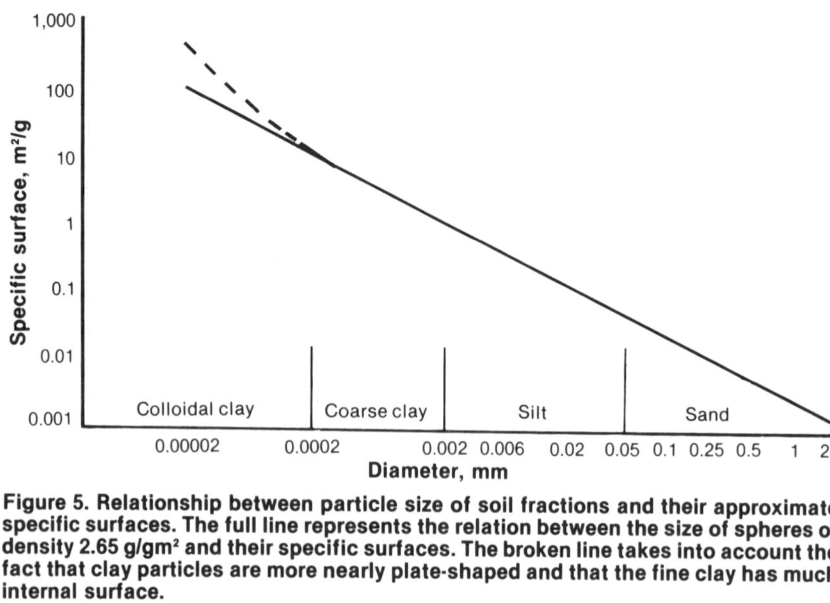

Figure 5. Relationship between particle size of soil fractions and their approximate specific surfaces. The full line represents the relation between the size of spheres of density 2.65 g/gm² and their specific surfaces. The broken line takes into account the fact that clay particles are more nearly plate-shaped and that the fine clay has much internal surface.

measures the negative charge associated with the particle surfaces. There is a close correlation between CEC and specific surface area of the particles.

CEC values for some common clay types include kalonitic clays, 3-15 me/100g; smectite (Montmorillonite) clays, 60-100 me/100 g; and illite clays, 20-40 me/100 g (*8*).

These differences in CEC values are due to variations in mineralogical constituents and arrangements and varying amounts of isomorphic substitution in their crystal structure.

Erosion

Soil erosion is a natural process that has proceeded at various rates over geologic times. Erosion occurs by water or wind action, the mechanisms of which have some similarities, yet many differences. Vegetative cover usually develops naturally over time, which helps to control erosion.

Man's activity can be helpful or detrimental in controlling erosion. Most drastically disturbed lands are highly susceptible to the erosive forces of nature unless reclamation management is well planned and executed. The greatest challenge is to provide a stable cover that will resist natural forces. In most situations, an adequate vegetative cover is the most economical and satisfactory management strategy to control erosion.

Particle size distribution can have a profound effect on erosion and on erosion control practices.

Water erosion. Soil erosion by water is the process of (1) detachment of soil particles from the soil surface, either from raindrop impact on the bare soil or where the force of overland flow exceeds the critical attractive force of the soil; (2) transport of soil particles; and (3) deposition of the sediment when the transport capabilities of the overland flow are reduced by ground cover, decreased slope steepness, or other factors (*6*).

Soil texture, the type and amount of clays, organic matter content, soil depth, and presence of an impervious layer influence erodibility by water. This is due to the effect of the above factors on infiltration rate, permeability, total water-holding capacity, dispersion, splashing, abrasion, and transporting forces of rainfall and runoff.

Soil erodibility increases with greater silt content and decrease with greater sand and organic matter content (*9*). The very fine sand fraction is comparable in erodibility to silt-sized particles (*10*). In general, the very fine sand and silt size fractions detach most easily. Large sand and clay particles are less easily detached because of the greater mass of the sand and the strong cohesive and adhesive forces of the clays. However, soils that are high in dispersed clays have low infiltration, high runoff, and high erosion.

Particle size is a factor used in estimating soil loss in the universal soil loss equation. Particle size is an integral part of the USLE's soil erodibility (K) factor. The K factor is determined using the percent fine sand and silt, percent sand, percent organic matter, soil structure code used in soil classification, and the soil profile permeability class (*10*).

Both running water and raindrop splash facilitate the detachment and transport of soil particles. In confined channels, most soil detachment is due to water flow. But on comparatively smooth surfaces, the force of raindrops has the greatest influence on detachment. On a soil subject to easy detachment, a torrential rain may splash as much as 100 tons of soil/acre, some of the drops rising as high as 2 feet and moving horizontally perhaps 4 to 5 feet (*2*). As with running water, very fine sand and silt are the most easily detached size fractions.

Raindrop splash is an important factor in creating a surface crust, thus increasing erosion by decreasing infiltration. Crust formation occurs when raindrops break down wet soil aggregates and finer soil materials are washed into the surface pores, reducing their volume. After breakdown of the surface aggregates, continued raindrop impact compacts the surface. Crusting effects are greater with fine-textured materials than coarse-textured materials. Crusting can impede seedling emergence, thereby directly affecting vegetation establishment on newly seeded areas.

Particle size affects the amount of water entering surface water and groundwater systems through runoff, infiltration, and percolation. Coarse-textured soils and overburden in general have high infiltration and percolation rates. Thus, surface runoff is relatively low. Water that percolates past the rooting zone eventually enters the groundwater system. Conversely, fine-textured soils generally have low infiltration and percolation rates. Therefore, much of the precipitation runs off, directly entering the surface water system. Also, because Na disperses the clays in the soil, surface runoff increases as the Na content of a soil increases.

Wind erosion. The physical character of soil or overburden affects the ease with which wind dislodges particles. The most important factor influencing wind erosion is moisture content. Wind erosion hazards are high where the mean annual rainfall is low—generally where annual rainfall is less than 30 cm (*4*).

Other soil characteristics influence wind erosion: particle size, mechanical stability of aggregates, the presence of a stable soil crust, and bulk density. Particles approximately 0.1 mm in diameter are most susceptible to wind erosion. Particles larger or smaller in size are less susceptible to movement. Because of this movement, particles approximately 0.1 mm in size can cause larger and smaller particles to move by dislodgement and bombardment (*2*). Three distinct types of wind erosion occur, depending upon particle size: suspension, creep, and saltation.

Suspension is the movement of very fine particles, mainly less than 0.1 mm in diameter (Figure 6). The falling speed of such small particles is so

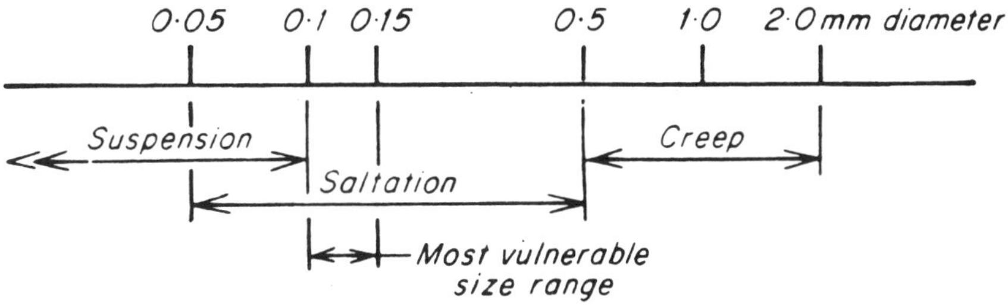

Figure 6. The form of soil movement by wind as a function of particle size (4).

low that once lifted, they remain suspended for long periods of time by air turbulence and eddy currents. Dust storms are fine particles in suspension that can result in large amounts of soil being moved long distances (*4*).

Creep is the movement of particles at the other end of the size scale. Pushed by the force of the wind and other particles moving with the wind, soil particles roll along the surface. Theoretically, there is no upper limit to the size of particle that could be rolled. But in practice, most rolling particles range in size from 0.5 mm to 1 or 2 mm diameter (*4*).

Saltation is the most important of the three forms of wind erosion. This process moves more soil than the other two processes, and neither creep nor suspension occur without saltation. Saltation is the movement of soil particles by a series of low bounces over the surface. It occurs among middle-size particles—those light enough to be sometimes lifted off the surface but too large to be suspended—particles from 0.05 mm to 0.5 mm diameter. Most saltation movement occurs among particles 0.1 to 0.15 mm in diameter. Particle size, thus, determines the type of movement a soil particle undergoes (Figure 6). However, because of many contributing factors, such as bulk density or wind velocity, the dividing lines are not precise (*4*).

In general, the wind or water erodibility of a particular soil or overburden is related to particle size gradation. This is because the most easily detached particles by either type of erosion are the very fine sand and silt-size particles.

The major area where wind and water erodibility differ is related to soil moisture. For a particular soil or overburden, wind erodibility increases with decreasing antecedent moisture content, while water erodibility decreases with decreasing antecedent moisture content.

Analytical procedures

Determining the particle size distribution or texture of soil or spoil materials can be done rather easily by "feel" when sufficient experience has been developed. It is a quick and simple procedure that may be adequate for some purposes, but lacks the precision required for other purposes. Any experienced soil scientist can use and describe the procedure because this method is used for field soil surveying.

Three laboratory techniques are more precise but more time consuming. The procedures are based on Stokes Law, which considers the viscous forces associated with a solid particle moving in a viscous liquid. Each of these methods require pretreatment of the soil sample to remove organic matter and disperse aggregates. Usually, organic matter is oxidized with hydrogen peroxide and aggregates dispersed with sodium metaphosphate. Physical agitation often is necessary to break up the aggregates or structural units to ensure that only individual particles exist in the soil water system. For precise analyses, the pipette method has been the procedure of choice among most soil scientists.

By controlling the temperature, size, and dimensions of the container and time periods, the particle size distribution for particles less than 2 mm in diameter can be determined quite accurately. The equipment necessary is elaborate, and the analysis time is extensive.

Another reasonably accurate method uses a hydrometer to measure suspension density instead of using a pipette to withdraw suspended particles at specific depths. This method is not as precise as the pipette method. However, if the same precautions and treatments of the soil water suspension are used, it can be accurate. The hydrometer method is less time consuming and requires less elaborate equipment. Both methods are discussed in detail by Day (*3*).

The third method, elutriation, generally is not used because soil scientists lack experience and familiarity with it. Elutriation determines particle size distribution by causing the soil water suspension to flow through various sized containers. The smaller the container, the more rapid is the rate of flow through the container. Only the larger particles settle out in the small containers. The smaller particles are carried in suspension to larger containers where the rate of flow is lower, thus allowing small particles to settle to the bottom. Based on this method, and with the use of specific sized containers, a size distribution can be determined. Considerations similar to those in the Stokes Law are functional in elutrication. But instead of particles flowing downward in a stationary body of water, as is the case in the pipette and hydrometer methods, the water flows at varying rates while the particles settle out of the moving water.

With adequate experience and calibration, each of these methods of determining particle size distribution can be used effectively. Because these methods deal primarily with sands, silts, and clays, sieves can be used for larger particles. Many overburden and spoil materials are composed mostly of pebbles, rocks, and boulders. However, when revegetation is the objective, concern for the sand, silt, and clay sized particles interspersed among the coarse fragments is paramount.

REFERENCES

1. Bernstein, Leon. 1974. *Crop growth and salinity.* In Jan Van Schilfgaarde [editor] *Agronomy.* No. 17. American Society of Agronomy, Madison, Wisconsin. pp. 39-54.
2. Brady, N. C. 1984. *The nature and properties of soils.* MacMillan Publishing Company, New York, New York. 750 pp.
3. Day, Paul R. 1965. *Particle fractionation and particle size analysis.* In C. A. Black [editor] Methods of Soil Analysis Part I. Agronomy Series No. 9. American Society of Agronomy, Madison, Wisconsin. pp. 545-567.
4. Hudson, N. 1971. *Soil conservation.* Cornell University Press, Ithaca, New York. 320 pp.
5. Kohnke, H. 1968. *Soil physics.* McGraw-Hill Book Company, New York, New York. 224 pp.
6. Meyer, L. D., and M. J. Romkens. 1976. *Erosion and sedimentation on reshaped land.* In Proceedings, Third Federal Inter-Agency Sedimentation Conference. Water Resources Council, Washington, D.C.
7. Richards, L. A., and L. R. Weaver 1944. *Moisture retention by some irrigated soils as related to soil moisture tension.* Journal of Agricultural Research 69: 215.
8. Thompson, L. M., and R. Troeh Frederick. 1978. *Soils and soil fertility.* McGraw-Hill Book Co., New York, New York. p. 516.
9. U.S. Department of Agriculture. 1977. *Preliminary guidance for estimating erosion on areas disturbed by surface mining activities in the interior western United States.* EPA-908/4-77-005. U.S. Environmental Protection Agency, Washington, D.C.
10. U.S. Department of Agriculture. 1978. *Predicting rainfall erosion losses.* Agricultural Handbook No. 537. Washington, D.C.

Chapter 5

Sodium

Mark G. Carlstrom, Francis A. Amendola, Dale A. Shay, and Douglas J. Dollhopf

Sodium (Na) influences soil material suitability, use, and management. With proper soil and overburden/spoil[1] evaluation and mine soil reconstruction, mine soils in Na-affected areas should support productive plant communities. The principles discussed herein should apply to Na-affected arid and semiarid areas throughout the western United States.

Physiochemical aspects of sodium pertinent to mine soils

Sodium indices. *Exchangeable-sodium percentage and sodium-adsorption ratio.* The exchangeable-Na percentage (ESP), which is the percentage of the cation exchange capacity (CEC) occupied by Na, is used as a primary measurement of Na hazard in soils (*37*). Exchangeable-Na percentage is calculated using the results of three analyses: extractable Na, soluble Na, and CEC. Exchangeable-Na percentage determinations are time-consuming, costly, and subject to error. Most of this error is the result of the CEC analysis, which, regardless of the procedure used, is suspect with regard to accuracy and reproducibility.

The ESP analytical technique proposed by Bower and associates (*16*) has been the standard used in many laboratories. After a study of the limitations of this and other methods used for the determination of CEC in calcareous and gypsiferous soils, Polemio and Rhoades (*65*) claimed that a reliable, generally applicable, and simple method for routinely determining the

[1]Key terms are defined in the glossary at the end of this chapter.

CEC of calcareous and gypsiferous soils was still lacking. Thus, they proposed a method for CEC determination that has special application to arid-land calcareous and gypsiferous soils and to soils containing zeolites and feldspathoids. They claimed that their method, which results in higher CEC values, eliminates steps that produce errors in conventional methods (*73*).

A later examination of both the Bower and associates method and the Polemio and Rhoades method resulted in the conclusion that both procedures overestimate the CEC of sodic soils (*34*). Yet another procedure was proposed for determining the CEC of calcareous sodic soils containing zeolites and feldspathoids. This new method eliminates the use of alcohol, which Gupta and colleagues claim eliminates the source of error in both of the earlier methods. They recommended the adoption of their method for sodic soils having free carbonates, zeolites, and feldspathoids. However, their method is not intended for gypsiferous or saline soils.

Apparently, the designation of a standard CEC method for arid and semiarid region soils analysis is debatable. Also in question is how the use of new CEC methods affect the applicability of established ESP guidelines. A standard, reproducible CEC method must be established so that consistent ESP determinations and interpretations can be made. New infiltration, permeability, and plant growth information may have to be derived from a new, standard CEC analytical method before concomitant ESP values can be properly evaluated.

The Na-adsorption ratio (SAR), an indirect measurement of Na hazard, has been correlated with ESP for agricultural soils and is often used to predict ESP because the chemical analysis for SAR is simpler and less costly than ESP analysis (*37*). Sodium, Ca, and Mg in the soil solution are in equilibrium with adsorbed cations and easier measurements can be made of soluble, saturation extract cations versus exchangeable cations. The SAR is calculated using the values of water-soluble Ca, Mg, and Na obtained from analysis of the saturation extract (*45, 46*). The main source of error in the SAR analysis is the determination of the saturation endpoint of the saturated soil paste. For a thorough discussion of saturation paste preparation, see the chapter herein on saturation percentage by Merrill and associates (*56*).

It is important to note that neither ESP nor SAR directly reflects the influence of salts, clay mineralogy, or texture on infiltration rate or hydraulic conductivity.

Exchangeable-sodium percentage—sodium-adsorption ratio relationships. Based on the analyses of 59 soils from the western United States, Richards (*76*) determined that SAR could be used to predict ESP by the following regression equation:

$$\mathrm{ESR} = 0.01475\ \mathrm{SAR} - 0.0126 \qquad r = 0.923$$

where ESR is the exchangeable-Na ratio. Exchangeable-Na percentage can then be calculated by:

$$\text{ESP} = \frac{100(\text{ESR})}{1 + (\text{ESR})}$$

Because the relationship in the former equation shows a high correlation, the SAR has been used extensively to estimate the sodicity of soil materials. It is commonly interpreted uncritically as a definitive measure of sodicity, something that was clearly not intended by its originators (*36*).

Discussions of ESR-SAR relationships (*36, 56*) show that the Richards' (*76*) correlation is not applicable to all soils, overburdens, or spoils. Deviations from the relationship in equation 1 may occur. These deviations may be attributable to experimental techniques and analytical methods, soil solution composition and concentration, texture, and/or clay mineralogy.

That these deviations exist for overburden has been affirmed (*37, 89*). Using the ESR-SAR regressions developed for overburden in these studies (*37, 89*), two teams of researchers showed that the ESP values estimated by SAR are lower than when equation 1 is used. Thus, by using the equation 1 relationship, the ESP of overburden materials would be overestimated by the SAR.

There are two ways that these deviations could be handled (*37*): (1) ESP could be determined on any soil or overburden/spoil samples when the SAR values of these samples approach guideline suspect levels, and/or (2) material-specific ESR-SAR regression analyses could be performed to develop satisfactory statistical relationships between ESP and SAR. The population to be represented by each ESR-SAR regression analysis should be largely homogeneous, especially with regard to salts, clay mineralogy, and texture.

These procedures could be used if the ESP analysis is reliable and concomitant ESP threshold values have been established for the mine-soil material in question. It may be more expedient to establish reliable SAR threshold values for the mine-soil material and then use SAR for direct assessment of that material.

Primary factors affecting mine soil suitability

Dispersion and swelling of clay particles. Dispersion and swelling of clays in the soil matrix may result in the deterioration of a soil's physical structure. This breakdown may have several ramifications: reduced aeration, water infiltration, and permeability and increased surface crusting, runoff and erosion, piping, surface instability, compaction, salt accumulation, and upward salt migration.

Dispersion and swelling are primarily influenced by a soil's water and exchangeable Na content, the electrolyte concentration of the water in the soil pores (the soil solution), and a soil's clay content and mineralogy. The adverse effects of dispersion and swelling tend to increase with increased soil water content and sodicity, reduced electrolyte concentration of the soil solution, increased clay content, and the presence of reactive clays (clays that have a high tendency to expand and disperse).

Because of its high exchange capacity and expanding nature, montmorillonite is the most reactive constituent among the soil mineral colloids. Ca-saturated montmorillonite clay particles commonly consist of weakly bonded packets (tactoids) of four to nine clay platelets arranged parallel to each other (*28*). These tactoids, with other soil particles, form soil aggregates.

Dispersion occurs when the ESP is about 10 to 20 (*27, 86*). In this ESP range Na will occupy the external surfaces of tactoids, resulting in the development of repulsive forces between adjacent tactoids and between tactoids and other adjacent soil particles. If the electrolyte concentration is sufficiently low (below the flocculation value of the clay), tactoids will become mobile (disperse), translocate, and plug conducting pores (*27, 28, 84, 86*). But soil hydraulic conductivity may increase under conditions of low soil bulk density and clay content and high flow rates. Under such conditions, dispersed particles may be removed from the soil profile, resulting in enlarged soil pores, especially along major flow channels (*27*).

Tactoid swelling occurs when ESP values exceed 15 to 25 (*27, 86*). At these exchangeable-Na concentrations, Na invades the interlayer positions between clay platelets. The presence of Na in these interlayer positions causes high repulsive forces or swelling pressures to develop. With repulsion between platelets, there is a volume increase of interlayer voids (swelling) and the tactoid expands like an accordian. The extent of swelling varies with the degree of hydration of interlayer cations. The osmotic attraction for external water (from the soil solution) is generally twice as high with monovalent cations than with divalent cations. Thus, swelling and repulsion will be greatest with monovalent cations, such as Na, and a low soil solution electrolyte concentration. Tactoids may swell to the point where attractive forces between platelets are no longer strong enough to oppose repulsive forces, and the tactoids begin to break apart. Complete tactoid breakdown occurs when the ESP reaches about 50 (*28*).

The swelling of clay particles closes off conducting pores and reduces infiltration and permeability (*19, 71*). Smaller tactoids and individual clay platelets, produced by tactoid swelling and breakdown, translocate and plug soil pores. But a swollen soil may shrink sufficiently upon drying to open cracks and channels that may enhance subsequent flow (*27*).

Swelling is a reversible process, while dispersion and particle migration are not. The addition of electrolytes or divalent cations to the soil system will cause swollen clay particles to shrink, thus opening soil pores. Mechanical manipulation may be the only means sufficient to improve the condition of a dispersed, clogged soil. The occurrence of both dispersion and swelling in Na-affected mine soils may be averted through the use of chemical amendments (see amendments).

Weathering. Soluble salts that occur in soils consist primarily of the cations calcium (Ca^{2+}), magnesium (Mg^{2+}), and sodium (Na^+) and the anions chloride (Cl^-) and sulfate (SO_4^{2-}). Potassium (K^+), carbonate (CO_3^{2-}), bicarbonate (HCO_3^-), nitrate (NO_3^-), and borate (BO_3^{3-}) are found in smaller quantities. The original sources of soluble salts are the primary minerals found in soils and exposed rocks. During the process of chemical weathering, which involves hydrolysis, hydration, solution, oxidation, and carbonation, salts are gradually released (*76*).

A soil's susceptibility to the adverse effects of Na when the soil is leached with a high quality solution (e.g., rainwater or snowmelt water), depends upon the soil's potential for releasing electrolytes (salts) through the dissolution of primary minerals and/or $CaCO_3$ (*86*). Soluble salts released by weathering may supply enough salt to maintain the soil solution concentration above the flocculation value of the soil's clay. In addition, it is probable that in many relatively unweathered soils the release of Ca and Mg upon weathering will reduce the SAR of the soil solution and lessen the Na hazard. Rowell and Shainberg (*79*) stated that through these release mechanisms some degree of soil self-protection may exist. They stated that under normal rainfall intensities, fresh loess in the surface soil horizon is likely to contribute significantly to the maintenance of soil structure because of the dissolution of minerals containing Ca.

Studies have been conducted to evaluate the amount of electrolyte that arid or semiarid region soils may release. In a soil column experiment, the dissolution of plagioclase, feldspars, hornblende, and other common mafic minerals accounted for a 3- to 5-meq/liter increase in effluent solution concentrations (*74*). Other researchers found that the release of cations from minerals increased lysimeter effluent concentrations 4.1 to 6.3 meq/liter (*15*). In another soil column experiment, Rowell and Shainberg (*79*) found that when a surface soil was leached with distilled water the effluent had an electrolyte concentration of 3 meq/liter. They hypothesized that easily weathered minerals were supplying salt to the soil solution.

Research has been conducted on sodic overburden to examine the effect of natural geochemical weathering on ESP and SAR. Hall and Berg (*37*) investigated the effect of weathering on 11 sodic overburden samples collected from mines in New Mexico, Colorado, North Dakota, and Montana. The samples were subjected to four different weathering treatments for periods of 16, 64, or 224 days. The weathering treatments caused increases in soluble Ca and significant decreases in sodicity to occur. Probable causes for the changes with weathering may have been one or a combination of the following: sulfide oxidation resulting in dissolution of $CaCO_3$; delamination of Ca-tactoids; and, for two samples, dissolution of gypsum.

A laboratory study of the weathering of sodic overburden from the Spring Creek Mine in Montana indicated that weathering can add a significant amount of Ca and Mg to the exchange plus solution phases (*41*). The researchers concluded that the importance of mineral weathering can be particularly significant in reducing the sodicity of overburden materials when dealing with slightly weathered calcareous overburden containing predominately nonswelling kaolinite, illite, and chlorite clay minerals.

Dollhopf and associates (*23*) found that the upper 30 cm (12 in) of an untreated sodic spoil underlying 70 cm (28 in) of good quality coversoil at the West Decker Mine in Montana showed an approximate 8-unit decline in SAR levels after 2 years. Lower SARs were largely attributed to an increase in Ca and Mg and, in part, to the leaching of Na. In another field study at the same mine, SAR values in a vegetated raw spoil declined from about 12 to 3 in the upper foot of spoil over a 7-year period (*77*). This decrease was thought to be the result of the leaching of Na by the percolation of rainwater and snowmelt water.

While the effect of mineral weathering on sodic overburden appears to be significant, additional research is needed to evaluate the factors and processes involved in the reduction of sodicity. The few studies that have been undertaken indicate that natural weathering processes will work to reduce the sodicity of overburden materials. Weathering rate should not be considered, however, in the classification of overburden because significant reductions in sodicity generally require prolonged, vigorous weathering (*37*).

Texture and clay mineralogy. Unlike swelling, dispersion does not necessarily require a high soil clay content or montmorillonite clay and may be more effective in soils of low clay content (*86*). Also, because the effect of Na on the swelling properties of clays is so pronounced, a coarse-textured soil may show a marked change in permeability provided the soil has an appreciable ESP (*70*). However, it has been shown that, in general, the physical properties of fine-textured soils are more adversely affected than the physical properties of coarse-textured soils under sodic conditions (*28, 48, 62, 76*). It is also recognized that under sodic conditions soils containing expanding clay minerals (montmorillonite) are more unstable than soils containing an equivalent amount of nonexpanding clay minerals (illite, kaolinite) (*20, 28, 49, 76, 96, 98*). But even soils with similar texture and clay mineralogy may differ greatly in their response to exchangeable Na (*86*).

The clay minerals in arid- and semiarid-region soils are predominately montmorillonite and illite (*44, 50, 76, 86*). It is more difficult to generalize about overburden clay mineralogy in these regions because of the variability of the overburden and the lack of overburden clay mineralogy data. It would be ideal to have clay mineralogy data for all soil, overburden, and spoil samples. But clay mineralogy analysis may not be practical or economically feasible for complete soil and overburden/spoil characterization. If information about clay mineralogy is unattainable, or if it is inadequate for the characterization of an area, saturation percentage can be used to screen mine-soil materials for reactive clays. Research to date supports the use of a saturation percentage suitability value of less than 90 (*24, 53*). Because this suitability value is based on information from only a few sites, however, further research is needed to confirm its utility.

Soil pH. Designation of sodic soils should not be based solely on soil pH. Sodic soils may be alkaline, neutral, or acid in reaction (*93*). In addition, soil pH is greatly influenced by several factors other than exchangeable Na, for example, other adsorbed cations, soil to water ratio, texture, CO_2 pressure, insoluble carbonates, gypsum, soluble salts, organic matter, and clay mineral type (*26*).

Richards (*76*), on the basis of experience and a statistical study (*26*), concluded that the following general statements about the interpretation of saturation-paste pH are possible: (1) a pH of 8.5 or higher almost invariably indicates an ESP of 15 or more, and (2) the ESP of soils having a pH less than 8.5 may or may not exceed 15. Abrol and associates (*1*) concluded that, based on their research and the data of others, most often an ESP of 15 to 20 is associated with a saturation-paste pH of 8.2. They suggested that a pH of 8.2 would be more appropriate than the commonly used value of 8.5 as a sodic soil diagnostic criterion.

Some studies have investigated the effect of pH on the physical properties of sodic soils. Findings in one case suggested that, even at constant ESP, any build up in the pH of the soil suspensions studied would enhance dispersion and the flocculation value (*35*). Arora and Coleman (*6*) found that the amount of salt required for clay flocculation increased with increasing pH for clay-size soil fractions.

In an investigation of the effect of pH on soil hydraulic conductivity, hydraulic conductivity values were lower at pH 9 than at pH 6 for a montmorillonitic and a kaolinitic soil (*92*). The adverse effect of high exchangeable Na on soil hydraulic conductivity is magnified by high soil and water pH in semiarid and arid regions. The researchers concluded that pH must be considered along with SAR and electrolyte concentration when assessing soil permeability potential. But they stated that, because of differences among soils, simple prediction of this pH effect is not possible at present.

Bulk density. Increasing the bulk density of a soil should increase that soil's sensitivity to conditions of high exchangeable Na and low electrolyte concentration. The smaller pores and increased tortuosity of a soil with an increased bulk density contributes to the effectiveness of dispersion and swelling in restricting and closing off transmitting pores (*28*).

Hydraulic conductivity and infiltration rate. Dispersion and swelling of soil clays tend to increase with increased soil sodicity and decreased electrolyte concentration of the percolating solution. Much research has been done on the effect of soil sodicity and soil solution concentration on soil hydraulic conductivity (*5, 25, 28, 40, 48, 49, 59, 69, 71, 75, 78, 79, 85, 95, 98*). Quirk and Schofield (*71*), who performed the first definitive work in this area, developed the "threshold concentration" concept. This was defined as the electrolyte concentration of the percolating solution that causes a

10-to 15-percent reduction in soil permeability at a given soil sodicity, that is, the concentration where factors that can cause a drastic reduction in permeability are becoming operative. This criterion had no absolute basis but was used as a convenient reference level.

Shainberg and Letey (*86*), upon examining the threshold concentration concept, concluded that a 50-percent reduction in soil permeability should be considered critical and the combination of soil solution concentration and ESP required to produce a 50-percent change in permeability be considered the threshold concentration. Based on a review of the literature, they concluded that the hydraulic conductivity of soils having an ESP below 10 is not affected by waters of salinity above 0.5 dS/m (5 meq/liter). An electrolyte concentration of 0.5 dS/m is enough to prevent the deleterious effect of low levels of exchangeable Na. Significant changes in soil hydraulic conductivity (>50 percent) occur only when soil ESP exceeds 15 to 20. It is interesting to note that in many of the hydraulic conductivity studies reviewed, ESP and SAR were found, or assumed, to be approximately equal and were used interchangeably in the discussions.

Dollhopf and associates (*24*) studied mine soil permeability relationships at the West Decker Mine in Montana. The spoil studied had an electrical conductivity of 2.8 dS/m, an ESP of 20 ($SAR = 23$), and was covered with 70 cm (28 inches) of good quality coversoil. The researchers concluded that when a mine soil is dominated by kaolinite and illite contains less than 35 percent clay or greater than 45 percent sand, and has a saturation percentage of less than 90, SAR values of 20 to 30 may not limit the site reclamation potential. A study of the dispersion of moderately saline (3.6 dS/m), montmorillonitic, clay loam sodic spoils in North Dakota revealed that the spoils readily dispersed and exhibited crusting and restricted water intake when SAR levels exceeded 20 (*8*). The researchers recommended that where a total of only 2.5 feet of coversoil are applied, the SAR of the underlying spoil material should be less than 20 and the clay content less than 35 percent in order for the spoil to be a suitable mine-soil component.

Most investigations concerning the effect of soluble salt concentration and composition on the transport of water and solutes in soils have focused on steady-state saturated flow conditions; however, in the field the transport of solutes and water takes place under transient, unsaturated conditions (*80*). Gardner and colleagues (*30*) measured soil water diffusivity to determine the effect of exchangeable Na and electrolyte concentration on unsaturated flow in sandy loam and loam soils. They found that in soils having an ESP less than 15 the weighted mean soil water diffusivity increased slightly as the soil solution concentration rose from 2 to 100 meq/liter. In contrast, soils high in exchangeable Na ($ESP > 25$) had diffusivities that increased by three orders of magnitude over the same concentration range. Gardner and associates (*30*) indicated that the same effects resulting from exchangeable Na and electrolyte concentration can be expected under both unsaturated and saturated flow conditions (*20*). Similarly, Russo and Bresler (*80*) found that when a loamy soil was subjected to mixed Na-Ca solutions, the unsaturated hydraulic conductivity declined as either the soil solution concentration decreased or the Na fraction in the soil solution increased. But they found that this effect becomes smaller as soil water content decreases.

Infiltration rate is more sensitive to soil exchangeable Na and the electrolyte concentration of the applied water than is hydraulic conductivity (*42, 61, 86, 87*). Under natural conditions, the soil solution concentration at the soil surface is controlled solely by the concentration of the applied water (i.e., high quality rainwater or snowmelt water). Dissolution of $CaCO_3$ and primary minerals is too slow to affect the surface soil solution concentration; therefore, particle dispersion may take place at low exchangeable Na levels. The effects of physical forces also contribute to the sensitivity of the soil surface by causing the mechanical breakdown of soil aggregates and by enhancing the kinetics of clay dispersion, both of which are related to crust formation and resulting lower infiltration rates.

Research has been conducted to determine the mechanisms of crust formation and the effect of high quality water application on the infiltration rates of soils with low ESP values. Agassi and colleagues (*2*) observed that crust formation in soils exposed to rain results from two mechanisms: (1) physical disintegration and compaction of soil aggregates caused by raindrop impact and (2) chemical dispersion of clay particles and their movement into a region of reduced porosity where they precipitate and clog the conducting pores [the "washed in" layer described by McIntyre (*47*)]. The relative importance of these two mechanisms depends upon the impact energy of the raindrops, the electrolyte concentration of the applied water, and the soil ESP (*3*). Other characteristics that make the soil surface especially susceptible to crusting are low organic matter and high silt content (*18*).

Several rainfall simulation experiments have shown that, when high quality water is applied to soil surfaces having ESP values greater than about 2, chemical dispersion, in addition to physical dispersion, became operative (*3, 29, 42*). Infiltration rates can be increased, and runoff reduced, by reducing the impact energy of raindrops with a mulch and/or by promoting flocculation by spreading an amendment, such as powdered phosphogypsum, on the soil surface (*3*).

Revegetation and reclamation as affected by sodium

Effects of sodium on vegetation. Excess Na indirectly affects plant growth through the deterioration of soil structure. This breakdown may result in restricted water movement, aeration, root elongation, and seedling emergence and development (*12, 18*). Sodium may also affect plant growth through nutritional effects. In nonsaline-sodic soils, increases in exchangeable Na are balanced by decreases in exchangeable Ca and Mg, leading to Ca and/or Mg deficiencies when the concentrations of these elements in solution become deficient (*10*). In saline-sodic soils at similar ESPs the concentrations of all elements in solution are greater and may be nutritionally adequate. Most crops are moderately tolerant of Na and, under nonsaline-sodic conditions, are affected nutritionally at ESPs of about 25. However, the poor physical condition of a sodic soil becomes the growth-limiting factor for most crop species, regardless of the varying abilities of plants to function nutritionally (*9*).

Bernstein and Pearson (*11, 64*) studied the effects of exchangeable Na on plant yields. The soils involved were treated with the soil conditioner vinyl acetate maleic acid to stabilize the soils' structure. By treating the soils thusly, the effect of exchangeable Na on plant yields could be determined without the interference of an unfavorable soil physical condition. Table 1 summarizes the results of these studies. Table 2 lists the ESP ranges at which several crops are affected nutritionally in the absence of an adverse soil physical condition (*63*). Table 3 presents suggested pH and ESP classes in relation to their sodicity hazard for crop growth (*1*).

Excess Na may adversely affect plant root growth and function in spoil. On a site where 58 cm (23 inches) of topsoil was placed over a sodic clay loam spoil (SAR = 37), it appeared to Gilley and associates (*31*) that crested wheatgrass root water extraction in the spoil was minimal. Power and colleagues (*66*) reported that when the soil contact over a sodic silty clay loam

Table 1. ESP values causing a 50 percent growth reduction in various plants (*11, 64*).

Species	ESP
Alfalfa	45
Clover	28
Barley	49
Oats	30
Tall fescue	34
Tall wheatgrass	64
Wheat	43

Table 2. ESP ranges at which the growth of certain crops is affected (*63*).

Species	ESP
Clover, oats, tall fescue	20-40
Wheat, alfalfa, barley	40-60
Crested and tall wheatgrass	>60

Table 3. Soil pH, exchangeable sodium, and sodicity hazard (*1*).

pH	Approximate ESP	Sodicity Hazard	Remarks
8.0-8.2	15	None to slight	The adverse effect of exchangeable Na on the growth and yield of crops in various classes occurs according to the relative crop tolerance to excess sodicity. The growth and yield of only sensitive crops are affected at ESP levels below 15. Only extremely tolerant native grasses grow at ESP levels above 70 to 80.
8.2-8.4	15-30	Slight to moderate	
8.4-8.6	30-50	Moderate to high	
8.6-8.8	50-70	High to very high	
>8.8	>70	Extremely high	

spoil (SAR = 25) was deeper in the mine soil profile, alfalfa, crested wheatgrass, blue grama, sideoats grama, and spring wheat all extracted water at greater depths. Barth (*7*) found that for plots with variable coversoil depths over sodic spoil (average ESP = 21, SAR's ≥ 25, average texture = clay loam), the penetration of crested, thickspike, and pubescent wheatgrasses and smooth brome roots appeared restricted to the upper 10 cm (4 inches) of spoil. Dollhopf and associates (*24*) studied root development in a stand of 11 commonly used reclamation species (predominately grasses) grown in 70 cm (28 inches) of soil over a sodic sandy clay loam to loam spoil (ESP = 20, SAR = 23). These researchers found a significant decline in root biomass across the soil-spoil interface. This decline was partially attributed to normal root distribution patterns of rangeland species, but the sodic condition of the spoil may have also had an effect.

Merrill and associates (*57*) studied the growth of crested wheatgrass on various coversoil depths [0.25 to 1.0 m (10 to 39 inches)] placed over a sodic silty clay loam spoil (SAR = 30). Coversoil profiles at each site consisted of 0.2 m (8 inches) of topsoil placed over varying thicknesses of subsoil. The purpose of this study was to develop an understanding of plant growth response in various mine soils by measuring water movement and use and root growth. The researchers found that total soil water potential and root weight densities at a given profile depth were similar whether the growth medium was subsoil or spoil; however, root water uptake was much less from the spoil than the subsoil. Low hydraulic conductivity emerged as the dominant factor limiting sodic spoil as a plant growth medium. The low hydraulic conductivity restricted the movement of stored spoil water to plant roots, thus restricting root water uptake. Power and associates (*67*), in a companion paper to that by Merrill and colleagues (*57*), showed that the constraint of low mine soil hydraulic conductivity on water uptake and movement affects N uptake in a qualitatively similar manner.

The studies discussed above considered the effects of Na on only a few grasses and legumes. Shrubs and trees are also important components of the reclaimed landscape, but little is known about the effects of Na on these plants. Additional research is needed to better define the effect of Na on all classes of reclamation vegetation.

Upward sodium migration. Mining disrupts established chemical gradients in natural soil systems. Because of this disruption and the replacement of coversoil over regraded spoil, new chemical gradients are established. These new chemical gradients can present a strong potential for the diffusion of solutes from spoil into coversoil.

Laboratory tests and field trials have been done to determine the mechanism(s) responsible for, and the extent of, Na movement in montmorillonitic mine soils (*53*). The finding: diffusion, not movement by convective flow, was the dominant salt transport mechanism over the coversoil-spoil interface. Diffusion is a significant transport mechanism when the spoil is highly dispersed, the necessary chemical gradient is present to drive the diffusion process, and the coversoil-spoil interface is far enough below the mine-soil surface so that the seasonal downward leaching of salts does not predominate. But the diffusion process is slow and self-limiting. The rate of diffusion is proportional to the square root of elapsed time. This rate declines as the concentration gradient that drives diffusion decreases.

The extent of Na movement and accumulation was observed at four North Dakota field sites over a four-year period (*53, 55*). Salt diffusion accounted for upward movement of Na in two mine-soil profiles having an average spoil SAR of 26, a saturation percentage of 133, and an electrical conductivity of 3.6 dS/m. In these mine soils, the lower halves of the 30-cm (12-inch) coversoils showed an average SAR increase of 14 after a period of 2 to 3 years. Apparently, downward convective salt movement (leaching) effectively abated diffusive upward Na movement in the two other mine-soil profiles, which had an average spoil SAR of 11, a saturation percentage of 87, and an electrical conductivity of 3.7 dS/m.

Given the semiarid conditions and characteristics of their study sites, Merrill and associates (*53*) concluded that significant Na accumulation will occur in the lower half of 30 cm of coversoil only if the associated spoil has very low hydraulic conductivity. This low hydraulic conductivity may result from dispersion and swelling due to sodicity and high clay content, severe compaction, or other reasons. Their calculations indicated that Na accumulation and the thickness of coversoil over which the accumulation would be significant would increase with increased coversoil depth.

Merrill and colleagues (*55*) singled out unstable, dispersed spoils with SAR values of 20 or more as having the greatest potential for degrading coversoils by chemical diffusion. But if a coversoil has sufficient salinity or can release sufficient salt through weathering, the hydraulic conductivity of the coversoil will be maintained, enhancing leaching, and coversoil sodication would be avoided.

Dollhopf and associates (*24*) studied salt movement under semiarid conditions at the West Decker Mine in Montana. Plots consisted of 70 cm (28 inches) of good quality sandy loam coversoil placed over kaolinitic sandy clay loam to loam spoil (kaolinite, 55 percent; smectite, 30 percent) having an average ESP of 20 (SAR = 23), a saturation percentage of 73, and an electrical conductivity of 2.8 dS/m. Over the 7-year study period, coversoil SAR levels declined significantly in the 0-to-23-cm (0-to-9-inch) depth, but increased significantly in the 23-to-70-cm (9-to-28-inch) depth (SAR increases of 1.5 to 2.5). Upward Na movement from spoil into coversoil occurred at a very slow rate. Linear regression was used to predict when SAR values in the coversoil would reach 15. At the 46-to-70-cm (18-to-28-inch) depth, SAR levels were predicted to reach 15 in 18 years in one replicate and in 36 years in another replicate. Use of a linear model assumes that the rate of upward salt migration is linear and constant; however, because the rate of movement of salts by diffusion is a function of the concentration gradient and the concentration gradient should decline with time, a reduction in the rate of salt migration would be expected to occur. In addition, leaching, especially in wet years, will retard the migration process. Therefore, these time predictions represent the worst-case scenario for coversoil degradation. Realistically it may take much longer, if ever, before Na levels become inhibitory (*24*).

Although salt migration occurred at the Decker Mine, it did not occur as rapidly or reach the levels reported in North Dakota mine soils. These differences in salt migration rates are attributable to mine-soil construction materials and techniques. In North Dakota, mine soils were often constructed of smectite-dominated, fine-textured materials having high soil water contents throughout the year. Because of the clay mineralogy and high soil water content of these materials, the clays tended to be dispersed, resulting in low hydraulic conductivities and very limited water movement across the coversoil-spoil interface. It is under these conditions of poor hydraulic conductivity and high soil water content that rapid upward migration of salts by diffusion will occur (*55*). At the Decker Mine the combination of sandy loam coversoil over sandy clay loam sodic spoil, the predominance of relatively nonswelling clay minerals throughout the profile, and a coversoil-spoil profile that was not seriously compacted increased hydraulic conductivity in the mine-soil profile. This higher mine-soil hydraulic conductivity and the resultant flux of water and Na salts through the profile were probably responsible for the greatly reduced rate of upward salt diffusion.

Miyamoto (*1983*) compiled published literature and unpublished data to assess potential Na and salt migration and its implication for vegetation stability at the San Juan Mine in New Mexico. Diffusional transfer of Na and salts was estimated to be much less than that reported in North Dakota, mainly due to the dryness of the New Mexico mine soils [average annual precipitation of about 15 cm (6 inches)]. In addition, high salts, which help to maintain hydraulic conductivity, and occasional heavy rains or winter precipitation promote downward salt leaching. Available data indicated that salinity and sodicity were decreasing with time under reclaimed range conditions; however, if the reclaimed areas were to be used for long-term irrigated pasture, a severe Na hazard may surface, especially in spoil derived from saline-sodic shale.

The movement of water and soluble salts in mine soils is a dynamic process. When precipitation or irrigation and hydraulic conductivity are sufficient for leaching, there is a net downward movement of water and salt. When leaching is not occurring and/or when mine-soil water is lost through evapotranspiration, the direction of salt movement may reverse, with salt being transported upward by diffusion or convective flow. Site-specific hydrologic cycles appear to control the distance salts will migrate upward into coversoil. When an equilibrium is achieved between downward leaching and the upward movement of salts, upward migration of salt should cease (*24*).

Coversoil depth requirements. The primary factor determining the depth of suitable coversoil needed to restore soil productivity is the quality of the underlying spoil (*22, 38*). When spoil has suitable characteristics for plant growth, it may be used in the root zone and a lesser coversoil thickness may be required. Sodium-affected spoil must be placed where it will not affect plant development.

Power and associates (*66*) determined optimum coversoil depth requirements for maximum plant production on disturbed sites in North Dakota. Coversoil materials were placed over a sodic spoil (SAR = 25, clay = 38 percent) in a wedge plot design. Subplots were seeded to four different crops: alfalfa, crested wheatgrass, blue grama and sideoats grama, and spring wheat. Yields of all crops increased as coversoil thickness increased to the 90- to 150-cm (35- to 59-inch) range. In most instances, more than 90 percent of maximum yields were obtained with 70 cm (28 inches) of subsoil plus 20 cm (8 inches) of topsoil—a total of 90 cm (35 inches) of coversoil. Considering the possibility of coversoil loss due to erosion and upward salt movement, the researchers recommended that more than 90 cm of coversoil be used to ensure adequate thickness.

Another study done to determine the depth of coversoil necessary to maximize production of cool-season perennial grasses involved the establishment of 15 coversoil wedge plots at mines in Montana, North Dakota, and Wyoming (*7*). These plots were grouped into four categories by spoil characteristics: generic spoil, sodic spoil, acid spoil, and soil-like spoil. The two sodic spoil plots, located in Montana and North Dakota, had an average texture of clay loam and an average ESP of 21. Optimum perennial plant production required 71 cm (28 inches) of coversoil over this spoil type. Maximum plant cover was obtained at much shallower coversoil depths.

In an extensive review of the literature on coversoil depth requirements for surface mine reclamation in semiarid regions of the western United States, Hargis and Redente (*38*) focused on coversoil thickness requirements for maximum plant production. Coversoil thickness, they concluded, should be 15 to 61 cm (6 to 24 inches) on spoils with no detrimental properties for plant growth and up to 102 cm (40 inches) on spoils with detrimental properties (such as high exchangeable Na).

Doll and associates (*22*) reviewed published and unpublished research on coversoil depth requirements for maximum plant production on North Dakota mine soils. They suggested guidelines for coversoil replacement based on spoil properties (Table 4). These coversoil depth suggestions are somewhat deeper than the optimum levels reported in the North Dakota experiments. But the researchers believe that, until more is known about changes occurring in unweathered spoil materials placed within the root zone and about Na movement, care must be taken to provide an adequate depth of coversoil to ensure the permanent restoration of maximum production levels. They also recommended that when topsoil or subsoil materials are sandy loam or coarser the subsoil replacement depth should be increased about 12 inches, thus providing a larger soil water reservoir to compensate for the low water-holding capacity of coarse-textured soils.

If an area has poorly developed soils, thus a limited soil resource, then the depths discussed above may not be applicable. An example of this condition exists in northwestern New Mexico where aeolian soils typically consist of 3 to 5 cm (1 to 2 inches) of A horizon over C horizon. W. L. Gould, retired professor of agronomy at New Mexico State University (personal communication, 1986), feels that intermixing 15 cm (6 inches) of aeolian material with about 15 cm of spoil and covering this mixture with 15 to 20 cm (6 to 8 inches) of aeolian material makes a better growth medium than 90 to 120 cm (35 to 47 inches) of aeolian material over spoil. Observations of native range condition support this plan; productivity and cover appear more favorable where 20 to 90 cm (8 to 35 inches) of native aeolian material overlies sodic

Table 4. Suggested guidelines for soil replacement based upon spoil properties in North Dakota (22).

Spoil Properties			Depth of Soil Replacement (cm)		
Texture	EC (dS/m)	SAR	Topsoil	Subsoil	Total
Coarse*	<6	<12	30	61-76	91-107
Medium†	<6	<12	30	30-46	61-76
N/A‡	N/A‡	12-20	30	61-91	91-122
N/A‡	N/A‡	>20	30	91-122	122-152

*Sandy loam or coarser.
†Loam or finer.
‡Not applicable, SAR dominant property.

shale than where the aeolian material is greater than 90 cm deep. In arid areas, a restrictive zone at 30 to 60 cm (12 to 24 inches) may make a site less droughty than a site with deeper soil.

Most coversoil depth research has focused on the coversoil depths needed to obtain maximum grass production. More limited is research regarding coversoil depth requirements that will ensure the establishment of "a diverse, effective, and permanent vegetative cover of the same seasonal variety native to the area of land to be affected and capable of self-regeneration and plant succession at least equal in extent of cover to the natural vegetation of the area" (Surface Mining Control and Reclamation Act of 1977, Public Law 95-87). Also, little is known about coversoil depth requirements for forbs, shrubs, and trees. Coversoil depths that provide for the maximum production of grasses may not ensure the vegetative cover, diversity, and permanence of grasses, forbs, shrubs, and trees that is required by law.

Plant community diversity, for example, may not be achievable in a system designed for maximum production. Species diversity relates inversely to biomass production, indicating that both may not be simultaneously attainable (*13, 91*). These results are consistent with predictions by various models (*33, 39, 94*) that suggest an inverse relationship between species diversity and production. Plant species diversity will probably be increased by reducing coversoil depth, but good plant establishment and rapid soil cover might be hampered and the potential resilience of the ecosystem reduced (*13*).

Determining correct coversoil depths must involve more than providing for maximum grass production. Comprehensive coversoil studies are needed of coversoil requirements for various plant morphological classes in the context of achieving the vegetation requirements of the law. Such studies would also help determine if these requirements are practical.

Amendments. The best amendment alternative for Na-affected spoils is soil. But adequate amounts of suitable soil are not always available for reclamation. Chemical or organic amendments must then be considered as an alternative or adjunct to soil replacement (*54*).

Chemical amendments are generally selected for their ability to supply soluble Ca for the replacement of Na on the exchange complex. This cation exchange reaction increases soluble Na, which must then be removed from the system by leaching. Chemical amendments may be of little value under natural conditions where low rainfall limits the leaching of Na and excess salts out of the root zone (*81*). Chemical amendments commonly used for sodic-soil reclamation include gypsum ($CaSO_4 \cdot 2H_2O$), calcium chloride ($CaCl_2 \cdot 2H_2O$), and sulfuric acid (H_2SO_4).

Gypsum is the most commonly used amendment, mainly because of its low cost. Surface-applied gypsum generally increases infiltration and reclaims sodic soils, but because gypsum has relatively low solubility, more time and water are required than with other amendments. Mixing gypsum with the soil usually accelerates the reclamation process (*17*).

Phosphogypsum, a by-product of the phosphate fertilizer industry, has received much attention (*29, 42, 43, 88*). The solubility product of gypsum is the same whether in the form of mined gypsum or phosphogypsum; however, the surface area (at a given fragment size) of phosphogypsum is larger than that of mined gypsum (for a given amount) and, therefore, the dissolution rate is higher (*43*). Because of this higher dissolution rate, phosphogypsum has proved more effective than mined gypsum in maintaining a high infiltration rate. This amendment is also inexpensive.

The high solubility of $CaCl_2$ produces a leaching solution of high electrolyte concentration, resulting in increased water intake rates (*4, 51*). The high electrolyte concentration is efficient for replacing exchangeable Na only in soils with high ESP (*68*). Under these conditions, the overall time and water requirements for reclamation are much less than for gypsum. But $CaCl_2$ can be expensive.

Sulfuric acid is an efficient amendment when used on sodic calcareous soils (*58*). It reacts with $CaCO_3$ to produce a source of soluble Ca and gypsum. It also promotes increased soil water intake due to increased electrolyte concentration and its ability to solubilize aluminum and iron-hydroxy compounds, which promote flocculation and stabilization of soil structure (*68*). Sulfuric acid does require special handling and equipment. It may also be expensive, depending upon its availability.

In addition to chemical amendments, organic matter can be added to soil materials to promote aggregation. Additions of organic matter and the subsequent improvement of structure in sodic soils have been recognized for many years (*81*). Organic matter can (1) reduce the interaction of water with inorganic colloids and (2) physically or chemically bind soil particles together (*18*). The application of a large quantity of organic matter is necessary to be effective.

Prather and associates (*68*) tested the effectiveness of three amendments—$CaSO_4$, $CaCl_2$, and H_2SO_4, individually and in combination—on two reclaimed sites. They found that combining either $CaCl_2$ or H_2SO_4 with gypsum appreciably reduced the time of leaching needed to improve reclamation, compared to using gypsum alone. Combinations of amendments, they concluded, were attractive because such combinations produced the fast-action, increased-permeability benefits of either $CaCl_2$ or H_2SO_4 and the cost-effectiveness of gypsum.

The application of amendments that supply Ca for exchange with Na on the exchange complex is an essential part of the reclamation process. The movement of these amendments through the soil to facilitate this exchange is also a prime consideration. Therefore, Reeve and Bower (*72*) proposed a method of reclaiming sodic soils that uses high-salt water as a flocculant and source of divalent cations. Doering and Willis (*21*) later performed laboratory studies on the use of this method in sodic spoil. They concluded that sodic spoil material can be reclaimed readily with solutions of $CaCl_2$ or any other Ca salt that is soluble enough in water to produce electrolyte concentrations greater than about 0.5 N (500 meq/liter). A cost analysis of this procedure resulted in the conclusion that chemical reclamation is not recommended for large-scale application to spoils. Use of soil on spoil was found to be a better alternative.

In a column study Weber and colleagues (*97*) determined infiltration rates on a saline-sodic spoil in New Mexico as a function of irrigation water quality, amendment type, and timing of amendment application. They found that gypsum improved infiltration only slightly because of the high sulfate levels in the spoil and water. Apparently, the common-ion effect retarded the dissolution of gypsum. The researchers concluded that both gypsum and H_2SO_4 can be effective amendments, but H_2SO_4 appears to be the best alternative if amendments are to be applied to spoils prior to irrigation with saline, high-sulfate water. Leaching that reduced sulfate and maintained electrolyte levels above the threshold level allowed either amendment to reclaim the spoil and promote adequate infiltration rates.

Stand and vigor of four native plants seeded and irrigated for two seasons on an amended sodic sandstone spoil were studied in New Mexico (*32*). The spoil was amended with gypsum, H_2SO_4, bottom ash, and soil, or a combination of these materials. Those field plots receiving 20 cm (8 inches) of soil showed the best stands with greatest vigor. Lowest plant populations were in plots receiving bottom ash. Besides soil, gypsum proved to be more effective than the other treatments; it was suggested as a supplemental treatment.

Jurinak and Wagenet (*41*) conducted column studies on sodic overburden from Montana to determine the relative reclamation efficiencies of $CaCl_2$, H_2SO_4, and 25 percent $CaCl_2$ mixed with 75 percent gypsum (equivalent weight basis). Two water management regimes, ponded and pulse applied, were compared in conjunction with the amendments. The results showed that $CaCl_2$ reduced the ESP of the overburden to a greater depth than H_2SO_4 or the $CaCl_2$-gypsum mix. Pulsed water applications proved to be more efficient than ponded water for reclaiming the sodic overburden. This study also indicated that when water application and/or leaching rates are limited, amendments must be applied in excess of calculated requirements.

Merrill and associates (*52*) studied the reclamation of nonirrigated, sodic spoil field plots in North Dakota using applied soil and gypsum. The application of 30 cm (12 inches) of good quality soil over a sodic, dispersed spoil (SAR = 27, saturation percentage = 119) increased crested wheatgrass forage yields 84 percent. Gypsum incorporated [to 15 cm (6 inches)] in the same spoil increased yields 19 percent. In this experiment soil proved to be the better amendment. Gypsum's effectiveness was hampered by a very low spoil hydraulic conductivity that limited downward leaching.

In laboratory experiments, Scholl and Miyamoto (*82*) evaluated the potential of using H_2SO_4, gypsum, and P, in combination with sand amendments, for revegetating sodic mine spoils. The spoils were of sandstone and shale origin from northwestern New Mexico. Results indicated that the following amendments show promise: for the sandstone-based spoil, H_2SO_4 treatment, gypsum-plus-P application, or P fertilization alone, depending upon water infiltration characteristics; for the shale-based spoil, H_2SO_4 treatment combined with either sand mulching, sand top-dressing, or sand incorporation. A gypsum-plus-P treatment in combination with sand top-dressing was also discussed as an alternative for the shale-based spoil.

Dollhopf and colleagues (*24*) examined the effects of irrigation and chemical amendments on Na migration and vegetation characteristics in a sodic mine soil in Montana. The amendments were applied and disked into the spoil surface, then 70 cm (28 inches) of coversoil were applied over the spoil. The 7-year investigation indicated that amendments will slow down, not prevent, the coversoil sodication process. Amendments provided Ca plus Mg for upward migration with Na, thereby reducing the negative impact of Na on the coversoil. Amendments also effectively reduced spoil SAR levels by 2 to 4 units compared to the check plots. Statistical analysis indicated that, among the four chemical amendments tested, the relative effectiveness of each after the first 2 years of the study was $CaCl_2 > Gypsum + CaCl_2 + NH_4NO_3 > gypsum > gypsum + (NH_4)_2SO_4 + CaCl_2$. Because natural leaching through the mine-soil profile occurred during the 2-year period, irrigation did not appear to be a significant management tool. Precipitation in the first and second year of the study was 63 percent above and 25 percent below, respectively, the average annual precipitation of 31 cm (12.2 inches).

Organic amendments have been examined recently as a treatment for sodic spoils, usually bentonite spoils. Bjugstad and associates (*14*) found that mulching, fertilization, gypsum application, and physical amendment application (sawdust, straw, perlite, and vermiculite) all helped in promoting plant establishment on a saline-sodic bentonite spoil. The researchers stated that the physical amendments provide rapid improvements in spoil physical conditions that often are unattainable with chemical amendment only.

Smith and colleagues (*90*) showed that wood residue and N fertilizer promoted vegetation establishment and growth on a bentonite spoil. Their data suggested that plant growth as well as establishment was stimulated by soil modifications induced by wood residue. These modifications can include increased water infiltration and storage, reduced crusting and bulk density, improved soil structure, and modification of soil temperature. Optimum vegetation density, cover, and biomass occurred on N-fertilized subplots.

Summary and recommendations

Sodicity hazard assessment. Exchangeable-Na percentage is that percentage of the CEC occupied by Na. This appears to be the ideal parameter for assessing directly the Na hazard. But the ESP determination is time-consuming, costly, and subject to error. This error is primarily the result of the CEC analysis, which, despite several attempts to improve it, is suspect with regard to accuracy and reproducibility. Until a generally accepted and reliable CEC analytical method is developed, the use of the Bower and associates method (*16*) is recommended. If another ESP analytical method is used, then a mine-soil ESP guideline should be developed on the basis of empirical tests (infiltration, permeability, plant growth) that incorporate the use of that method.

The SAR analysis is more convenient, less expensive, and useful for determining the general sodicity of mine-soil materials. If reliable ESP threshold values have been established, then the SAR can be used as a screening tool for Na hazard assessment. When mine-soil SAR values approach guideline suspect values, ESP analyses would be performed. Alternatively, material-specific SAR-ESR statistical relationships could be developed so that SAR values could be properly evaluated. Again, this is assuming that credible mine-soil specific ESP threshold values have been established. To have confidence in such regression equations, the CEC and exchangeable-Na determinations should be made under carefully controlled laboratory conditions using the same analytical method used to establish the ESP threshold values. The regression equation developed by Richards (*76*) is not recommended for specific application to mine soils.

An option to these procedures would be to correlate mine-soil SAR values directly with the infiltration, permeability, and plant growth medium characteristics of that mine soil. Then SAR could be used for direct assessment of the mine-soil material.

Sodic spoil weathering. Evidence suggests that as sodic spoils weather, sodicity will decrease because of the increased availability of soluble Ca and Mg. At this time it is not recommended that either mining companies or regulatory agencies account for this weathering phenomenon in determining the suitability of sodic materials. But if site-specific weathering criteria can be developed, then the concept may have application.

Texture and clay mineralogy. Generally, soils with high clay and/or expanding clay contents are more adversely affected by high sodicity than soils with low clay and/or nonexpanding clay contents. Problems associated with fine-textured, Na-affected mine soils may include crusting; low infiltration, permeability, available water content, and aeration; upward salt migration; and increased compaction, runoff, and erosion. Use of loamy mine-soil materials (<35 percent clay) is recommended to avoid these problems.

A complete soil and overburden/spoil clay mineralogy analysis would identify those soil and overburden materials that contain reactive clays. If complete clay mineralogy analysis is not practical or economically feasible, alternatives, including mineralogical analysis of representative samples and/or saturation percentage analysis of all samples, could be used. A saturation percentage equal to or greater than 90 can be used to screen for materials that may contain reactive clays.

Effects on vegetation. The adverse effects of Na on plant growth are usually a function of a deteriorated soil physical condition. Sodium can be toxic to reclamation species, but deterioration of the soil physical condition and hampered plant performance usually occur before toxic effects are evident. Limited information about the effects of Na on vegetation implies that, barring an adverse soil physical condition, reclamation species should function properly in mine soils with ESP values less than about 20. Root growth and function have been shown to be inhibited by spoils with ESPs/SARs greater than about 20. Therefore, a sufficient depth of suitable mine soil (ESP/SAR$<$20) should be provided to ensure proper plant development.

Upward sodium migration. Upward Na migration may occur when the underlying spoil is dispersive, has a low hydraulic conductivity, and an SAR greater than 20. A sufficient chemical gradient must exist between the spoil and mine soil to drive the diffusion process. The diffusion process is slow and self-limiting, and the rate of diffusion will decline as the concentration gradient decreases. Seasonal leaching will retard the diffusion process, especially if mine-soil hydraulic conductivity is maintained. Also, it appears that when mine-soil moisture is lower, diffusive sodication has a lower impact. When an equilibrium is achieved between downward leaching and upward diffusion, upward salt migration should cease.

Sodium-affected spoil should be placed below the mine-soil profile. The addition of amendments to the surface of a Na-affected spoil before mine-soil placement may help slow upward salt movement. Also, an additional thickness of suitable spoil material (a buffer) between a Na-affected spoil and the overlying mine soil should mitigate upward salt movement. Mine-soil profiles should be constructed to allow for the development of the highest possible permeability.

Coversoil depth requirements. If maximum grass production is the reclamation goal, mine operators in semiarid regions should attempt to supply at least 90 cm (35 inches) of coversoil over a Na-affected spoil. Coversoil thickness may be reduced if the mine soil includes a thickness of suitable spoil. Sodium-affected spoil must be placed where it will not affect plant development. The total mine-soil depth may have to be increased to accommodate deeper rooting species or to buffer against mine-soil loss due to upward Na migration and erosion.

The depth criteria discussed above may not be applicable in arid areas with poorly developed soils. Mulching, top-dressing, soil-spoil mixing, amendment use, or the use of restrictive zones (to stop downward water movement) may have greater application in these areas.

Little is known about how mine-soil thickness affects vegetative cover, diversity, and permanence or the growth of forbs, shrubs, and trees. More information is needed about these effects so that mining companies and regulatory agencies will know what mine-soil depths should be used, under various circumstances, to ensure reclamation success and compliance with the law.

Amendments. The best amendment for Na-affected spoils is soil, but adequate amounts of suitable soil are not always available for reclamation. Chemical amendments can supplement coversoil replacement. Chemical amendments provide added divalent ions (usually Ca), thereby reducing spoil sodicity and its negative impacts. Chemical amendments are most successful when Na and excessive salts can be leached out of the root zone by irrigation or natural precipitation.

Combining chemical amendments can result in effective reclamation and optimize the cost:benefit ratio. Gypsum, which has relatively low solubility, should compose the bulk of the treatment for a low-cost, long-term solution. A rapid reacting amendment, such as $CaCl_2$ or H_2SO_4, should make up the balance of the treatment to provide immediate alleviation of a sodicity problem.

Use of amendments on spoil is recommended when there is not enough suitable mine soil available to provide for adequate root zone thickness, or when the possibility exists that a portion of the mine soil may be lost to upward Na movement. Surface amendments may be required when the infiltration rate of the topsoil prohibits an adequate supply of water or air from moving into the mine-soil profile, or when crusting interferes with seedling emergence and growth. Mulches, cover crops, organic matter additions, or phosphogypsum can be used at the mine-soil surface to protect or enhance beneficial soil surface characteristics.

Suitability guidelines. The dispersion and swelling of soil clays tend to increase with increased water content and sodicity and reduced electrolyte concentration. Sodium usually has a more adverse effect on fine-textured mine soils, mine soils with a high-expanding clay content, and compacted mine soils. The adverse effects of high Na also appear to be enhanced by high mine-soil pH.

An increase in a mine-soil component's solution concentration should reduce the adverse effects of Na. Topsoil should be the most labile mine-soil component at a given Na concentration because of the direct effects of high quality rainwater and snowmelt water ($\sim$0.0 meq/liter) and physical impacts on surface soil structure. The subsoil component should be more protected from the adverse effects of a given concentration of Na because of the probability that the subsoil solution concentration would be higher than the topsoil solution concentration. This higher concentration would be attributable to an inherently higher salt concentration in the subsoil material, the opportunity for the weathering of $CaCO_3$ and primary minerals to contribute soluble salts, and the accumulation of soluble salts leached from the topsoil. These factors should keep the subsoil solution concentration above the flocculation value for subsoils with ESPs/SARs up to about 15 to 20.

The spoil component of the mine soil should be the least labile of all the mine-soil components at a given Na concentration. Spoil solution concentrations should be the highest of all components because of the following:

► The probability of having the highest inherent salt concentrations.

► Water retention by the coversoil, combined with the low precipitation and high evapotranspiration of arid and semiarid areas, should provide little, if any, water for the leaching of salts beyond the spoil component of the mine soil.

► Salts elluviated from the overlying coversoil and salts released by in-situ weathering should contribute additional soluble salts to the spoil component.

The spoil component should, therefore, maintain a solution concentration with a flocculation value that corresponds to ESP/SAR values exceeding 15 to 20.

Spoil infiltration, permeability, and Na migration studies indicate that the spoil component of the mine soil should have a SAR less than 20, a saturation percentage less than 90, and less than 35 percent clay to maintain suitable mine-soil physical and chemical characteristics. Also, information on the effects of Na on vegetation suggests that reclamation species should perform adequately in mine soil with ESP's/SAR's up to about 20.

Given the research to date, there is enough information available to form a general Na guideline for mine soils (Table 5). If possible, empirical tests should be performed to define material-specific relationships involving Na and to establish the utility of the analytical methods and parameters used.

Table 5. Recommended suitable ESP/SAR values for topsoil, subsoil, and overburden/spoil used as a portion of the mine soil.*

Topsoil	Subsoil	Overburden/Spoil
<10†	<15	<20

*Suitable texture: <35% clay. Suitable saturation percentage: <90.

†Value depends on surface soil condition and management.

REFERENCES

1. Abrol, I. P., R. Chhabra, and R. K. Gupta. 1980. *A fresh look at the diagnostic criteria for sodic soils. In Salt-Affected Soils.* Central Soil Salinity Research Institute, Karnal, India. pp. 142-147.
2. Agassi, M., I. Shainberg, and J. Morin. 1981. *Effect of electrolyte concentration and soil sodicity on infiltration rate and crust formation.* Soil Science Society of America Journal 45: 848-851.
3. Agassi, M., J. Morin, and I. Shainberg. 1985. *Effect of raindrop impact energy and water salinity on infiltration rates of sodic soils.* Soil Science Society of America Journal 49: 186-190.
4. Alperovitch, N., and I. Shainberg. 1973. *Reclamation of alkali soils with $CaCl_2$ solutions.* In A. Hadas et al. [editors] *Physical Aspects of Soil Water and Salts in Ecosystems* (vol. 4). Springer-Verlag, Berlin, West Germany. pp. 431-440.
5. Alperovitch, N., I. Shainberg, and R. Keren. 1981. *Specific effect of magnesium on the hydraulic conductivity of sodic soils.* Journal of Soil Science 32: 543-554.
6. Arora, H. S., and N. T. Coleman. 1979. *The influence of electrolyte concentration on flocculation of clay suspensions.* Soil Science 127: 134-139.
7. Barth, R. C. 1984. *Soil-depth requirements to reestablish perennial grasses on surface-mined areas in the Northern Great Plains.* Mineral and Energy Resources 27: 1-20.
8. Bauer, A., G. W. Gee, and J. E. Gilley. 1976. *Physical, chemical, and biological aspects of reclamation of strip-mined lands in Western North Dakota.* OWRC Grant No. 10470016. North Dakota Agricultural Experiment Station, North Dakota State University, Fargo.
9. Bernstein, L. 1974. *Crop growth and salinity.* In Jan Van Schilfgaarde [editor] *Drainage for Agriculture.* American Society of Agronomy, Madison, Wisconsin.
10. Bernstein, L. 1975. *Effects of salinity and sodicity on plant growth.* Annual Review of Phytopathology 13: 295-312.
11. Bernstein, L., and G. A. Pearson. 1956. *Influence of exchangeable sodium on yield and chemical composition of plants. I. Greenbeans, garden beets, clover, and alfalfa.* Soil Science 82: 247-258.
12. Bernstein, L., and H. E. Hayward. 1958. *Physiology of salt tolerance.* Annual Review of Plant Physiology 9: 25-46.
13. Biondini, M. E., and E. F. Redente. 1986. *Interactive effect of stimulus and stress on plant community diversity in reclaimed lands.* Reclamation and Revegetation Research 4: 211-222.
14. Bjugstad, A. J., T. Yamamoto, and D. W. Uresk. 1981. *Shrub establishment on coal and bentonite clay mine soils.* In *Shrub Establishment on Disturbed Arid and Semi-arid Lands.* Wyoming Game and Fish Department, Laramie. pp. 104-122.
15. Bower, C. A., G. Ogata, and J. M. Tucker. 1968. *Sodium hazard of irrigation waters as influenced by leaching fraction and by precipitation or solution of calcium carbonate.* Soil Science 106: 29-34.
16. Bower, C. A., R. F. Reitemeier, and M. Fireman. 1952. *Exchangeable cation analysis of saline and alkali soils.* Soil Science 73: 251-261.
17. Bresler, E., B. L. McNeal, and D. L. Carter. 1982. *Saline and sodic soils: Principles-dynamics-modeling.* Springer-Verlag, Berlin, West Germany.
18. Cary, J. W., and D. D. Evans, editors. 1974. *Soil crusts.* Tech. Bull. No. 214. Arizona Agricultural Experiment Station, University of Arizona, Tucson.
19. Chen, Y., and A. Banin. 1975. *Scanning electron microscope (SEM) observations of soil structure changes induced by sodium-calcium exchange in relation to hydraulic conductivity.* Soil Science 120: 428-436.

20. Christenson, D. R., and H. Ferguson. 1966. *The effect of interactions of salts and clays on unsaturated water flow.* Soil Science Society of America Proceedings 30: 549-553.
21. Doering, E. J., and W. O. Willis. 1975. *Chemical reclamation for sodic strip-mine spoils.* ARS-NC-20. U.S. Department of Agriculture, Mandan, North Dakota.
22. Doll, E. C., S. D. Merrill, and G. A. Halvorson. 1984. *Soil replacement for reclamation of strip-mined lands in North Dakota.* Bull. 514. North Dakota Agricultural Experiment Station, North Dakota State University, Fargo.
23. Dollhopf, D. J., E. J. DePuit, and M. G. Klages. 1980. *Chemical amendment and irrigation effects on sodium migration and vegetation characteristics in sodic minesoils in Montana.* Bull. 736. Montana Agricultural Experiment Station, Montana State University, Bozeman.
24. Dollhopf, D. J., R. C. Postle, R. B. Rennick, and S. A. Young. 1985. *Chemical amendment and irrigation effects on sodium migration and vegetative characteristics in 7-year old sodic minesoils.* Montana Agricultural Experiment Station, Montana State University, Bozeman.
25. Felhendler, R., I. Shainberg, and H. Frenkel. 1974. *Dispersion and the hydraulic conductivity of soils in mixed solutions.* International Congress of Soil Science Transactions 1: 103-112.
26. Fireman, M., and C. H. Wadleigh. 1951. *A statistical study of the relation between pH and the exchangeable-sodium-percentage of western soils.* Soil Science 71: 273-285.
27. Frenkel, H., and J. D. Rhoades. 1978. *Effects of dispersion and swelling on soil hydraulic conductivity.* Journal of Testing and Evaluation 6: 60-65.
28. Frenkel, H., J. Q. Goertzen, and J. D. Rhoades. 1978. *Effects of clay type and content, exchangeable sodium percentage, and electrolyte concentration on clay dispersion and soil hydraulic conductivity.* Soil Science Society of America Journal 42: 32-39.
29. Gal, M., L. Arcan, I. Shainberg, and R. Keren. 1984. *Effect of exchangeable sodium and phosphogypsum on crust structure—scanning electron microscope observations.* Soil Science Society of America Journal 48: 872-878.
30. Gardner, W. R., M. S. Mayhugh, J. O. Goertzen, and C. A. Bower. 1959. *Effect of electrolyte concentration and exchangeable sodium percentage on diffusivity of water in soils.* Soil Science 88: 270-274.
31. Gilley, J. E., S. A. Schroeder, and D. D. Schlenker. 1984. *In situ saturated hydraulic conductivity and water-holding capacity of undisturbed and surface-mined sites.* Research Report No. 102. North Dakota Agricultural Experiment Station, North Dakota State University, Fargo.
32. Gould, W. L. 1982. *Amendments on sodic spoils for revegetation in the Southwest.* In E. F. Aldon and W. R. Oaks [editors] *Reclamation of Mined Lands in the Southwest: A Symposium.* Rocky Mountain Forest and Range Experiment Station, Forest Service, U.S. Department of Agriculture, Fort Collins, Colorado.
33. Grime, J. P. 1979. *Plant strategies and vegetation processes.* Wiley and Sons, Ltd., New York, New York.
34. Gupta, R. K., C. P. Singh, and I. P. Abrol. 1985. *Determining cation exchange capacity and exchangeable sodium in alkali soils.* Soil Science 139: 326-332.
35. Gupta, R. K., D. K. Bhumbla, and I. P. Abrol. 1984. *Effects of sodicity, pH, organic matter, and calcium carbonate on the dispersion behavior of soils.* Soil Science 137: 245-251.
36. Hall, A. 1979. *Weathering of sodic coal overburden.* M. S. thesis. Colorado State University, Fort Collins.
37. Hall, A., and W. A. Berg. 1983. *Prediction of sodium hazard in coal mine over-*

burden. Reclamation and Revegetation Research 2: 191-204.

38. Hargis, N. E., and E. F. Redente. 1984. *Soil handling for surface mine reclamation*. Journal of Soil and Water Conservation 39: 300-305.
39. Huston, M. 1979. *A general hypothesis of species diversity*. American Naturalist 113: 81-101.
40. Johnston, M. A. 1975. *The effects of different levels of exchangeable sodium on soil hydraulic conductivity*. Proceedings, South African Sugar Technical Association 49: 142-147.
41. Jurinak, J. J., and R. J. Wagenet. 1982. *Spring Creek Coal: Factors affecting the sodicity and reclamation of overburden materials*. Annual Report 2. Utah Agricultural Experiment Station, Utah State University, Logan.
42. Kazman, Z., I. Shainberg, and M. Gal. 1983. *Effect of low levels of exchangeable sodium and applied phosphogypsum on the infiltration rate of various soils*. Soil Science 135: 184-192.
43. Keren, R., and I. Shainberg. 1981. *Effects of dissolution rate on the efficiency of industrial and mined gypsum in improving infiltration of a sodic soil*. Soil Science Society of America Journal 45: 103-107.
44. Klages, M. G. 1974. *Clay minerals of Montana soils*. Proceedings, Montana Academy of Science 34: 12-18.
45. Knudsen, D., G. A. Peterson, and P. F. Pratt. 1982. *Lithium, sodium, and potassium*. In A. L. Page et al. [editors] *Methods of Soil Analysis: Part 2*. American Society of Agronomy, Madison, Wisconsin. pp. 225-246.
46. Lanyon, L. E., and W. R. Heald. 1982. *Magnesium, calcium, strontium, and barium*. In A. L. Page, et al. [editors] *Methods of Soil Analysis: Part 2*. American Society of Agronomy, Madison, Wisconsin. pp. 225-246.
47. McIntyre, D. S. 1958. *Permeability measurements of soil crusts formed by raindrop impact*. Soil Science 85: 185-189.
48. McNeal, B. L., D. A. Layfield, W. A. Norvell, and J. D. Rhoades. 1968. *Factors influencing hydraulic conductivity of soils in the presence of mixed-salt solutions*. Soil Science Society of America Proceedings 32: 187-190.
49. McNeal, B. L., and N. T. Coleman. 1966. *Effect of solution composition on soil hydraulic conductivity*. Soil Science Society of America Proceedings 30: 308-312.
50. McNeal, B. L., and T. Sansoterra. 1964. *Mineralogical examination of arid-land soils*. Soil Science 97: 367-375.
51. Magdoff, F., and E. Bresler. 1973. *Evaluation of methods for reclaiming sodic soils with $CaCl_2$*. In A. Hadas et al. [editors] *Physical Aspects of Soil Water and Salts in Ecosystems* (vol. 4). Springer-Verlag, Berlin, West Germany. pp. 441-452.
52. Merrill, S. D., E. J. Doering, and F. M. Sandoval. 1983. *Reclamation of sodic minespoils with topsoiling and gypsum*. ASAE Paper No. 83-2141. American Society of Agricultural Engineers, St. Joseph, Michigan.
53. Merrill, S. D., E. J. Doering, J. F. Power, and F. M. Sandoval. 1983. *Sodium movement in soil-minespoil profiles: Diffusion and convection*. Soil Science 136: 308-316.
54. Merrill, S. D., F. M. Sandoval, E. J. Doering, and J. F. Power. 1981. *The role of gypsum and other amendments in the reclamation of strip-mined lands in semi-arid environments*. In D. P. Borris and P. W. Boody [editors] *Phosphogypsum*. Florida Institute of Phosphate Research, Bartow. pp. 218-242.
55. Merrill, S. D., F. M. Sandoval, J. F. Power, and E. J. Doering. 1980. *Salinity and sodicity factors affecting suitability of materials for mined-land reclamation*. In *Adequate Reclamation of Mined Lands?* Soil Conservation Society of America and WRCC-21 Committee, Billings, Montana. pp. 3-1 to 3-25.
56. Merrill, S. D., J. R. Deutsch, and M. W. Pole. 1986. *Saturation percentage*. In *Reclaiming Mine Soils and Overburden in the Western United States, Analytic*

Parameters and Procedures. Soil Conservation Society of America, Ankeny, Iowa. pp. 35-58.
57. Merrill, S. D., S. J. Smith, and J. F. Power. 1985. *Effect of disturbed soil thickness on soil water use and movement under perennial grass.* Soil Science Society of America Journal 49: 196-202.
58. Miyamoto, S., R. J. Prather, and J. L. Stroehleim. 1975. *Sulfuric acid and leaching requirements for reclaiming sodium-affected calcareous soils.* Plant Science 43: 573-585.
59. Naghshineh-Pour, G., G. W. Kunze, and C. D. Carson. 1970. *The effect of electrolyte composition on hydraulic conductivity of certain Texas soils.* Soil Science 110: 124-127.
60. Omodt, H. W., F. W. Schroer, and D. D. Patterson. 1975. *The properties of important agricultural soils as criteria for mined land reclamation.* Bulletin 492. North Dakota Agricultural Experiment Station, North Dakota State University, Fargo.
61. Oster, J. D., and F. W. Schroer. 1979. *Infiltration as influenced by irrigation water quality.* Soil Science Society of America Journal 43: 444-447.
62. Park, C. S., and G. A. O'Conner. 1980. *Salinity effects on hydraulic properties of soils.* Soil Science 130: 167-174.
63. Pearson, G. A. 1960. *Tolerance of crops to exchangeable sodium.* Agricultural Information Bulletin No. 216. U.S. Department of Agriculture, Washington, D.C.
64. Pearson, G. A., and L. Bernstein. 1958. *Influence of exchangeable sodium on yield and chemical composition of plants. II. Wheat, barley, oats, rice, tall fescue, and tall wheatgrass.* Soil Science 86: 254-261.
65. Polemio, M., and J. D. Rhoades. 1977. *Determining cation exchange capacity: A new procedure for calcareous and gypsiferous soils.* Soil Science Society of America Journal 41: 524-528.
66. Power, J. F., F. M. Sandoval, R. E. Ries, and S. D. Merrill. 1981. *Effects of topsoil and subsoil thickness on soil water content and crop production on a disturbed soil.* Soil Science Society of America Journal 15: 124-129.
67. Power, J. F., S. D. Merrill, and S. J. Smith. 1985. *Effect of soil thickness on nitrogen distribution and use by crested wheatgrass.* Soil Science Society of America Journal 49: 1,461-1,465.
68. Prather, R. J., J. O. Goertzen, J. D. Rhoades, and H. Frenkel. 1978. *Efficient amendment use in sodic soil reclamation.* Soil Science Society of America Journal 42: 782-786.
69. Pupisky, H., and I. Shainberg. 1979. *Salt effects on the hydraulic conductivity of a sandy soil.* Soil Science Society of America Journal 43: 429-433.
70. Quirk, J. P. 1971. *Chemistry of saline soils and their physical properties.* In T. Talsma and J. R. Phillip [editors] *Salinity and Water Use.* John Wiley and Sons, Inc., New York, New York.
71. Quirk, J. P., and R. K. Schofield. 1955. *The effect of electrolyte concentration on soil permeability.* Journal of Soil Science 6: 163-178.
72. Reeve, R. C., and C. A. Bower. 1960. *Use of high-salt waters as a flocculant and source of divalent cations for reclaiming sodic soils.* Soil Science 90: 139-144.
73. Rhoades, J. D. 1982. *Cation exchange capacity.* In A. L. Page et al. [editors] *Methods of Soil Analysis: Part 2.* American Society of Agronomy, Madison, Wisconsin. pp. 149-157.
74. Rhoades, J. D., D. B. Krueger, and M. J. Reed. 1968. *The effect of soil mineral weathering on the sodium hazard of irrigation waters.* Soil Science Society of America Proceedings 32: 643-647.
75. Rhoades, J. D., and R. D. Ingvalson. 1969. *Macroscopic swelling and hydraulic*

conductivity properties of four vermiculitic soils. Soil Science Society of America Proceedings 33: 364-369.
76. Richards, L. A., editor. 1954. *Diagnosis and improvement of saline and alkali soil.* Agricultural Handbook No. 60. U.S. Department of Agriculture, Washington, D.C. 160 pp.
77. Richardson, B. Z., and E. E. Farmer. 1982. *Changes in sodium adsorption ratios following revegetation of coal mine spoils in southeastern Montana.* INT-87. Intermountain Forest and Range Experiment Station, Forest Service, U.S. Department of Agriculture, Ogden, Utah.
78. Rowell, D. L., D. Payne, and N. Ahmad. 1969. *The effect of the concentration and movement of solutions on the swelling, dispersion, and movement of clay in saline and alkali soils.* Journal of Soil Science 20: 176-188.
79. Rowell, D. L., and I. Shainberg. 1979. *The influence of magnesium and of easily weathered minerals on hydraulic conductivity changes in a sodic soil.* Journal of Soil Science 30: 719-726.
80. Russo, D., and E. Bresler. 1977. *Effect of mixed Na/Ca solutions on the hydraulic properties of unsaturated soils.* Soil Science Society of America Journal 41: 713-717.
81. Sandoval, F. M., and W. L. Gould. 1978. *Improvement of saline-and sodium-affected disturbed lands.* In F. W. Schaller et al. [editors] *Reclamation of Drastically Disturbed Lands.* American Society of Agronomy, Madison, Wisconsin. pp. 485-504.
82. Scholl, D. G., and S. Miyamoto. 1984. *Response of alkali sacaton and fourwing saltbush to various amendments on coal mine spoils from northwestern New Mexico, II. Sodic spoil.* Reclamation and Revegetation Research 2: 243-252.
83. Schuman, G. E., T. A. Sedbrook. 1984. *Sawmill wood residue for reclaiming bentonite spoils.* Forest Products Journal 34: 65-68.
84. Shainberg, I., and A. Caiserman. 1971. *Studies on Na/Ca montmorillonite systems. II. The hydraulic conductivity.* Soil Science 111: 276-281.
85. Shainberg, I., J. D. Rhoades, and R. J. Prather. 1981. *Effect of low electrolyte concentration on clay dispersion and hydraulic conductivity of a sodic soil.* Soil Science Society of America Journal 45: 273-277.
86. Shainberg, I., and J. Letey. 1984. *Response of soils to sodic and saline conditions.* Hilgardia 52: 1-57.
87. Shainberg, I., and M. Gal. 1982. *The effect of lime on the response of soils to sodic conditions.* Journal of Soil Science 33: 489-498.
88. Shainberg, I., and M. J. Singer. 1985. *Effect of electrolyte concentration on the hydraulic properties of depositional crust.* Soil Science Society of America Journal 49: 1,260-1,263.
89. Shay, D. A., and F. E. Parady. 1983. *Significance of refinement in the state-of-the-art of predicting sodic hazard in soils and overburden.* In E. J. DePuit and G. E. Schuman [editors] *Soils and Overburden in Reclamation of Arid/Semiarid Mined Lands.* Wyoming Agricultural Experiment Station, University of Wyoming, Laramie.
90. Smith, J. A., G. E. Schuman, E. J. DePuit, and T. A. Sedbrook. 1985. *Wood residue and fertilizer amendment of bentonite mine spoils: I. Spoil and general vegetation responses.* Journal of Environmental Quality 14: 575-580.
91. Stark, J. M., and E. F. Redente 1985. *Soil-plant diversity relationships on a disturbed site in northwestern Colorado.* Soil Science Society of America Journal 49: 1,028-1,034.
92. Suarez, D. L., J. D. Rhoades, R. Lavado, and C. M. Grieve. 1984. *Effect of pH on saturated hydraulic conductivity and soil dispersion.* Soil Science Society of

America Journal 48: 50-55.
93. Tan, K. H. 1982. *Principles of soil chemistry.* Marcel Decker, Inc., New York, New York.
94. Tilman, D. 1982. *Resource competition and community structure.* Princeton University Press, Princeton, New Jersey.
95. Vander Pluym, H.S.A., J. A. Toogood, and R. A. Milne. 1973. *Reclamation of saline-sodic soil by the high saltwater dilution method.* Canadian Journal of Soil Science 53: 473-480.
96. Velasco-Molina, H. A., and A. R. Swoboda. 1971. *Dispersion of soils of different mineralogy in relation to sodium adsorption ratio and electrolyte concentration.* Soil Science 111: 282-287.
97. Weber, S. J., M. E. Essington, G. A. O'Connor, and W. L. Gould. 1979. *Infiltration studies with sodic mine spoil material.* Soil Science 128: 312-318.
98. Yaron, B., and G. W. Thomas. 1968. *Soil hydraulic conductivity as affected by sodic water.* Water Resources Research. 4: 545-552.

ADDENDUM

Western states' sodium suitability guidelines

Colorado

The Mined Land Reclamation Division (MLRD) of the Colorado Department of Natural Resources established a written guideline for overburden/spoil in 1982. MLRD does not have a formal guideline for soil. However, unsuitability criteria for soil and overburden/spoil are similar. The MLRD guidelines, according to J. Herron, an MLRD soil scientist, are based on a review of the literature and modeled after the Montana and Wyoming guidelines.

The MLRD considers SAR values of 10 to 15 and saturation percentages of 60 to 80 to be marginal; SAR values greater than 15 and saturation percentages greater than 80 are unsuitable. Texture is considered in the evaluation of Na levels on a case-by-case basis. All of these parameters are required in the analysis of soil and overburden/spoil.

Montana

In 1983 the Montana Department of State Lands (MDSL) drafted revised soil and overburden guidelines. The following SAR and ESP suspect levels and related guidelines were developed on the basis of a literature review and a review of comments received:

Soil and Soil Substitutes		Overburden and Regraded Spoil	Suitable Textures for Both Soil and Overburden/Spoil
SAR	ESP	ESP	
>11	>15	>15	cl (≤35%c), scl, sicl (≤35%c, ≥15%s), sil (≥20%c, ≥15%s), 1(≥20%c)
>14	>18	>18	sl, 1(<20%c), sil(<20%c, ≥15%s)

The MDSL recommended that an SAR screening technique be used for overburden/spoil characterization whereby overburden/spoil ESP analysis would be run only when overburden/spoil SAR values exceed the suitability value for soil. An unsuitability limit for saturation percentage of greater than 85 applies to all soil and overburden/spoil materials.

New Mexico

In 1985 the New Mexico Energy and Minerals Department developed informal "Overburden and Soils Inventory and Handling Guidelines." The guidelines are based mainly on the literature, but also on in-state experimental work, experience of mine reclamation specialists, and the Montana and Wyoming guidelines (personal communication in 1986 with E. Kelly, director, Mining and Minerals Division, New Mexico Energy and Minerals Department).

Sodium hazard is assessed for soil and overburden using the SAR, modified by texture:

Suitable SAR	Texture
≤18	Sandy loam or coarser
≤16	Loams and clay loams
≤14	Clays (>40% clay)

A saturation percentage limit of greater than 80 applies to all soil and overburden materials.

North Dakota

The North Dakota Public Service Commission's (NDPSC) critical Na values for topsoil and subsoil are based on the work of Omodt and associates (*60*). These researchers reviewed the data on, and the properties of, soils important to farming and ranching in North Dakota. As a result of this review, the following critical SAR and ESP values were developed for soils:

	SAR	ESP
Topsoil	≥4	≥5
Subsoil	≥10	≥12

As a general policy, NDPSC considers any overburden/spoil materials with an SAR of greater than 14 to be unsuitable (personal communication in 1986 with J. R. Deutsch, senior environment scientist, Reclamation Division, NDPSC). This level was arrived at through a general consensus among research scientists in North Dakota and field research performed in the state; it is applied universally to all mined land situations in the state (personal communication in 1983 with E. C. Doll, superintendent, Land Reclamation Research Center, North Dakota Agricultural Experiment Station, Fargo).

NDPSC requires particle-size analysis on a routine basis for soil and spoil, but no suitability criteria have been established. Saturation percentage determination is not required by NDPSC.

Utah

The Division of Oil, Gas, and Mining (DOGM) in the Utah Department of Natural Resources recently drafted soil and overburden guidelines (personal communication in 1986 with J. S. Leatherwood, reclamation soils specialist, DOGM). These guidelines are based on the literature and guidelines of other western states. DOGM has set its SAR unsuitability criteria for soil and overburden at greater than 12 for fine-textured soil materials and greater than 15 for coarse-textured soil materials. If native materials exceed these values and are unavoidable, companies may test these materials to demonstrate their suitability.

DOGM uses a saturation percentage criterion of greater than 80 to help identify suspect materials.

Wyoming

The Wyoming Department of Environmental Quality (WDEQ) revised its soil and overburden guidelines in 1984. WDEQ's revision is based on the literature, comments received, and practical experience within the state. The flexibility of the new guidelines allows for their application on a more site-specific basis (personal communication in 1986 with D. Boon, soil scientist,

Land Quality Division, WDEQ). WDEQ's SAR guidelines for soil and overburden/spoil are as follows:

	Suitable	Marginal*	Unsuitable
SAR†	0-10	10-12‡	>12‡
		10-15§	>15§

*Evaluated on an individual basis for suitability.
†As an alternative to SAR calculations, ESP can be determined. ESP should be determined if suitable SAR values are exceeded.
‡For fine-textured soils (clay ≥ 40%).
§For medium- and coarse-textured soils (clay < 40%).

WDEQ considers materials with saturation percentages of greater than 80 to be marginal, but saturation percentage analysis is no longer required by WDEQ for mine soil characterization.

GLOSSARY

Arid: A term applied to regions or climates that lack sufficient moisture for crop production without irrigation. Contrast with *Semiarid.*

Bentonite: A sedimentary rock formed from the alteration in place of volcanic ash. It is largely composed of smectite. Sodium bentonite has great ability to absorb water and swell.

Bulk Density: The mass of dry soil per unit bulk volume.

Cation Exchange Capacity (CEC): The total quantity of cations that a soil can absorb by cation exchange (the interchange of a cation in solution with another cation on a surface-active material), usually expressed as meq/100 g. Measured values of CEC vary depending on the method used for determination.

Convective Flow: In the context of this paper, it is meant to be the physical movement of solutes by water. Contrast with *Diffusion.*

Coversoil: That portion of the mine-soil profile that consists of topsoil and subsoil.

Diffusion: The independent or random movement of ions or molecules that tends to bring about their uniform distribution within a continuous system. The movement of solutes in response to a concentration gradient. Contrast with *Convective Flow.*

Electrical Conductivity: A measure of salinity. A physical quantity that measures the readiness with which a medium conducts electricity. Electrical conductivity can be related to the electrolyte concentration of soil waters and aqueous extracts of soils by the equation: electrical conductivity (dS/m) $\times$ 10 = sum of cations or anions (meq/liter), which is applicable to solutions in the EC range of 0.1 to 5 dS/m.

Electrolyte Concentration: See *Electrical Conductivity.*

Exchangeable Cation: A cation that is adsorbed on the exchange complex and that is capable of exchange with other cations.

Exchangeable Sodium: See *Exchangeable Cation.*

Exchangeable-Sodium Percentage (ESP): A direct measure of the percentage of the soil cation exchange capacity occupied by sodium. It may be calculated by the formula:

$$ESP = \frac{ES(100)}{CEC}$$

where the units are expressed as meq/100 g of soil.

Exchangeable-Sodium Ratio (ESR): Exchangeable-sodium percentage (ESP)-sodium-adsorption ratio (SAR) relationships are determined by correlating ESR and SAR using empirical data where:

$$ESR = \frac{ES}{CEC - ES}$$

$$ESP = \frac{(100)\ ESR}{1 + ESR} = \frac{ES(100)}{CEC}$$

where the units are expressed as meq/100 g of soil.

Flocculation Value: The minimum electrolyte concentration that causes the flocculation (aggregation) of clay particles.

Hydraulic Conductivity: The rate of flow of water through a unit cross section of soil per unit of hydraulic gradient under specified temperature and hydraulic conditions, usually expressed as cm/h.

Illite (hydrous mica): A 2:1 clay mineral. Such properties as hydration, cation absorption, swelling, shrinkage, and plasticity are less intensively developed in hydrous mica than in montmorillonite; however, hydrous mica exceeds kaolinite in respect to these characteristics. There is a disagreement in the naming of hydrous mica. It has been called illite, mica, hydromica, and sericite.

Infiltration: The downward entry of water (and air) into the soil.

Infiltration Rate: A soil characteristic determining or describing the maximum rate at which water can enter the soil under specified conditions.

Kaolinite: A 1:1, relatively nonexpanding clay mineral. See *Illite.*

Mine Soil: The reconstructed plant root zone consisting of coversoil and suitable spoil.

Montmorillonite: A 2:1, expanding clay mineral. See *Smectite* and *Illite.*

Overburden: Undisturbed, consolidated or unconsolidated materials overlying a coal seam to be mined (this excludes the suitable soil that would be salvaged).

Permeability: The ease with which water can pass through a bulk mass of soil or layer of soil. This can be either a qualitative term, or a quantitative term if the rate of movement is specified. See *Hydraulic Conductivity.*

pH, Soil: The degree of acidity or alkalinity expressed by the negative logarithm of the hydrogen-ion activity of a soil. The neutral point is pH 7.0. All pH values below 7.0 are acid, and all above 7.0 are alkaline.

Primary Mineral: A mineral that has not been altered chemically since deposition and crystallization from molten lava.

Root Zone: That part of the natural soil or mine soil that is invaded by plant roots.

Saturation, Soil: The point at which a soil will no longer absorb any amount of water without losing an equal amount.

Saturation Extract: The solution extracted from a saturated soil paste prepared by using distilled water.

Saturation Percentage: The moisture percentage of a saturated soil paste, expressed on a dry-weight basis.

Semiarid: A term applied to regions or climates where moisture is normally greater than under arid conditions but still definitely limits the growth of most crops. Contrast with *Arid.*

Smectite: The smectite group includes all the clay minerals formerly classified in the montmorillonite group of expansible layer silicates. In soil environments three smectitites—montmorillonite, beidellite, and nontronite—are important. See *Montmorillonite.*

Sodium-Adsorption Ratio (SAR): A ratio used to express the relative activity of sodium in a saturation extract, as given in the following equation:

$$SAR = Na^{+}/\sqrt{(Ca^{2+} + Mg^{2+})/2}$$

in which ionic concentrations are expressed in meq/liter.

Sodium-Affected: Earthen materials that contain excess sodium in regards to reclamation success, especially plant growth.

Soil Solution: The aqueous phase of the soil and its solutes consisting of ions dissociated from the surfaces of the soil particles and of other soluble materials.

Spoil: Overburden that has been disturbed by mining activities.

Subsoil: Soil materials typically consisting of the lower B and upper C horizons; synonymous with second-lift soil materials.

Threshold Concentration: The relationship between soil ESP or solution SAR and the concentration of the percolating solution that distinquishes stable pore structure from unstable structure.

Topsoil: Soil materials typically consisting of the A and upper B horizons; synonymous with first-lift soil materials.

Chapter 6

Selenium

Scott E. Fisher, Jr., Frank F. Munshower, and Fred Parady

Selenium (Se) was discovered by the Swedish Chemist Baron Jons Jakob Berzelius in 1817 (*1*). The element was isolated from deposits in the lead chambers of a sulfuric acid plant at Gripsholm, Sweden (*85*). It belongs to the group-VI A elements—S, Se, and Te—that possess similar chemical and biochemical properties. Like S, Se occurs in several allotropic forms (*85*). These include (a) metallic hexagonal, crystalline-stable form, lustrous gray to black in color; (b) red Se, monoclinic crystals consisting of Se8 milecules; and (c) amorphous Se, which may exist as black, amorphous red, or colloidal Se.

The element is widely distributed in the earth's crust. It frequently is found in mineralogical association and with sulfide minerals, primarily Pb, Zn, Hg, Ag and Cu (*60*), with some 40 microminerals known at present. Selenium is a rare, chalcophyllous (showing strong affinity for sulfur) element (*45*). The *Handbook of Chemistry and Physics* (*21*) gives the average crustal abundance of Se as 0.09 mg/kg, but Taylor (*75*) suggested a lower concentration of 0.05 mg/kg was more appropriate (both figures based on igneous rocks). Goldschmidt (*33*) put the Se content of the earth's crust at 0.091 ppm.

Selenium rarely occurs naturally in concentrations exceeding a few hundred parts per million, and it tends to concentrate in sedimentary materials. In such deposits Se primarily occurs as oxidates of vanadium-uranium ores or in hydrolysates, such as shales (*27*). Selenium is most frequently found in the natural world as a contaminant element in a series of sulfide, uranyl-V,

Mo, phosphorite, and S deposits (*69*). The most common inorganic forms of Se are selenides, selenites, selenates, hydrogen selenides, selenium dioxide, selenious and selenic acids, and selenium halides and sulfides (*85*).

Selenium is produced commercially in several countries; these include Canada, the United States, the Soviet Union, Japan, and Sweden. The element frequently is a by-product of electrolytic Cu refining (*1*). Six companies produced Se in the United States in 1967; five of the six produced the element as a by-product of Cu refining (*17*). Industrially, the element is widely used in the chemical industry as a catalyst and in other situations to improve alloys. It is also used in electrical manufacturing of semiconductors and thermo and photo elements and as a decolorizer or colorant in glass.

Selenium is an element closely related to S—both chemically and biologically. It has an atomic weight of 78.96 and an atomic number of 34 while S has an atomic weight of 32.06 and an atomic number of 16 (*68*). Selenium reacts with mineral acids to produce hydrogen selenide, which is a stronger reducing agent than the related hydrogen sulfide (*64*). It is closely related to S in its geochemistry and occurs in sulfide minerals. The ionic radii of the two elements are similar [$S(2^-) = 1.85A$ and $Se(2^-) = 2.0A$] (*28*). Selenium occurs as -2, 0, $+4$ and $+6$ valence states; binding in these states is primarily covalent (*85*).

Much of the interest in this element is based on animal toxicities that have long been associated with Se-induced acute livestock mortalities (selenosis) and chronic general loss of thriftiness (*62*). More recently, Se has been found deficient in some livestock forages, resulting in lower production and an increase in offspring mortality. Concentrations of Se in surface water and groundwater, elevated to the point of potential toxicity, have not received much attention until recent events at the Kesterson Reservoir in California (*50*).

Studies and the impacts of Se have an interesting history. Illness and death among horses in Marco Polo's caravan during 1295 in China are thought to have been the result of selenosis (*60*). A similar condition was reported in the Great Plains of the United States in 1856; the condition was subsequently called "alkali disease" or "blind staggers." The Wyoming Agricultural Experiment Station's Third Annual Report (*88*) discussed illness and mortality among horses eating irrigated hay from the Shirley Basin area of Carbon County. Another major loss of livestock—sheep in this case—took place in the Shirley Basin in 1907 (*89*). An understanding of the cause was not achieved until the 1933 to 1936 era and the publication of early works by Beath and associates (*10, 11*), Byers and Knight (*19*), and Byers (*18*). These papers described the resource and environmental factors involved in the disease selenosis and means of possibly alleviating the problem. Selenium toxicities or selenosis have been reported from many parts of the world, but problems with Se deficiency are more widespread (*27*).

Selenium in surface mine reclamation[1] and regulation

The importance of Se in surface mine reclamation generally lies in two relatively distinct areas. First is the potential contamination of minesoils through the addition of Se-bearing materials that may ultimately be taken up by crops, forages, or native plants, resulting in toxicities to animals and/or the related capacity of the reclaimed landscape to meet the postmine land use. Second is the quality of both groundwater and surface water after reclamation. The forage produced after reclamation must have the ability to support livestock grazing where such use is identified as one of the final land uses. Similar requirements are necessary for meeting wildlife land use goals and habitat/ecosystem requirements. If the forage produced from the revegetated soils (mine soils) contains sufficient quantities of Se to be toxic to livestock, it cannot be considered capable of meeting the defined postmine land use. In summary, Se is of concern in successful reclamation because it can directly influence the value and utility of reclaimed land.

State and federal regulations control the quality of water draining from and/or being recharged through mined land. To accomplish successful reclamation of surface-mined land, the concentration of Se in both surface water and groundwater must be in compliance with these regulations and standards.

An additional but more limited Se-related concern in surface coal mining and reclamation involves the burial of fly ash in reclaimed landscapes. The potential for both groundwater and surface water contamination and the production of toxic plant material or forage exist if such wastes are not properly disposed of (isolated) during the backfilling and grading portions of reclamation operations.

Metabolic role of selenium

Selenium is not considered an essential element for plant growth and development, although growth of some plant species (primarily accumulators) has been enhanced when the element was added to the plant growth medium (*77*). The element is not known to retard plant growth at any concentrations generally encountered in soils, but Se toxicity to certain plant species has occurred in greenhouse studies.

The element is essential to animal nutrition. Deficiencies produce a char-

[1]The term reclamation, as used here, follows the definition of Munshower (*51*), which includes rehabilitation of all aspects of the environment or ecosystem. Such a definition must be carefully differentiated from the concepts of reclamation as ecosystem restoration or revegetation.

acteristic syndrome—white muscle disease in livestock (*23, 39*). Early work with Se appeared to suggest carcinogenic activities associated with elevated dietary levels, but more recent studies suggest the opposite—some potential for retardation in the development of some types of tumors. Underwood (*80*) indicated that a dietary intake of 0.1 ppm in dry feeds is adequate for grazing cattle and sheep. Vitamin E and Se are closely related to the development of white muscle disease, and symptoms as developed in this myopathy can be produced by a vitamin E deficiency (*29*).

Toxicities resulting from the ingestion of excessive amounts of Se have occurred widely throughout the world, but forage threshold levels capable of producing chronic or acute toxicity cannot be precisely defined. Table 1 provides some data on a species-specific level. Differences in individual and species responses, differing reactions to various chemical forms of Se, and associated dietary and environmental conditions affect animal response to Se in foodstuffs. The literature indirectly suggests that animals can develop some tolerance to elevated dietary levels of Se. Underwood (*80*) suggested that the minimum Se level in plant tissue that would produce toxicity in grazing livestock is approximately 5 ppm. Byers (*18*) reported 4 ppm as the tolerance limit for Se concentrations in livestock forage; he also suggested that 5 ppm should be considered potentially hazardous. The National Ecological Research Laboratory (*85*) suggests that 0.2 ppm Se in foodstuffs is required to prevent deficiencies, that animals generally have a tolerance to 4 or 5 ppm, and that 8 to 10 ppm Se will be toxic. Ganje (*30*) suggested that livestock consumption of water with Se concentrations greater than 0.5 ppm should be considered dangerous.

All animals are thought to be susceptible to Se toxicosis, although herbivores directly consuming seleniferous vegetation are believed to be the most susceptible. The factors influencing the toxicity of the Se ingested include the total amount and frequency of Se ingestion, associated dietary protein level, complexing or synergistic dietary substances, type of Se compound in-

Table 1. Selenium requirements and toxicity in livestock, laboratory animals, and poultry feeds.*

Livestock Type	Requirement Level	Tolerance Level	Toxic Level
	ppm		
Cattle	0.10	2.0	8.0
Sheep	0.10	4.0	10.0
Swine	0.10	2.5	7.0
Horses	?	2.0	10.0
Poultry	0.15	5.0	15.0
Turkey	0.20	?	?
Rat	0.08	4.0	10.0

*From National Ecological Research Laboratory (*85*) as adapted from Schwarz and Foltz (*66*).

Table 2. Estimates of livestock response to dietary selenium content.

Physiological Condition	Dietary Se Intake
Deficiency	0.1 ppm
Adequacy	0.1 - 1.0 ppm
Chronic toxicity	3 - 20 ppm
Acute toxicity	100 ppm

gested—organic or inorganic, etc.—and the effectiveness of the animal in eliminating Se when ingested (*85*).

Rotruck and associates (*63*) described the isolation of the selenoenzyme glutathione peroxidase from mammalian tissue. The enzyme serves to break down hydrogen peroxide in the presence of reduced glutathione to water and oxidize the glutathione. The specific mechanism of action of the enzyme and its chemical structure have not been determined.

The majority of analyses of plant tissue indicate that Se concentrations generally do not exceed 1 ppm (*22, 24*). Acute toxicity occurs when animals consume vegetation containing several hundred ppm Se. There is some evidence that livestock animals that do not routinely consume forage containing elevated levels of Se are particularly susceptible to Se-enriched forages. Chronic toxicity occurs only when animals consume plants containing 3 to 20 ppm over a long period (*52*). Table 2 summarizes this dietary information.

Koval'skii and Ermakov (*45*) summarized the roles Se plays in biological systems: (a) Selenium has an effect on cell growth and mitosis of organisms; (b) functional antagonism exists between S and Se—under some conditions Se replaces S in organic compounds; (c) inorganic Se compounds undergo methylation in the developmental processes of several microorganisms—selenomethionine can substitute for methionine in transmethylation reactions; (d) many biochemical reactions of Se compounds are not specific because they also can be carried out by their S analogs; (e) mechanisms and roles for Se-containing compounds remain relatively unclear; (f) additions of Se to the diet of animals increase the total activity of the alpha ketoglutaric acid oxidase system; (g) compounds of Se influence the process of oxidative decarboxylation of pyruvate; and (h) a link exists between Se and the processes of phosphorylation—altering the rate of formation of adenosine triphosphate (ATP) in animal tissue.

Distribution and geochemical relationships of selenium

Geologic materials. Selenium is found in highest concentration in sedimentary materials or rocks. Koval'skii and Ermakov (*45*) provided the rock-type Se contents found in table 3. Brimhall (*14*) reported the highest Se concentrations in shales that were dark colored, carbonaceous, calcareous, soft, usually gypsiferous, and contained significant iron oxide or pyrite. Boon and Smith (*13*) found enrichment of Se, as well as other elements, in carbonaceous materials in Wyoming. Savelev (*65*) reported above normal concentrations of Se in Upper Cretaceous sedimentary rocks in Central Asia; highest concentrations (1,800 ppm) occurred in parent rocks containing high levels of iron sulfide and organic carbon. Wells (*83*) reported average Se concentrations (ppm) in parent rocks of New Zealand as follows: rhyolitic pumice, 0.16; andesite ash and andesite, 1.19; basalt, 0.61; granite, 0.15; schist, 0.22; gneiss, 0.25; ultrabasic rocks, 0.20; greywacke, 0.36; argillite, 1.80; mudstone, 0.33; and limestone, 0.20.

Higher Se contents occur in the Pierre and Niobrara Formations of south central Colorado in more clastic units that include abundant volcanic materials. Most Se in these unweathered shales is within the diagenetic pyrite as opposed to being found adsorbed on hydrous ferric oxides or as elemental Se in the weathered shales (*40, 41, 42*). Sindeeva and associates (*69*) reported 900 to 51,000 mg/kg Se in volcanic S. The Se content in ppm of various minerals in ferrous Fe-Cu ores is as follows: galena, 440-1160; sphalerite, 198-940; chalcopyrite, 50-512; arsenopyrite, 70; pyrite, 41 (maximum 600); pyrrhotite, 22; ankerite, 6-16; and quartz-dolomite-melanterite-chalcanthite, 2-4 (*6*). The Se content in pyrite depends upon the type of deposit and usually is related to Cu-Mo types, with the highest Se content in pyrite found at the medium stage of Cu-Mo mineralization (*16*). The average ratio of S:Se in the lithosphere is 6,000:1, with the plutonic rock reaching 7,000:1 and larger concentrations of Se occurring in the schist/clays, lowering the ratio to 4,000:1 (*45*).

Lakin (*48*) identified three processes for concentration of Se in sedimentary rocks: (a) precipitation of Se from volcanic emanations by rain; (b) deposition of erosional products from volcanic sulfur, seleniferous tuffs, and sulfide deposits; and (c) precipitation of Se from streams or groundwater carrying unusual quantities of Se from older seleniferous sediments.

Table 3. Average content of selenium reported by researchers in various rock types (45).

Rock Type	Selenium (ppm)	Source
Volcanic	9.1	(*32*)
Volcanic	140	(*68*)
Limestones	110	
Limestones and dolomites	110 to 1,100	
Schists	61	
Schists	51 to 110	
Schists and clays	61	
Basalts, gabbros, granites and granodiorites	51	

Table 4. Selected overburden data from the Hanna Basin, Wyoming (*14*).

Depth (feet)	Lithology	Se (μg/g)
46-52	sandstone	0.05
52-62	shale-carbonaceous shale	0.30
62-83	siltstone	0.03
83-85	shale	0.02
91-94	siltstone-coal	0.12
94-97	shale-carbonaceous shale	0.35
97-101	siltstone	0.08
101-105	shale	0.06

In the western United States much of the area west of the current Mississippi River was inundated by water from Devonian to Miocene times. Periodic volcanism in the area of the existing west coast deposited significant amounts of gaseous and particulate Se into the prevailing westerly winds. The element was deposited with the sedimentary materials being formed in this enormous basin. Under anaerobic conditions, generally characterizing the depositional environment of these sediments, Se is frequently found associated with volcanic S and sulfides (*3, 47*). Furthermore, it is generally found in highest concentrations in shales with significant accumulations of organic material (*27*). Overburden data in table 4 provide evidence of these differential lithologic types of concentrations in Wyoming (*13*).

Sedimentary materials of Cretaceous age that have been identified as commonly seleniferous include the Pierre, Niobrara, and Greenhorn formations. Various Jurassic and Triassic formations support mildly or occasionally highly seleniferous vegetation, but they are not as widespread or as potentially toxic as the Cretaceous deposits. Other formations of non-Cretaceous age containing significant amounts of Se include the DeSmet and Arlington tills (South Dakota), Eocene tuffs (Wyoming), Morrison (South Dakota), Gunnison (Colorado), Phosphoria (Wyoming), and Manning Canyon (Utah) (*47*). Additional information about geologic formations containing elevated concentrations of Se can be found in Trelease and Beath (*78*), Lakin (*46, 47*) or Rosenfeld and Beath (*62*).

Elevated concentrations of Se have been identified in parent materials and soils of the western United States (*46, 47*). Known seleniferous formations and soils derived from these materials are distributed from Canada to Mexico. These materials exhibit wide variation in both lateral and vertical distribution of Se over relatively short distances. To better understand the erratic distribution of Se in Great Plains vegetation, emphasis must be placed on characteristics of the soil parent materials/rock, pedogenic processes by which the soil was formed, and plant communities, including those currently on the site and those that occupied the landscape during soil formation.

Overburden generally is found in a reduced state prior to mining. But mining exposes overburden to the atmosphere and portions of it become oxidized. Of primary concern to reclamation programs are the oxidation of nonavailable (reduced) forms of overburden Se to more readily available forms during mining and the potential placement of the latter within the near surface or resaturated zones. Mine soils generally contain some spoil materials within the active plant root zone of 1.5 to 2.0 meters. Such materials generally will become oxidized through pedogenic activity, resulting in Se increasingly becoming available for plant uptake. Unfortunately, no analytical method or predictive model exists to estimate the rate of oxidation of nonavailable Se to the more soluble forms following surface mining. Estimates of potential plant Se concentration must be made on evaluation of the total pool of Se in the material and its position in the mine soil or regraded spoil. The plant species expected to be present in the postmine landscape, both from seeding and plant invasion, are equally important considerations in determining suitability levels for Se in mine soil materials.

Fly ash materials. Elemental Se is the principal form of Se in fly ash (*5*). The forms of Se in settling pond materials include selenite, selenate, and, in some instances, a form such as $SeSO_3^{-2}$. Palmer (*59*), using Gutermann and associates (*36*) data, suggests that about 3.25 percent of the total Se in fly ash is soluble. All types of power plant wastes at the Navajo power plant in the Four Corners area of New Mexico show levels of water-soluble Se less than the detection limits or well below established toxic standards (personal communication, John Thompson, Utah International). Goodson and associates (*34*) reported the leachate Se content from fly ash (Tables 5 and 6). Their studies and modelling efforts on the feasibility of using fly ash for backfilling abandoned coal mines in Colorado Springs, Colorado, suggest a potential for groundwater contamination with Se.

Soils. The Se content of soil reflects Se concentration in the parent materials/rock from which the soil is derived, secondary deposition or redistribution of Se within parent materials/rocks, accumulation or deposition by Se-accumulating plants, and erosion and deposition of Se from other parent rocks or volcanic sources. Most soils contain 0.1 to 2.0 ppm total Se, although soils with concentrations of more than 1,000 ppm have been reported (*28*). Hance (*38*) reported several samples of volcanic materials from Hawaii exceeding 6,000 ppm. Table 7 provides the Se content of some of the world's soils.

Soil materials containing Se may develop through in situ weathering of Se rocks (*78*). Soils formed from such materials are not uniformly seleniferous because the outcrops of Se-bearing sedimentary materials are frequently discontinuous in Se concentration with depth. Furthermore, weathering, erosion, and deposition of Se-bearing materials have occurred through pedogenic activity (Table 8). Soils developing in transported sediments, such as alluvial fans and floodplains, will frequently exhibit even wider ranges in Se with depth. While water-soluble Se usually is leached from such deposits during or after transport, erosion of residual materials can lead to the development of soils in the depositional zone with horizons high in Se. Other horizons in the same soil from different sediment sources may contain relatively small amounts of Se. Allway (*2*) describes the fate of Se in soil formation and weathering as being controlled by oxidation/reduction reactions and by the solubility of the combinations formed by differential oxidation states of the element with other products of the soil and parent

Table 5. Analytical data from three fly ash leachate analyses.*

Parameter	Sample 1	Sample 2	Sample 3	EPA Maximum
TDS (mg/liter)	11,000	12,000	12,000	500
pH	9.0	8.9	9.0	6.5-8.5
As (mg/liter)	0.073	0.100	0.100	0.050
B (mg/liter)	57.0	55.0	51.0	NA
Cr (mg/liter)	0.22	0.31	0.31	0.05
F (mg/liter)	25.0	25.0	23.0	1.4-2.4
Se (mg/liter)	0.82	0.79	0.67	0.01
SO_4 (mg/liter)	6,900	7,800	7,000	250

*Abstracted from Goodson and associates (*34*), detailed characterization studies on three initial fly ash samples.

Table 6. Mean sample values of analytical parameters for the 25 fly ash leachate samples (*34*).

Parameter	EPA Limit	Sample Mean
TDS (mg/liter)	500	10,264
pH	8.5	9.43
As (mg/liter)	0.05	0.039
Cr (mg/liter)	0.05	0.321
F (mg/liter)	2.40	14.95
Se (mg/liter)	0.01	0.49
So_4 (mg/liter)	250	6,648

Table 7. Selenium content of soils (*45*).

Soil and Region	Selenium (ppm)	Source
Regional soils of Russian lowlands	1.1	(*81*)
Other Russian soils	1.2 to 37	(*44*)
Scottish soils	11 to 210	(*73*)
New Zealand soils	71 to 168	(*82*)
Kyogo soils Japan	41 to 110	(*79*)
USA soils	1.1 to 10	(*48*)

material/rock complex. The National Ecological Research Laboratory (*51*) prepared a map (Figure 1) of the United States providing patterns of plant Se concentrations that, in many respects, reflects the properties and characteristics of soils as identified earlier by Allaway (*2*).

Studies of the solubility of Se in seven soils by Geering and associates (*31*) indicated that Se concentration in the soil solution was governed primarily by ferric oxide-selenite-adsorption complex (Se oxidation state +4). Oxidation states are governed by redox potential, soil pH, and ions with which Se might combine. Olson and Moxon (*57*) demonstrated the presence of the relatively soluble selenate ion in soils. Vinogradov (*81*) suggested that the selenate anion would be most stable under the high oxidation potential and alkaline conditions characteristic of arid soils. Koval'skii and Ermakov (*45*) provided data supporting the presence of the selenite anion in aqueous extracts from Chestnut soils of the Tuva region of the Soviet Union.

Accumulation of Se in plants and soils of arid lands is related to its solubility and supergene (ores or minerals formed by generally downward enrichment) transport characteristics. Selenium is soluble in waters under oxidized conditions (Eh $>$ +0.241). Selenium is readily precipitated by Fe in limonites or as selenates of heavy metals. Whereas Se salts of alkaline metals are soluble, under reducing conditions these salts are readily reduced to elemental Se, which is often codeposited with S and replaces that element isomorphously. Selenium salts of alkaline metals are soluble. Selenium is not concentrated in solonetz (dark strongly alkaline) soils.

Accumulation in soils formed in arid regions takes place when the soil/parent material/rock system contains Se-forming strata or when conditions exist for the development of increased Se concentration and supergene Se transport (*67*).

Because the analysis of soil Se provides a poor index of potential toxicity, the definition of seleniferous soils by total or water-soluble soil Se concentrations historically has not been practiced widely. Instead, suspect soils have been located and mapped by the presence of Se accumulator species (*78*). Identification of soil regions thought to contain suspect seleniferous soils has and can be based on the knowledge of potential Se contents of geologic materials. Only after vegetation identified as accumulator species have been mapped are soil samples collected and Se concentrations determined.

Table 8. Distribution of selenium within a Wyoming soil profile (*70*).

Horizon	Depth (cm)	pH	EC (mmhos/cm)	SAR	Se (ppm)	Texture
A1	0-5	6.7	0.3	0.5	0.02	loam
C1	5-28	7.2	2.4	0.1	0.02	sandy loam
C2cs	28-40	7.6	4.5	0.5	0.03	sandy loam
C3rcs	40-75	7.2	6.2	0.5	0.22	silt loam
C4r	75+	6.5	4.2	0.7	0.50	silt loam

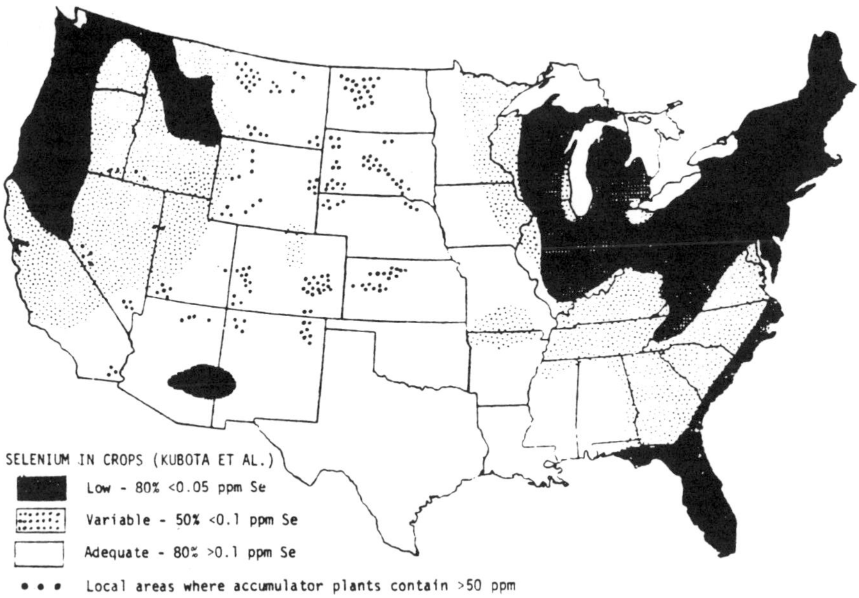

Figure 1. Regional plant tissue Se concentrations.

Seleniferous materials have higher concentrations of total Se ranging upward from 5 ppm. Such soils may be toxic or nontoxic depending upon the amount of available Se. Toxic seleniferous soils and spoils usually contain water-soluble or extractable Se in concentrations >0.1 ppm. Nontoxic seleniferous and nonseleniferous soils also may contain small quantities (as much as 0.1 ppm) of the element in a water-soluble form.

Redistribution and/or secondary deposition. Selenium redistribution from subsurface strata to surface soils through plant accumulation and subsequent decomposition of vegetative material also has occurred and is a continuing process where such plant species exist (*78*). Maximum leaching of soil Se takes place when the weathering processes are relatively weak and do not significantly alter the chemical and physical properties of the iron fraction of the parent material and when oxidation of the Se minerals can take place without the products being complexed with Fe. In reviewing Se movement in New Zealand soils, Watkinson (*82*) suggested that a large portion of the selenides are weathered to form selenites, which are rapidly complexed with the Fe oxides and clays. Redistribution of Se is limited severely in soils of more humid regions where the soil environment is generally much more acid and Fe oxides are prevalent.

Increased Se in soil materials via uptake and decomposition of terrestrial and/or aquatic vegetation may have contributed historically to the formation of toxic seleniferous soils and parent materials in the Great Plains. Similar mineral-cycling processes take place with the presence of black greasewood (*Sarcobatus vermiculatus*) plants and Na (*61*).

Selenium-accumulating plant species may convert nontoxic soils to toxic seleniferous plant growth media (soils). The primary accumulator plants (*62*) are those species capable of accumulating large quantities of Se. In the process of Se uptake and storage these plants may convert selenites or other less available forms of Se to more readily available forms, such as selenates or organic Se (*43, 78*). The more available forms may be subsequently absorbed by nonaccumulator grasses and forbs. Conversion of nontoxic soils to seleniferous materials is of concern on mined land where Se-accumulating species are allowed to invade or are planted in nontoxic seleniferous mine soils where the root zone contains potentially toxic concentrations of Se in a nonavailable form. Many primary accumulator species, *Astragalus* and *Stanleya* genera in particular, invade drastically disturbed areas. An important shrub being successfully established on mine soils in the arid and semiarid West is fourwing saltbush (*Atriplex canescens*), which is a secondary Se-accumulator species.

Availability and plant uptake

Relationships between soil characteristics and plant Se concentrations are complex. Generalizations about toxic soils must be made with caution. Differences in total, water-soluble, and base-extractable Se indicate that the element exists in different forms in soils and geologic materials. Various investigators have reported elemental Se and nonavailable selenides present in soils as well as selenites, selenates, and organic Se (*62*). Of these various complexes, the very soluble selenates and organic Se are the forms most readily available to plants. However, Thornton and Alloway (*76*) stated, "Despite the complex factors affecting the relationships between trace elements in rock, parent materials, soil, plants, and animals, problems of deficiency and excess in livestock forage frequently reflect the chemical composition of soils and geochemical nature of the underlying rock or parent material."

It is generally accepted that Se uptake by most plant species occurs in ionic forms. Allaway (*2*) developed a schematic representation for soil Se (Figure 2).

Workman (*87*) stated that between pe + pH 14.5 and 7.5 $HSeO_3^-$ and SeO_3^{-2} are the predominant Se ion species while below pe + pH 7.5, HSe^- becomes the dominant ion. Selenium speciation is significantly affected by redox, which is strongly different than S where SO_4^{-2} is the major species at all but the very low redox levels ($pe + pH < 3.3$). Acid soils, cold temperatures, and the application of super phosphate fertilizers reduce the amount of Se taken up by plants (*85*). Using oxidation potentials of the various oxidative states of Se as a function of pH and the solubility of ferric selenite, Workman (*87*) drew two conclusions concerning Se solubilities and plant uptake. First, in regions of alkaline soils, particularly in regions of low rainfall, some of the soil Se would occur as selenate. This form would be quite soluble and readily available to plants. Second, in acid soils Se would be found in ferric selenite precipitates. The solubility and availability to plants would be quite low.

The literature on soil Se uptake provides additional insights into the plant/edaphic factors influencing the uptake of Se. First, plants growing in soils with higher clay content generally exhibit reduced Se concentrations (*12*). Second, S reduces the capacity of plants to absorb Se from the soil system and increases the amount of Se loss to percolating water (*17, 84*). Third, the Se content of oats declined significantly during summer months with temperatures below normal, compared to months with temperatures above normal (*49*). Fourth, Oldfield (*55*) found reduced concentrations of Se in plant tissue under high levels of irrigation and related increased amounts of Se being leached from the soil.

In acid or reduced environments Se is found in the elemental form or as selenides. Neither is readily available to plants. In more alkaline and oxidized environments, Se is converted to selenites or selenates, with the selenates being more soluble and readily available to plants. The selenites readily form insoluble ferric oxide complexes, thereby reducing the poten-

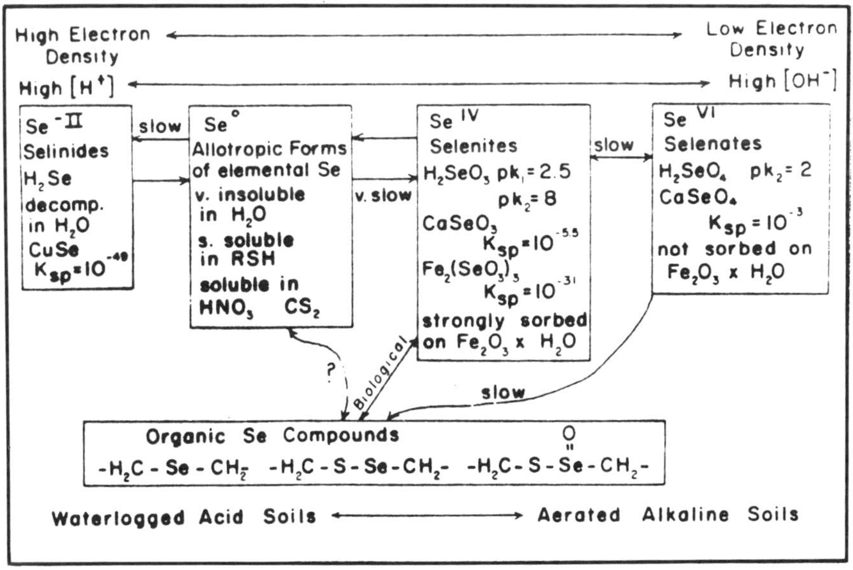

Figure 2. Allaway's (2) generalized chemistry of Se in soils and weathering sediments.

tially plant-available Se. This phenomenon is common in many western areas (*85*).

Plant tissues may contain elevated Se concentrations although the Se source may not occur in the surface soil horizon(s) or epipedons. Plants may accumulate Se from deeper soil horizons, from paralithic materials, or from Se leached from up-slope materials and deposited in the lower portion of the soil/root zone of the vegetation analyzed. The distribution of Se-bearing geologic strata overlying the coal are of major importance in planning the reclamation of drastically disturbed lands. These materials may be transported during mining and backfilling into root zones and remain as future contaminants at the site of deposit. Spoil materials constitute a major portion of the potential root zone of most mine soils.

Selenium deficiencies in animal diets are widespread in the United States and Canada (Figure 1). It is important under such conditions to either increase the Se content of the grain or forage being produced or directly supplement the diets of herbivores. One option is to apply Se to the soil system. This has been accomplished in a variety of experiments with selenite and/or selenate as well as elemental Se. Selenium applied in the form of selenate generally is 10 times more effective in terms of plant uptake than applications of selenite. Cary and Allaway (*20*) found that the Se content in crops stayed at levels sufficient to prevent animal deficiencies for 4 years after treating agricultural soils with Se. Table 9 illustrates the increases in Se concentrations in various crop species after soil treatment. Note the differences in concentration among plant species and within various tissues of the same species.

Not only is the distribution of soil Se highly variable, but plant species also reveal differing capacities to absorb the element. The ability of certain plants to accumulate large quantities of Se has contributed to the history of

Table 9. Plant tissue concentration of selenium after soil application of the element.*

Plant Species/ Tissues	Se Concentration by Se Treatment		
	0 kg/ha	2.24 kg/ha	4.48 kg/ha
	ppb		
Oats			
Whole plant	39	1,260	3,540
Grain	29	1,480	2,490
Straw	31	1,000	1,910
Corn			
Grain	15	375	1,040
Stover	53	353	804
Alfalfa	53	355	742
Timothy	39	345	889

*Modified from National Ecological Research Laboratory (*85*) and Cary and Allaway (*20*).

livestock poisoning and the identification of the agent causing the poisonings. In several papers, Beath and associates (*7, 8, 9*) developed the concept of the accumulation of Se by indicator plants and the recognition of seleniferous areas by the presence of plants with high Se-accumulating capacity.

Plants exhibit differing capacities to absorb Se. Two-grooved milkvetch (*Astragalus bisulcatus*) may accumulate large quantities of the element, up to several thousand ppm. Common western wheatgrass (*Agropyron smithii*) may accumulate concentrations of a few hundred ppm. But the vast majority of plants do not accumulate more than a few ppm Se. Wide variability in the uptake and accumulating capacity of different plant species makes it difficult to define precisely the relationships between Se content in plant tissue and the soil environment. Therefore, relatively precise and accurate soil-plant Se ratios have not been developed for native soils, mine soils, or spoil, nor are they likely to be.

Because species differ radically in their ability to accumulate Se, Beath and his associates (*7, 8, 9*) proposed three categories of plant species based on the species' capacities to accumulate Se. The categories were (a) plant species with very high Se-accumulating power—primary selenium accumulators, (b) plant species with moderate accumulating power—secondary selenium accumulators, and (c) plant species with very low Se-accumulating capacities.

Mentioned in the first group were species of the vetches (*Astragalus*), goldenweed (*Oonopsis*), woody aster (*Xylorrhiza*), and prince's plume (*Stanleya*). These plants exhibit Se concentrations exceeding 1,000 ppm and ranging upward to 15,000 ppm. The second group included species of *Aster*, saltbush (*Atriplex*), gumweed (*Grindelia*), broom snakeweed (*Gutierrezia*), *Penstemon*, and others. Generally, these plants accumulate a few hundred ppm if available Se is present in the soil environment. The group with low Se-accumulating power (usually up to 10 ppm) included most cultivated crops, many native grasses, forbs, and even species of the genus *Astragalus*.

Plant species that accumulate large amounts of Se (Beath's indicator species) may even require the element for maximum growth. There are indications that the species can compete better in seleniferous soils than nonaccumulating vegetation (*78*). These species also contribute to the increase in plant-available Se in surface soils through the decomposition of plant tissues high in organic Se. Thus, these species not only indicate the presence of seleniferous soils, they also cause further degradation of rangeland by increasing the soil pool of Se available to nonaccumulating plant species.

Analytical methods for selenium determination

Selenium compounds are generally volatile. The odor of Se is obvious when seleniferous vegetation is collected or brought into the laboratory.

Therefore, care must be exercised to prevent loss of the element from vegetation and perhaps soil samples. The loss of the element from samples being heated has been reported by several investigators (*10, 32*). Artifically drying vegetation or soils, if the samples are destined for Se analysis, should be avoided. Analysis of vegetation should be on a wet-weight basis and soils on an as-received basis. Under some circumstances the plant tissue Se concentration may be below detection levels for many instruments, and in those cases air drying of the sample would be required. Air drying of soil samples at room temperature is appropriate. While there are reports to the contrary, plant tissues at or near maturity do not appear to lose significant amounts of Se with moderate oven drying (Fisher and Keil, work in progress).

Soils. Analyses of soils for Se can be separated into three categories. These categories include methods that determine (a) total soil/overburden Se concentration (b), some index of plant-available Se in the soil/overburden, and (c) specific forms of the element present in the soil/overburden.

Byers studied the Se content of several South Dakota soils (*18, 19*). Beath and his co-workers studied the Se content of soils in Wyoming (*10*). The work of these scientists and others indicated that total and extractable soil Se had little relationship to Se concentrations in plant tissues. This lack of consistent correlation is due to the forms of the element in the soil, interactions between Se and other elements in the soil system, isolation of Se-supplying horizons from the soil surface, and plant species' specific absorptive capacities.

While the determination of total Se in soils or overburden has little relevance to plant Se levels, the analysis of the elemental pool of Se is of value as an indicator of the absolute amount of Se in the sample. Part or all of the pool of total Se may be oxidized or converted to a plant-available form in the mining/reclamation process. It may be important to know the quantities and forms of Se when overburden materials are to be deposited within plant root zones or substituted for surface soils.

Identification of the quantity of Se in various valence states or chemical forms in mine soils has value in describing the long-term potential maturation of the available soil Se pool, but it should not be uniformly addressed at all mines. Ultimately, the derivation of an index of available soil/mine soil Se content and its relationship to plant levels that may pose a threat to grazing animals would be useful to regulatory agencies and the surface mining industry as well as to agricultural interests. But as has been pointed out, the discontinuous nature of seleniferous materials and the ability of Se-accumulating plants to absorb Se from sources deep within the soil profile indicate that surface soil elemental analyses have serious limitations as a means of forecasting toxicity problems in potential mine soils.

Nevertheless, soil analytical techniques have been recommended for the

Table 10. Selenium in soils and vegetation (*58*).

Soil Depth (inches)	Soluble Se (ppm)	Se Content in Western Wheatgrass (ppm)
0-12	0.12	18
12-24	2.76	
24-36	0.58	
0-12	0.08	2
12-24	0.10	
24-36	0.12	

evaluation of Se in soils and overburden. The two techniques most commonly used are hot-water-soluble Se (*26*) and Soltanpour's ammonium bicarbonate DTPA extraction (*71, 72*). Soltanpur's work indicated that the two techniques extract closely correlated amounts of Se ($r^2 = 0.99$).

Either of the extraction techniques may indicate a seleniferous area if the soil sample selected for analysis represents the existing seleniferous soil zone. However, the seleniferous material may be located at the soil surface or in subsurface horizons. If this material is located beneath the surface, it may escape collection and analysis. The data in table 10 illustrates this point. In the first example the grass revealed potentially toxic Se concentrations, but the surface 12 inches of soil were "unsuitable" only if the soil Se level for unacceptable root zone materials was 0.1 ppm. The second sample did not produce Se-enriched vegetation, but contained almost the same Se concentration as the first soil. The conflict is clear. To identify adequately toxic soils, as illustrated in Olson and associates (*58*), analyses of successive 1-foot increments would be necessary at this site. Many soils would require sampling at increments much less than 1-foot intervals and to depths extending below the soil solum, an obviously impossible sampling regime.

Analyses of overburden for hot-water-soluble Se will reveal elevated levels only if the Se has been oxidized to the selenate form. Elevated levels of Se and related *potential toxicity,* when in the form of elemental Se, selenides, or selenites, will largely not be detected by either the ammonium bicarbonate-DTPA or hot-water-soluble extracts. If overburden or resulting spoil high in insoluble Se is deposited within plant root zones or the zone of oxidation, however, these materials may be oxidized and thereby produce seleniferous soils and vegetation.

Conclusions and recommendations

Caution is the order of the day with respect to routine agronomic soil analysis for Se. Significant research remains to be done before relatively precise correlations can be developed between soil/overburden analytical results and the concentration of Se in plant tissue *on a species-specific level.* Given the requirements of Public Law 95-87 (1977) and its associated 30

CFR regulations (1983), as well as the need for quality land use/reclamation planning, decisions and approaches concerning the identification, handling, and mitigation of potential toxicity from Se introduced into the biosphere by surface coal mining must now be addressed. Decisions must be made concerning many of the problems related to reclamation without an adequate information or research framework.

The following two recommendations have been developed with input from a variety of individuals ranging from industry and regulatory technical staff people to academic and research staff people. First, provide sufficient data to allow development of a plan whereby Se contamination of mine soils and surface water and groundwater will not occur during or after the reclamation process. Second, develop the most cost-effective means by which the baseline data can be generated and evaluated.

Several options exist in meeting the above goals: (a) grid sampling of premine soil mapping units and analysis of all samples for both total and extractable Se; (b) analysis of all overburden samples for both total and extractable Se; and (c) sampling of native vegetation for Se analysis, achieving sample adequacy in sampling and analysis of all plant species represented on the minesite.

Following is an outline for evaluation of the existing and potential toxicities related to Se and surface coal mine reclamation planning. We believe it is the best data collection program for defining potentially limiting Se problems (accepting some risk) while limiting costs. The procedure does not negate in any respect the critical need for research to resolve several of the primary issues relative to Se and its role in resource evaluation and reclamation planning.

Vegetation inventory (arid and semiarid lands). If a sufficiently detailed vegetation inventory was prepared as part of the premine environmental data collection program, it can become an important part of the Se evaluation system. A review of the plant species list for any of the primary or secondary Se-accumulator species described earlier should be undertaken. If present, the distribution of the primary and secondary accumulator species must be prepared. In new permit applications, in areas where warranted, the Se/plant inventory effort should become an integral part of the baseline data collection program. After identification and delineation of the species, plant tissue samples from both primary and secondary accumulator species should be collected and analyzed. Quality control is of paramount importance when working with Se. Duplicate analyses of 10 percent randomly selected samples should be analyzed by a second independent laboratory. An additional 10 percent of the samples should be analyzed in duplicate within the principal laboratory. National Bureau of Standards samples should be submitted for analysis to both laboratories as part of the quality control program. Plant tissue representing a single species should be collected from a single plant with a total of three plants (samples) from the same species being sampled within a delineated stand. Tissue analysis for Se would define Se content in existing plant communities or individuals, which is probably the best definition of potentially toxic soil and parent material/rock. Sample collection should take place only during the most active growing period because a major decline in Se content in plant tissue follows maturation. All species representative of primary and secondary accumulators within the delineation should be sampled. Samples must be kept at temperatures less than 10 °C from the time of collection until delivery to the laboratory, preferably in coolers with dry ice. Samples should be analyzed on a fresh-weight basis, as mentioned earlier, with appropriate split sample drying to permit conversion of fresh-weight to dry-weight concentrations.

Soil investigations (arid and semiarid lands). If the plant tissue analyses indicate the presence of elevated concentrations of Se (>4 ppm dry weight), a soil sampling and analysis program should be undertaken. Detailed soil profile descriptions should be prepared for each soil unit in the area delineated by the Se-accumulator species. Soil units should be restricted to series and phase, but also should include depth increments, such as subdivisions of 0 to 10 and 10 to 20 inches on the shallow Entisols and 20 to 30 and 30 to 40 inches on the moderately deep soils, for example. Soil sampling should be based on the horizons identified while preparing the soil profile description. However, the epidedon samples should have samples taken at intervals of 5 cm or less, and the sulfate enrichment or zone of concentration, if present, should have a similar sampling interval. Sampling should continue below the soil solum to a minimum depth of 5 feet or lithic contact. If the pedon is underlain with paralithic material, sampling should continue to the point where no root penetration has taken place. Samples should be placed in sealed jars, kept at temperatures less than 10 °C, and analyzed promptly. Samples should be screened to 2 mm and mechanically split where subsamples are required. They should be analyzed on an as-received basis with subsample drying for moisture correction. Either the hot water or the ammonium bicarbonate-DTPA extracts should be run on all samples.

Total Se analysis should be run on 50 percent of the samples taken from the C horizons and paralithic materials. Some reduced forms of Se may be present in such materials. As a data base is developed, the number of Se analyses can be reduced to some extent.

Overburden (all western climatic regimes). The analyses here must serve a twofold purpose: describing potential impacts on spoil materials incorporated into mine soils or near potentially oxidized surface zones and describing similar impacts on groundwater systems. No single analytical method can provide direct answers concerning the potential toxicity of an overburden or spoil sample. The use of extractants, such as ammonium bicarbonate-DTPA or hot water, on overburden samples has limited value in reduced samples (i.e., limited or no extraction of the more insoluble forms of Se). The overburden samples should be identified by lithologic type and depositional environment and samples selected from each of these categories. The largest concentration of sampling should take place in the lignitic or carbonaceous shale types of materials, but all types should be represented. Total Se analyses should be run on 50 percent of the samples collected and additional samples can be analyzed by a hot-water-soluble or ammonium bicarbonate-DTPA extraction technique. Reduction in the number of samples collected and analyzed can be accomplished as a data base is developed for each of the lithologic and/or depositional environments.

Research needs

With respect to development and testing of analytical methods, studies are needed on:

► The solubility of Se in spoil and fly ash materials under simulated weathering conditions.

► Evaluation of a wide variety of analytical techniques using a large, diverse population of soil and spoil samples, including several forms of Se in spike and recovery operations.

► Evaluation of Se-extraction procedures in defining potential plant uptake of Se from seleniferous and nonseleniferous soil, spoil, and fly ash materials. This would require a wide range of chemical and physical conditions in each type of sample.

► Review and evaluation of the process of addition of concentrated hydrogen peroxide after extraction of samples with ammonium bicarbonate-DTPA.

► Attempted correlation of column leach and simulated weathering chamber analytical data with total analysis results for prediction of Se contributions to groundwater systems.

Among field and related soil/plant research needs are the following:

► Determination of Se concentration in native plant species and forage and crop plants growing on seleniferous soils, including Se concentration in specific tissues and temporal changes through the growing season.

► Long-term studies correlating plant tissue concentration of Se with extractable soil Se in mine soils.

► Greenhouse studies assessing native plant uptake of Se from naturally seleniferous soils, mine soils, and soil materials spiked with a variety of Se forms and concentrations, to be followed by selected field studies.

► Evaluation of edaphic factors and their respective impacts on plant uptake of Se and the relationship to results from several extractable Se methods.

► Determination of the effect of moisture stress on the uptake and concentration of Se in plant tissue.

► Study of naturally occurring seleniferous ecosystems in the western United States, including geology, soils, plant species, and herbivores.

► Greenhouse and long-term field studies of the uptake of Se from various types of fly ash, ash, and soil mixes and types/form of burial, including development of several burial and disposal options to prevent incorporation of toxic levels into forage and crops.

► Construction and testing of models to define the potential for contamination of groundwaters via burial of fly ash and spoils containing Se.

► Development and testing of grazing management systems for reclaimed land to minimize hazards to livestock and wildlife.

► Evaluation and definition of foodchain effects—biomagnification potentials—on naturally seleniferous and man-created seleniferous soils/plants.

► Definition of the forms of Se at various depths in naturally occurring seleniferous soils.

REFERENCES

1. Ageton, R. W. 1970. *Selenium.* In *Mineral Facts and Problems.* Bureau of Mines, U.S. Department of Interior, Washington, D.C. 1,291 pp.
2. Allaway, W. H. 1968. *Control of the environmental levels of selenium.* Proceedings, Trace Substances in Environmental Health. University of Missouri, Columbia.
3. Allaway, W. H. 1970. *Sulphur—selenium relationships in soils and plants.* The Sulphur Institute Journal X: 3-5.
4. Allaway, W. H., and E. E. Cary. 1964. *Determination of submicron amounts of selenium in biological materials.* Analytical Chemistry 36: 1,359-1,362.
5. Andren, A. W., Y. Talmi, D. H. Klein, and N. E. Bolton. 1975. *Physical and chemical characterization of selenium in coal fired steam plant emissions.* Environmental Sciences Division, Oak Ridge National Laboratory, Oak Ridge, Tennessee.
6. Babcan, J., and J. Ilavsky. 1966. *Geochemistry from the stratiform pyrite deposits of ferrous sulfide—upper ores in Smolnik.* Sb. Geol. Vied, Zapado Karpaty, No. 6: 85-106. In C. A. Gent (1976) *Annotated Bibliography of the Geology of Selenium, 1958-1974.* Bulletin No. 1419. U.S. Geological Survey, Washington, D.C. p. 49.
7. Beath, O. A., C. S. Gilbert, and H. F. Eppson. 1939. *The use of indicator plants in locating seleniferous areas in the western United States, I. General.* American Journal of Botany 26: 257-269.
8. Beath, O. A., C. S. Gilbert, and H. F. Eppson. 1939. *The use of indicator plants in locating seleniferous areas in the western United States, II. Correlation studies by states.* American Journal of Botany 26: 296-315.
9. Beath, O. A., C. S. Gilbert, and H. F. Eppson. 1940. *The use of indicator plants in locating seleniferous areas in the western United States, III. Further studies.* American Journal of Botany 27: 564-573.
10. Beath, O. A., H. F. Eppson, and C. S. Gilbert. 1935. *Selenium and other toxic minerals in soils and vegetation.* Bulletin No. 206. Wyoming Agricultural Experiment Station, University of Wyoming, Laramie. 55 pp.
11. Beath, O. A., J. H. Draie, H. F. Eppson, C. S. Gilbert, and O. C. McCreary.

1934. *Certain poisonous plants of Wyoming activated by selenium and their association with respect to soil types.* American Pharmaceutical Journal 23: 95-97.
12. Bisbjerg, B., and G. Gissel-Nielsen. 1973. *The uptake of applied selenite and selenate by barley and red clover as influenced by sulphur.* Journal of Science Food Agriculture 24: 649.
13. Boon, D. Y., and P. J. Smith. 1985. *Carbonaceous materials—problems associated with reclamation.* In Proceedings, Second Annual Meeting, American Society for Surface Mining and Reclamation. American Society for Surface Mining and Reclamation, Princeton, West Virginia. pp. 248-256.
14. Brimhall, W. H. 1963. *Progress report on selenium in the Manning Canyon shale, central Utah.* Geology Studies. Brigham Young University, Provo, Utah. pp. 104-120.
15. Brown, M. J., and D. L. Carter. 1969. *Leaching of added selenium from alkaline soils as influenced by sulfate.* Soil Science Society of America Proceedings 33: 563.
16. Budalov, S. T., and S. R. Ruzmatov. 1960. *Contributions to the geochemistry of selenium and tellerium in the Almalyk ore deposit.* Akad Nauk. SSSR Inst. Mineralogie Geokhimii Kristallokhimii Red. Elementov, Trudy. 4: 24-27.
17. Butterfield, J. A. 1968. *Selenium—1967.* Engineering and Mining Journal 169: 131.
18. Byers, H. G. 1935. *Selenium occurrence in certain soils in the United States with a discussion of related topics.* Technical Bulletin No. 482. U.S. Department of Agriculture, Washington, D.C. pp. 1-47.
19. Byers, H. G., and H. G. Knight. 1935. *Selenium in soils in relation to its presence in vegetation.* Industrial and Engineering Chemistry 27: 902-904.
20. Cary, E. E., and W. H. Allaway. 1973. *Selenium content of field crops grown on selenite treated soils.* Agronomy Journal 65: 633.
21. Chemical Rubber Company. 1970. *Handbook of chemistry and physics* (51st edition). Cleveland, Ohio.
22. Davis, E. B., and J. H. Watkinson. 1966. *Uptake of native and applied selenium by pasture species: I. Uptake of selenium by browntop, ryegrass, cocksfoot and white clover from Atiamuri sand.* New Zealand Journal of Agricultural Research 9: 317-327.
23. Demayo, A., M. C. Taylor, and S. W. Reeder. 1979. *Guidelines for surface water quality. Volume 1: Inorganic Substances—Selenium.* Water Quality Branch, Inland Waters Directorate, Environment Canada, Ottawa, Ontario. 21 pp.
24. Ehlig, C. F., W. H. Allaway, E. E. Cary, and J. Kubota. 1968. *Differences among plant species in selenium accumulation from soils low in available selenium.* Agronomy Journal 60: 43-47.
25. Ewan, R. C., C. A. Baumann, and A. L. Pope. 1968. *Determination of selenium in biological materials.* Agricultural and Food Chemistry 16: 212-215.
26. Fine, L. O. 1965. *Selenium.* In C. A. Black [editor] *Methods of Soil Analysis, Part II.* Agronomy No. 9. American Society of Agronomy, Madison, Wisconsin.
27. Fleming, G. A. 1980. *Essential micronutrients II: Iodine and Selenium.* In B. E. Davies [Editor] *Applied Soil Trace Elements.* John Wiley and Sons, Ltd., New York, New York.
28. Fleming, G. A., and T. Walsh. 1957. *Selenium occurrence in certain Irish soils and its toxic effects on animals.* Royal Irish Academy Proceedings 58, Section B, 7: 151-165.
29. Frost, D. V. 1972. *The two faces of Selenium—can selenophobia be cured.* CRC

Critical Reviews of Technology I: 467-514.

30. Ganje, T. J. 1966. *Selenium.* In H. L. Chapman [editor] *Diagnostic Criteria for Plants and Soils.* Division of Agriculture Science, University of California, Riverside. pp. 394-404.
31. Geering, H. R., E. E. Cary, L.P.H. Jones, and W. H. Allaway. 1968. *Solubility and redox criteria for the possible forms of selenium in soils.* Soil Science Society of America Proceedings 32: 35-40.
32. Gissel-Nielsen, G. 1970. *Loss of selenium in drying and storage of plant species.* Plant and Soil 32: 242-245.
33. Goldschmidt, V. M. 1939. *Geochemische Verteilungsgesetze der Elemente.* Dywab in Komm., Oslo, Norwey.
34. Goodson and Associates, Inc. 1986. *Feasibility of using fly ash for backfilling abandoned coal mines in Colorado Springs, Colorado.* Draft Final Report. Office of Surface Mining, Western Technical Center, Denver, Colorado.
35. Gorsuch, T. T. 1970. *The destruction of organic matter.* Pergamon Press, New York, New York. 151 pp.
36. Gutermann, W. H., C. A. Bache, W. D. Youngs, and D. J. Lisk. 1976. *Selenium in fly ash.* Science 191: 966-967.
37. Hamilton, J. W., and O. A. Beath. 1963. *Uptake of available selenium by certain range plants.* Journal of Range Management 16: 261-264.
38. Hance, F. E. 1939. *Report on selenium chemistry studies.* In Proceedings, Hawaiian Sugar Planter's Association Report. Committee in Charge of Experiment Station, Honolulu, Hawaii. p. 88.
39. Hartley, W. J., and A. B. Grant. 1961. *A review of selenium responsive disease of New Zealand livestock.* Federation Proceedings 20: 679-688.
40. Howard J. H. III. 1969. *Selenium in the Upper Cretaceous Niobrara Formation and Pierre shale of south central Colorado.* Abstracts for 1968, Special Paper 121. Geological Society of America, Boulder, Colorado. p. 606.
41. Howard J. H. III. 1970. *Geochemical behavior of selenium in earth surfce environments.* Dissertation Abstracts International, Section B, Science and Engineering 30: 5554B-5555B.
42. Howard, J. H. III. 1973. *Sources of selenium in sediments of western Great Plains: Late Cretaceous centers of west central Montana.* Geological Society of America Abstracts 5: 485.
43. Kolm, K. E. 1975. *Special report: Selenium in soils of the lower Wasatch Formation, Campbell County, Wyoming—geochemistry, distribution and environmental hazards.* NASA 5-21799. Goddard Space Flight Center, National Aeronautics and Space Administrator, Greenbelt, Maryland.
44. Koval'skii, V. V., and V. V. Ermakov. 1967. *Biochemical selenium regions of Tuva.* Geokhimiya.
45. Koval'skii, V. V., and V. V. Ermakov. 1970. *The biological importance of selenium.* Translated by S. J. Wilson, edited by Raymond C. T. Powell. National Lending Library for Science and Technology, Boton Spa, Yorkshire, England.
46. Lakin, H. W. 1961. *Selenium content of soils.* In M. S. Anderson et al. [editor] *Selenium in Agriculture.* Agricultural Handbook No. 200. Agricultural Research Service, U.S. Department of Agriculture, Washington, D.C. pp. 27-34.
47. Lakin, H. W. 1961. *Vertical and lateral distribution of selenium in sedimentary rocks of western United States.* In M. S. Anderson et al. [editor] *Selenium in Agriculture.* Agricultural Handbook No. 200. Agricultural Research Service, U.S. Department of Agriculture, Washington, D.C. pp. 12-24.
48. Lakin, H. W., and H. G. Byers. 1948. *Selenium occurrence in certain soils in the United States, with a discussion of related topics.* Technical Bulletin No. 950.

U.S. Department of Agriculture, Washington, D.C.

49. Lindberg, P., and N. Lannek. 1970. *Amounts of selenium in Swedish forages, soils and animal tissues.* In C. F. Mills and E. and S. Livingstone [editors] *Trace Element Metabolism in Animals.* WAAP-IBP, Edinburgh, Scotland.
50. Marshall, Eliot. 1985. *Selenium poisons refuge, California politics.* Science 229: 144-146.
51. Munshower, F. F. 1983. *Problems in reclamation planning and design.* In S. E. Fisher, Jr. [project coordinator] *Coal Development: Collected Papers* (volume 2). Bureau of Land Management, Denver Service Center, Denver, Colorado. pp. 1,287-1,307.
52. National Research Council. 1971. *Selenium in nutrition.* National Academy of Sciences, Washington, D.C. 79 pp.
53. National Research Council. 1976. *Selenium.* National Academy of Science, Washington, D.C. 203 pp.
54. Office of the Federal Register, National Archives and Records Service. 1983. *Title 30—Mineral Resources. Chapter VII. Office of Surface Mining Reclamation and Enforcement. Revised as of October 1, 1983.* In *Code of Federal Regulations.* General Services Administration, Washington, D.C.
55. Oldfield, J. E. 1972. *Selenium deficiency in soils and its effect on animal health.* Geology Society American Bulletin 83: 173.
56. Olson, O. E., and A. L. Moxon. 1939. *The availability to crop plants of different forms of selenium in the soil.* Soil Science 47: 305-311.
57. Olson, O. E., and A. L. Moxon. 1942. *The availability to crop plants of different forms of selenium in soils.* Soil Science 54: 47.
58. Olson, O. E., E. A. Whitehead, and A. L. Moxon. 1942. *Occurrence of soluble selenium in soils and its availability to plants.* Soil Science 54: 47-53.
59. Palmer, I. S. 1986. *A study of the effects of fly ash burial on the selenium content of vegetation at the San Juan Mine and Navajo Mine at Farmington, N.M.* Utah International, Waterflow, New Mexico.
60. Raptis, S. E., G. Kaiser, and G. Tolg. 1983. *A survey of selenium in the environment and a critical review of its determination at trace levels.* Zeitschrift fuer Analytische Chemie 316: 105-123.
61. Rickard, W. H. 1965. *Sodium and potassium accumulation by greasewood and hopsage leaves.* Botanical Gazette 126: 116-119.
62. Rosenfeld, I., and O. A. Beath. 1964. *Selenium.* Academic Press, Inc., New York, New York. 411 pp.
63. Rotruck, J. T., A. L. Pope, H. E. Ganther, A. B. Swanson, D. G. Hafeman, and W. G. Hoekstra. 1973. *Selenium: Biochemical role as a component of glutathione peroxidase.* Science 179: 588.
64. Sauchelli, V. 1969. *Trace elements in agriculture.* Van Nostrand-Reinhold Co., New York, New York. pp. 248.
65. Savelev, V. F. 1963. *Selenium containing minerals of sedimentary rock of central Asia.* Byulleten Nauchno-Tekhnicheskoi Informatsii.
66. Schwar, K., and C. M. Foltz. 1957. *Selenium as an integral part of factor 3 against dietary necrotic liver degeneration.* Journal of American Chemical Society 79: 3,292.
67. Sidel'nikova, V. D. 1973. *Geological occurrence of selenium in arid regions.* In S. G. Batulin [editor] *Geokhimiya Landshaftov Protsessy Aipergeneza.* Moscow, USSR. pp. 64-92.
68. Sienko, M. J., and R. A. Plane. 1966. *Chemistry* (third edition). McGraw-Hill Book Co., New York, New York. 654 pp.
69. Sindeeva, N. D., Kurbanova, and N. Z. Clarke. 1959. *Selenium in several rocks of the USSR.* Doklady Academii Nauk 20:(82).

70. Smith, P. J., and D. T. Boon. 1985. *Plant analysis: A measure of successful reclamation.* In Proceedings, Second Annual Meeting, American Society for Surface Mining and Reclamation. American Society for Surface Mining and Reclamation, Princeton, West Virginia. pp. 95-101.
71. Soltanpour, P. N. 1982. *Use of the ammonium bicarbonate-DTPA soil test to assess elemental availability and toxicity to plants.* In Proceedings, Symposium on Trace Element Mobilization in Western Energy Regions. U.S. Geological Survey and Bureau of Land Management, Denver, Colorado.
72. Soltanpour, P. N., and S. M. Workman. 1980. *Use of ammonium bicarbonate-DTPA soil test to assess availability and toxicity of selenium to alfalfa plants.* Communications in Soil Science and Plant Analysis 11: 1,147-1,156.
73. Swaine, D. J. 1955. *The trace element content of soils.* Technical Communications No. 48. Commonwealth Bureau of Soil Science, Rothamsted Experiment Station, Harpenden, England.
74. Swaine, D. J. 1969. *Trace element content of soils.* Technical Communication No. 48. Commonwealth Bureau of Soil Science, Rothamsted Experiment Station, Harpenden, England. 157 pp.
75. Taylor, S. R. 1964. *Abundance of chemical elements in the continental crust: A new table.* Geochimica Et Chosmochimica Acta 28: 1,273-1,285.
76. Thornton, I., and B. J. Alloway. 1974. *Geochemical aspects of the soil-plant-animal relationship in the development of trace element deficiency and excess.* Proceedings, Nutritional Society 33: 257-266.
77. Trelease, S. F., and H. M. Trelease. 1938. *Selenium as a stimulating and possibly essential element for indicator plants.* American Journal of Botany 25: 372-380.
78. Trelease, S. F., and O. A. Beath. 1949. *Selenium.* Champlain Printers, Burlington, Vermont. 292 pp.
79. Tsuge, T., and S. Terada. 1949. *The selenium content in the pyrite soils in Japan.* Journal Agricultural Chemical Society Japan 23: 421.
80. Underwood, E. J. 1971. *Trace elements in human and animal nutrition* (third edition). Academic Press, New York, New York. 543 pp.
81. Vinogradov, A. P. 1957. *Geokhimiya redkikh i rasseyannykh khimicheskikh elementov v pochvakh* (The chemistry of rare and common chemical elements in the soil). Akademii Nauk SSR.
82. Watkinson, J. H. 1962. *Soil selenium and animal health.* In *Transactions, Joint Meeting, Committees IV and V, International Soil Science Society.* Wageningen, The Netherlands. p. 149.
83. Wells, N. 1967. *Selenium content of soil-forming rocks.* New Zealand Journal of Geology and Geophysics 10: 198-208.
84. Westerman, D. T., and C. W. Robbins. 1974. *Effect of sulfate sulfur fertilization on selenium concentration of alfalfa.* Agronomy Journal 66: 207.
85. Whanger, T. D. 1974. *Bioenvironmental impact of selenium.* National Ecological Research Laboratory, Corvallis, Oregon. 71 pp.
86. Williams, K. T., and H. G. Byers. 1936. *Selenium compounds in soils.* Industrial and Engineering Chemistry 26: 912-914.
87. Workman, S. M. 1983. *Selenium equilibrium in soils.* Paper prepared for AG 680. Colorado State University, Fort Collins.
88. Wyoming Agricultural Experiment Station. 1893. *Third annual report.* In O. A. Beath [editor] *The Story of Selenium in Wyoming (1982).* Bulletin 774. Agricultural Experiment Station, University of Wyoming, Laramie.
89. Wyoming State Board of Sheep Commissioners. 1908. *Tenth annual report—December 10, 1908.* In O. A. Beath [editor] *The Story of Selenium in Wyoming (1982).* Bulletin 774. Agricultural Experiment Station, University of Wyoming, Laramie.

Chapter 7

Boron

R. C. Barth, R. C. Severson, and G. Weiler

The trace element boron (B) is of interest in reclamation situations for several reasons. It plays an essential though largely unidentified role in the growth of higher plants. In agronomic situations B deficiencies are common, and deficiencies in reclamation situations have been suggested but not documented. Among micronutrients, B is unique because the range from deficient concentrations to toxic concentrations either in the soil solution or in plant tissue is narrower than for any other micronutrient. In reclamation situations excessive amounts of B can occur in the soil or in near-surface mining wastes and thus interfere with reclamation objectives, especially in arid and semiarid regions. Also, B is mobile and appears subject to both upward transport (and possible contamination of overlying material) and downward transport (and possible contamination of surface water and groundwater).

Boron in surficial materials

Boron in geologic materials. Generalizations about the average B content in rocks cannot be made with any certainty (Table 1). In igneous and metamorphic rocks B usually occurs as high-temperature borosilicates, which are resistant to weathering and not readily available to plants. In these rock types B is usually of little significance in reclamation situations. Sedimentary rocks contain, in addition to the resistant borosilicates, hydrous borates that are easily weathered, thus supplying B to plants.

Minerals containing B are subdivided into three main groups: cyclosilicates, anhydrous borates, and hydrous borates (*42*). The common cyclosilicates and the anhydrous borates are formed at high temperatures. They occur mainly in granite pegmatites and at the borders of large, intrusive rock bodies. These high-temperature minerals are relatively insoluble. As a result, they do not contribute to the B nutrition of plants directly. However, weathering of these minerals over long periods of time is a source of B for sedimentary basins. The common hydrous borates are low-temperature minerals found in evaporite deposits of salt lakes or sea waters. There are at least 150 known B-containing minerals (*23*). No solubility data exist for a majority of these minerals; therefore, dissolution-precipitation reactions cannot be calculated to predict the availability of B from many of the minerals (*44*).

Boron in soils. In soils B occurs primarily in the form of silicates, adsorbed forms, and organic matter. The B-containing silicate minerals are generally trace constituents. Because this group of minerals resists weathering, the minerals continually supply small amounts of B to plants over long periods of time. These minerals are important in meeting plant B requirements. Adsorbed forms of B are essentially restricted to alkaline soils. Adsorption is highly pH-dependent, with maximum adsorption occurring in a pH range of 7 to 9. At pH 7, $B(OH)_3$ is the dominant species in soil solution. As pH increases toward 9, $B(OH)_4^-$ beomes dominant. Organic matter is the main source of plant-available B in acid soils. In alkaline soils, minerals rather than organic matter appear to be the important source of plant-available B.

Surface mining often produces waste carbonaceous material that can be an important B source. Carbonaceous shales, coal lenses, leonardite, and coal slack are dominated by biologically inert organic matter that often contains B. Researchers believe this B originates from the living plants that were the progenitors of the carbonaceous material. Mining removes these materials from their depositional environment, usually an acid-reducing environment, and places them in an oxidizing environment. This activity can

Table 1. Average boron content in rocks, from selected studies.

Rock Type	B Content (ppm) in Three Studies		
	Krauskopf (*42*)	Aubert and Pinta (*3*)	Norrish (*49*)
Earth's crust	10	50	3
Igneous rocks			
Granite (acidic)	15	3-10	3
Basalt (basic)	5	1-5	1-2
Metamorphic rocks	No data	5-12	No data
Sedimentary rocks			
Limestone	20	8-10	100
Sandstone	35	No data	100
Shale	100	No data	100

result in the release of substantial quantities of water-soluble B. Highest concentrations are often found in thinly bedded carbonaceous material immediately above or below a major coal seam.[1]

Nonionized $B(OH)_3$ predominates in soil solutions, and it is most likely the form absorbed by plants. Plants absorb B passively, and B has limited mobility within the plant. A continuous supply of B is thus necessary. The amount of plant uptake can be related to, among other factors, the amount of water transpired and the B concentration in the soil solution (*61*). Ionized $B(OH)_4^-$ predominates in soil solution as soil pH approaches 9.2. A soil pH this high is uncommon, and acid production at root surfaces probably converts the ionized species to the unionized species.

Total B in soils ranges from less than 5 ppm to about 300 ppm, and average concentrations generally range from 10 ppm to 60 ppm (Table 2). Water-soluble B appears to range from less than 1 ppm to 97 ppm in soil and to average less than 2 ppm. In revegetated spoil to which topsoil had been applied, hot-water-soluble B ranges from less than 0.5 ppm to 26 ppm.

Factors affecting B availability. Several researchers have suggested that light, temperature, aeration, and moisture affect B uptake by plants (*33, 43*). Because plants absorb B passively as a nonionized species, the way environmental factors affect plant growth rates, transpiration rates, and soil moisture supplies will affect B uptake in a similar manner. Increasing temperature, light, and moisture or decreasing aeration generally increases B uptake.

Boron fixation is another important factor in determining the element's availability. Boron in the soil solution, rather than adsorbed or fixed B, is the primary form available to plants (*38*). The equilibrium between solution B and fixed B determines the instantaneous supply of B to the plant. The most important factors influencing fixation and removal of B from soil appear to be pH, type of clay mineral, and exchangeable ions (*1, 11, 37, 38*).

Organic matter, pH, texture, mineralogy, and moisture are the soil properties that most influence B availability. Studies relating these soil properties to B availability are numerous and sometimes conflicting. One researcher concluded that a single soil property cannot be isolated and used to estimate B availability or uptake (*25*). Other researchers found that recovery of B following leaching declined with increasing soil organic matter, clay, and free iron and alumina oxide content, but statistical separation of the effects of individual components was complicated by intercorrelations among these soil properties (*54*).

In humid areas with acid soils, the effects of organic matter and pH on B

[1]Personal communication with D. Boon , Wyoming Department of Environmental Quality, Cheyenne, and D. Boon, P. Smith, and R. Lawton, "Carbonaceous Materials, Problems Associated with Successful Reclamation," a paper presented at the second annual meeting, American Society for Surface Mining and Reclamation, October 8-10, 1985, Denver, Colorado.

availability are probably dominant because moisture supplies are generally adequate and B adsorption is minimal at a pH below neutrality. Texture may also have some influence under these conditions by physically retarding moisture flow and leaching. In arid regions with alkaline soils, moisture, texture, and mineralogy are probably the most important factors in determining B availability and plant uptake because pH is generally between 7 and 9 and organic matter is low.

A number of cation interactions involving B can increase or decrease its availability (*1*). On acid soils the addition of limestone reduces B availability (*33*). Other studies have shown that Ca and K accentuate B deficiencies in low-B soils, while in high-B soils Ca reduces toxicity and K and possibly other soluble salts accentuate toxicity (*25, 53*). One research team suggested that, unlike the case with some cations, B adsorption is site specific and

Table 2. Abundance of boron in soils and other materials.

Method of Analysis	B Content (ppm) Average Content	B Content (ppm) Range in Content	Comments	Reference
Total	41	18-96	Topsoil; western coal regions	*72*
Total	43	16-115	Subsoil; western coal regions	*72*
Total	10	2-100	United States	*14*
Total	50	5-150	Agricultural soils	*16*
Total	34	<20-300	Surficial material	*73*
Total	29	<20-70	Topsoil	*17*
Total	26	<20-70	Subsoil	*17*
Total	61	25-102	Western soils	*63*
Total	-	1-270	World soils	*3*
Total	113	13-337	Arid soils	*22*
Total	-	16-83	Western soils	*72*
Total	16	< 5-41	A horizon; New Mexico	*69*
Total	15	< 5-43	C horizon; New Mexico	*69*
Total	6.9	< 5-23	New Mexico mine soil; topsoil	*69*
Total	13	7.3-25	New Mexico mine soil; spoil	*69*
Total	32	4-202	Western coal	*
Hot-water-soluble	0.1-2	-	World soils	*3*
Hot-water-soluble	0.6	0.5-1.9	Western soils	*6*
Water-saturation extract	1.3	0.3-3.4	Arid soils	*22*
Water-saturation extract	2.2	-	Carbonaceous spoil	*
Water-saturation extract	-	2-97	Central Nevada soils	*64*
Hot-water-soluble	0.2-0.4	-	Western soils	*69*
Hot-water-soluble	1.0	<0.5-9.5	Western mine soil-topsoil	*70*
Hot-water-soluble	1.1	<0.5-26	Western mine soil-spoil	*70*
Hot-water-soluble	0.3	<0.5-4.0	A horizon; New Mexico	*69*
Hot-water-soluble	0.5	<0.5-3.5	C horizon; New Mexico	*69*
Hot-water-soluble	0.6	<0.5-3.5	New Mexico mine soil; topsoil	*69*
Hot-water-soluble	1.8	0.5-11	New Mexico mine soil; spoil	*69*

*Communication with D. Boon, Wyoming Department of Environmental Quality, Cheyenne.

other anions do not compete for these adsorption sites (*12*). As a result, anions apparently have little influence on B availability and uptake.

Boron in plants

Physiological role. Unlike other micronutrients, B is not part of the plant enzyme system, and it plays a poorly understood role in a number of other plant systems. B is essential for the elongation of growing tips, both shoots and roots. When normally growing plants are transferred to a B-free growth medium, elongation of growing tips ceases within hours. Other research suggests that B is necessary for the production of nucleic aids, synthesis of cell wall components, translocation of sucrose, P absorption, and carbohydrate metobolism (*26, 33, 35*).

The amount of B in plants varies with age, species, plant part, growth stage, and the concentration and availability of B in the growth medium. Numerous variables affect B uptake and use. Any attempt to classify plants by their B requirements is, therefore, tenuous. Generally, dicotyledonous plants require about four times as much B as monoctyledonous plants (*25*). Grasses thus have a low B requirement, while many legumes and shrubs need somewhat more B.

Table 3 displays the concentrations of B that are essential to support normal growth of plant species useful in reclamation. The table also presents deficient and toxic levels of B in cases where these levels have been established. The range of B concentration is probably more useful in defining the B status of a plant than the mean concentration.

Deficiency and toxicity. Insufficient B causes degeneration of meristematic tissues in plants, including a breakdown of cell walls and feeble development of vascular tissues. External symptoms include dieback or abnormal growth of terminal shoots; thick, brittle, or curling leaves; thickened or crooked stems; and chlorotic spotting on leaves and stems (*15, 33*). Because B is relatively immobile in plants, younger leaves and stems may display deficiency symptoms while older tissues contain sufficient or even toxic amounts of B. These symptoms are characteristic of several other nutritional disorders as well; therefore, tissue analysis usually is needed to confirm suspected B deficiencies. Symptoms of B deficiency are most common in plants grown in light-colored acid sands and silt loams in humid regions.

Boron toxicity is characterized by chlorosis of the leaf margins and tips of older leaves, followed by necrosis or death of the affected area (*15, 47*). B is transported in the xylem but becomes immobile in leaves, especially along the margins. B levels greater than 1,500 ppm have been recorded in chlorotic leaf tissue, compared to values of 40 to 100 ppm in healthy portions of plants (*25*). However, the B loss from guttating leaves can amount to 80 percent of the previously absorbed B (*50*). This indicates that B immobilization in leaves does not occur in all cases. As the toxicity continues, the entire leaf will have a burned or scorched appearance and will drop prematurely. Moreover, plants suffering from B toxicity often display retarded growth (*75*). One researcher found B leaf chlorosis in older plant parts and deficiency symptoms on the younger parts (*20*). Beneficial and injurious effects can thus overlap within the same plant. Precise diagnosis of B toxicity is difficult, and such diagnosis should be based on visual symptoms, tissue analysis, and the soil B level.

Plant tolerance to excessive B appears related to a slower accumulation rate for tolerant plants rather than a difference in tissue sensitivity to B (*25, 51*). In the case of some species older plants tolerate excessive B better than

Table 3. Boron levels in plant tissue of commonly used reclamation species.

	B in Dry Matter (ppm)					
	Deficient		Sufficient*			
Plant Species	Mean	Range	Mean	Range	Toxic Range	Reference
Grasses						
Green needle grass	-	-	7.5	6.8-8.7	-	*77*
Needle-and-thread	-	-	14.9	12.0-17.7	-	*77*
Indian ricegrass	-	-	14.3	-	-	*68*
Prairie junegrass	-	-	16.1	11.3-22.7	-	*77*
Japanese brome	-	-	4.0	3.2-4.5	-	*77*
Bluegrass	-	-	16.9	3.4-30.0	-	*77*
Smooth brome	-	-	6.8	6.5-7.0	-	*77*
	-	-	14.0	-		*8*
	-	-	24.0	14-36	-	*31*
Sixweeks fescue	-	-	25.6	-	-	*77*
Alkali sacaton	-	-	9.1	7.8-12.0	-	*30*
	-	-	-	-	>262	*76*
Blue grama	-	-	27.5	-	-	*77*
Pubescent wheatgrass	-	-	9.8	3.6-17.7	-	*77*
Crested wheatgrass	-	-	10.1	11.9-21.5	-	*77*
	-	-	16.0	11.0-48.0	-	*24*
	-	-	10.4	3.1-24.0	-	*30*
	-	-	8.0	-	-	*8*
	-	-	17.0	10.1-22.0	-	*7*
Slender wheatgrass	-	-	4.9	2.5-7.0	-	*77*
	-	-	7.7	1.4-36	-	*31*
Tall wheatgrass	-	-	-	-	>366	*76*
Western wheatgrass	-	-	13.1	4.1-37.2	-	*77*
	-	-	9.0	4.0-15.0	>202	*40*
	-	-	12.7	-	-	*68*
	-	-	-	-	>569	*76*
	-	-	-	10.0-24.0	-	*41*
Thickspike wheatgrass	-	-	17.7	8.9-26.5	-	*77*
Intermediate wheatgrass	-	-	6.5	5.9-7.1	-	*77*
	-	-	11.0	-	-	*8*
	-	-	8.3	1.6-20	-	*31*
Galleta grass	-	-	5.0	3.2-16	-	*30*
Timothy	-	-	-	3-93	>102	*33*

Continued

Table 3, continued

Plant Species	B in Dry Matter (ppm)					Reference
	Deficient		Sufficient*		Toxic	
	Mean	Range	Mean	Range	Range	
Legumes						
Alfalfa	-	<20	-	31-80	>100	*33*
	-	<30	-	30-80	> 80	*76*
	-	-	65.2	46-144	-	*21*
	-	-	32.0	-	-	*8*
	-	-	49.0	19-130	-	*31*
Sanfoin	-	-	20.0	11-43	-	*31*
Red clover	-	12-20	-	20-52	> 60	*33*
Sweet clover	-	-	28.0	-	-	*8*
Birdsfoot trefoil	14	-	-	30-45	> 68	*33*
Shrubs						
Winterfat	-	-	30.0	-	-	*68*
Fourwing saltbush	-	-	57.0	40->65	-	*30*
	-	-	-	-	> 90	*76*
	-	-	47.7	-	-	*68*
	-	-	125.0	18-610	-	*31*
	-	-	13.8	17-770	-	*2*
Big sagebrush	-	-	24.0	9.6-78	-	*30*
Broom snakeweed	-	-	21.0	11->42	-	*30*
Trees						
Red pine	-	-	-	21-34	-	*78*
White pine	-	-	-	21-34	-	*78*
Radiata pine	-	-	-	13-70	>200	*75*
Digger pine	-	-	-	-	140	*29*
Madrone pine	-	-	-	-	20	*29*
California laurel	-	-	-	-	100	*29*
Bigleaf maple	-	-	-	-	7	*29*

*Plant levels reported in the literature without mention of deficient or toxic symptoms were assumed to represent the sufficient range.

younger plants (*29*). Some species may prematurely drop leaves as a means of removing excess B.

Boron sampling and analysis

In plants. Because of the wide range in B concentrations within individual plants and because there may be both beneficial and injurious effects within the same plant, the entire above-ground portion of grasses or legumes should be harvested for analysis. For woody plants, leaves produced throughout the growing season should be collected and combined for analysis. Plants should be collected during active growth and immediately washed with distilled water—either dipped in water or placed in a sonicator—to remove any soil particles that might adhere to the plants. Care should be taken during washing to minimize ion removal by minimizing immersion time. After washing, plant samples should be oven dried at a temperature between 50 and 70 °C. The dried plants should be ground or cut to pass a 40-mesh sieve and stored in air-tight containers. Storage time is unknown, but plant samples thus prepared probably could be stored for several years without degradation.

Wet or dry ashing followed by the azomethine-H colorimetric method is probably the most common technique used to determine B levels in plants. This method is reproducible, simple, does not require concentrated acids, and is convenient for use on automatic systems (*33*). However, wet ashing with nitric acid can introduce nitrate compounds that interfere with B determination using the azomethine-H method (*28*). This problem can be avoided by using inductively coupled plasma spectrometry or by using a dry ashing method. One analyst found that B was lost during wet ashing and recommended use of dry ashing (*36*). This problem was attributed to the volatilization of boric acid from the acid solution during digestion (*34*). Apparently, then, the most desirable method of B determination in plant tissue is dry ashing, followed by a colorimetric procedure, preferably the azomethine-H method, or inductively coupled plasma spectrometry.

In soils and overburden. The particle size distribution, including coarse fragments, of natural soils developed from sedimentary rocks is generally much different than mine soils developed from fresh, fractured sedimentary rocks. Because the purpose of collecting samples is to evaluate B status over both the short term and long term, the samples should be collected to represent the present B status and to predict the future B status due to physical and chemical rock weathering. Overburden initially low in B—less than 5 ppm, hot-water-soluble—significantly increased in B after 2 to 3 years of weathering under field conditions and values as high as 50 ppm (hot-water-soluble) have been recorded (personal communication with P. J. Smith, Wyoming Department of Environmental Quality). However, when Lance-formation samples of overburden containing up to 138 ppm total B (39.2 ppm hot-water-soluble B and 28.1 ppm saturated-paste-extract B) were weathered, there was no increase in the B concentration of the leachate with time (*48*). Therefore, the release of B during weathering of spoil is difficult to predict.

Sampling the less-than-2-mm fraction of the A horizon, a common agricultural practice, is not practical for mine soils because genetic horizons do not exist. The coarse fragments may be important in long-term B availability. Sampling depth is also important. Because B is accumulated by plants in a passive process that depends upon the moisture availability in the growth medium, the rooting character of the plant and the depth of moisture penetration should be considered when deciding which soil (spoil) zones to sample.

Different methods of sample preparation can either dilute or concentrate the B-containing fraction. If the samples are disaggregated but not ground and the coarse fraction is discarded, then a dilution of B could occur if the discarded coarse fraction was B-enriched, for example, with a highly organic or fine-textured sedimentary rock. Conversely, if the discarded fraction was low in B, for example, in the case of a coarse-textured sedimentary rock, then a B-concentrating effect could be expected. The initial mechanical sample preparation procedure should (a) be appropriate for the objectives of the study, (b) consider B release through weathering, and (c) reflect anticipated field conditions. In summary, that portion of the sample expected to contribute to B availability should be analyzed.

Common methods of detecting total B in soils and overburden have been reviewed (*1, 12*). The inductively coupled plasma spectrometry process is a relatively new technique to determine B levels in soil extracts (*19, 27*). Various colorimetric, potentiometric, selective ion electrode, and spectroscopic methods have certain merits (*12, 19, 27*). Generally, measures of total B in soils and rocks are poor indications of plant-available B and the B-supplying capacity of the soil.

Table 4 lists extraction methods that have been used to assess B availability in soils. The common method for overburden and mine soils is refluxing a 1:2 soil-to-water suspension to determine hot-water-soluble B (*66*). However, some researchers believe that the method produces erratic and nonre-

Table 4. Extraction methods used to assess available boron in soils.

Method	Comments of Referenced Author(s)	Reference
Hot-water-soluble	Widely used method; appears to be a good predictor of deficient levels.	*12, 66*
Reflex 1:2 soil-water for 5 minutes	Better than total or acid-extractable for assessing plant-available boron.	*10*
Reflux with aqueous $CaCl_2$	Similar results as hot-water-soluble, but provides a clarified solution.	*5*
Water-saturation-extract	Values relate well to plant uptake; appears to be a good predictor of toxic levels.	*12, 37, 80*
Soxhlet extraction with water	Measures the capacity factor, includes all forms of available boron including adsorbed.	*45*
Soxhlet extraction with water	The ratio of Soxhlet B to hot-water-soluble B ranged from 1.25 to 12.5 and generally followed the sequence of acid < neutral < alkaline soils.	*4*
Mannitol-$CaCl_2$ extraction for 16 hours	Assesses regenerative power of soil for boron.	*12, 62*
Acid extraction with final pH < 2.8	Estimates reserve supply; roots can use some acid-extractable boron.	*58*
"Acid" extractions	Do not correlate well with plant levels of boron for calcareous soils.	*18*
Ammonium acetate extractant	Simple and effective method for both calcareous and noncalcareous materials.	*32*

producible results, while the water-saturation-extract method provides stable and reliable results (*9, 12, 37*). In one test, results with the hot-water-soluble method were significantly more variable than results with the saturated-paste-extraction (*48*). Another test found that 1 N ammonium acetate (pH 4.8) was a more reliable extractant of soil B, compared to the hot-water-extractant, especially in evaluating potentially B-deficient soils (*32*).

The extraction method should be based on an evaluation of soil, climatic, and plant variables, and the study's objective. Extraction methods that evaluate plant-available B at one point in time (intensity factor) will be different than methods that evaluate a soil-supplying power (capacity factor). For the western United States the intensity factor could be estimated by either hot-water-soluble B or the water-saturation-extract method. The capacity factor could be estimated by either the mannitol-$CaCl_2$ or the Soxhlet-extraction method (Table 4). Research is needed to determine the best method or to develop a new method that accounts for the soil, climate, and plant conditions in revegetated areas of the western United States.

Sample storage should have little effect on increasing or decreasing B availability if soil drying is the only change that occurs (*39*). However, if B is adsorbed on fresh Al or Fe oxyhydroxides or the pH of the soil changes during sample storage, determination of B availability may be affected. Weathering of Fe and Al oxyhydroxides can reduce B extractability (*74*). A reduction in available B may also result from drying the samples at temperatures exceeding normal field temperatures (*25*). Oxidation of pyrite-containing samples and the resulting drop in sample pH can increase B availability. Acidifying spoil samples may nearly double the B in a saturation extract (*60*). Similarly, if microbial decay of organic matter in stored samples occurs, B availability tends to increase. The magnitude of such increases or decreases in stored samples cannot be generalized because of variability in the character of the sample and storage conditions.

Contamination sources. There are several sources of B contamination for both surficial materials and plants. Use of high-B water in drilling mud can contaminate overburden core samples. Glue in paper bags, which are often used to collect and store samples, can also interfere with B analysis (*66*). Borosilicate glassware used in wet digestion is a source of contamination as well. This source of contamination can be eliminated by using nalgene bottles. Muffle furnaces used in dry combustion can also be a source of contamination (*28*). Such furnaces should be evaluated carefully prior to use.

Boron in the reclamation situation

Boron in the growth medium in reclaimed areas is generally viewed from the standpoint of B toxicity rather than from the standpoint of B deficiency.

While the probability of B toxicity is much higher than the probability of B deficiency, B deficiency in reclaimed areas should not be ignored. Borderline B deficiency conditions have been found at a number of western surface mines (*71*). Potential deficiencies are best detected through plant and soil sampling. Deficiencies are easily corrected by applying a variety of fertilizers containing B (*1*).

Recognition of potential problem areas. Although procedures for sampling premining overburden vary from state to state and are often site specific, most sampling efforts are of sufficient intensity to delineate areas high in B. Sedimentary rocks, shales, and particularly carbonaceous material have a higher probability of containing undesirable levels of B than other types of overburden. Sampling of shale and especially carbonaceous material should be more intense than sampling of other overburden materials to identify problem areas. If areas of high B are detected in the premining overburden assessment, additional drilling and sampling in the suspected areas may be necessary to define accurately the volume of B-affected material. In such situations, B should be included in the postmining analysis of recontoured overburden.

Soil sampling and analysis should identify soil horizons with excessive B. If such horizons are found, additional sampling should be conducted. Reclamationists should not assume that soil horizons high in B are a trait of the particular soil series sampled. If the B content of soil is high enough to render it unsuitable for salvage, as based on state standards, the material should be treated as undesirable overburden and handled accordingly.

Determining hot-water-soluble B in soil and overburden is expensive and analytically difficult. When soluble salts accumulate, B often accumulates, and there appears to be a correlation between electrical conductivity and water-soluble B. The relatively simple and inexpensive determination of electrical conductivity of a saturation extraction could be performed as a first step to determine if B toxicity potentially exists. Some researchers have found a log/linear relationship between hot-water-soluble B and electrical conductivity (*70*). When problem areas are indicated by electrical conductivity, subsequent analyses for B should be conducted. In a related study, a high correlation ($r = 0.90$) was found between increasing levels of hot-water-soluble B and increasing amounts of total organic carbon. This high correlation suggests that carbonaceous material should be routinely analyzed for B (personal communication with D. Boon, Wyoming Department of Environmental Quality, Cheyenne).

Acceptable B levels in soil and overburden. For a number of years, researchers believed that soil and overburden containing more than 5 ppm of hot-water-soluble B were unsuitable for use as a plant growth medium. This limit was based on the response of agronomic crops to B. But this limit is not an accurate reflection of the plant species and growing conditions in arid and semiarid reclamation situations. Many native species are well adapted to B levels in excess of 5 ppm. About 45 percent of the plant varieties in one study grew better with 5 ppm of plant-available B in the growth medium than with less than 5 ppm B (*20*). One research team found that B toxicity to plants is most likely to occur at hot-water-soluble levels exceeding 5 ppm (*61*). Because of the overlapping in beneficial and injurious levels of B in the growth medium and because of the numerous factors affecting B uptake and use, setting an optimum B concentration in the growth medium becomes theoretical and relative to numerous conditions.

In western reclamation situations a wide range of B can be tolerated in the growth medium. One test found that 7.2 to 10.2 ppm of 1 N ammonium-acetate-extractable B reduced yields by only 10 percent in wheatgrasses and alkali sacaton used to revegetate retorted oil shale (*76*). In another test tall wheatgrass tolerated 35 ppm B in a saturated-paste-extract without significantly reducing yields (*67*). Basin wildrye can tolerate 37 ppm B in a water-saturated extract before above-ground yields are reduced 50 percent (*64*).

While present knowledge is insufficient to define precisely the acceptable B levels in a growth medium, soil and overburden containing more than 5 to 8 ppm of hot-water-soluble B may require special revegetation considerations. Boron limits must be applied with flexibility, and consideration must be given to individual situations.

Boron migration. Leaching experiments suggest that soluble B in the growth medium is relatively mobile. Applying 120 cm of water in one test removed 58 percent of the B in the top 30 cm of soil (*65*). Other researchers found that applying 120 cm of water reduced an initial B concentration of 54 ppm to 6.9 ppm (*80*). However, still other tests found that, after leaching, some soils may regenerate soluble B to its original level (*55, 62*). Mechanisms for B regeneration have been described (*56*).

Because of the mobility of soluble B, soil placed on top of overburden containing elevated levels of B could be contaminated by upward migration of B under textural and moisture conditions conducive to capillary movement. Evidence of upward migration of B from spoil high in B into overlying soil was found in southwestern Wyoming (personal communication, D. Boon, Wyoming Department of Environmental Quality, Cheyenne). Also, B could leach from overburden placed in groundwater zones. This possibility should be considered when overburden high in B is encountered and when groundwater moves through a portion of the overburden.

Reclamation strategies

Following are some reclamation strategies that may be appropriate in mitigating high-B materials. One word of caution must be noted with respect to these strategies: their cost-effectiveness has not been evaluated, and such an evaluation should be made before any remedial actions are taken.

Amendment application. Acidic materials high in B can be amended with lime, and presumably with other neutralizing agents, to reduce plant uptake of B (*33*). If sewage sludge is used as an amendment to increase the fertility of the growth medium, the B content of the sludge should be measured prior to its use. One study found that fiberglass in sewage sludge released as much as 56.5 mg hot-water-soluble B/kg of sludge (*81*). This B was toxic to some plant species.

Overburden placement. Materials containing excessive B, such as carbonaceous spoil, can be selectively placed to limit both upward and downward migration of B. Such placement depends upon the depth of water penetration, patterns of groundwater movement, topography, physical and chemical traits of surrounding materials, and other factors. Material high in B should be covered by at least 70 cm of suitable material. Cover depths for undesirable material containing soluble constituents that can migrate upward have yet to be established with any precision. Depths depend upon substrate conditions, such as ionic species, physical traits, and patterns of water movement, and upon reclamation considerations, including species and cultural practices. Recent laboratory and field research have provided some insight into burial depths. Laboratory studies suggest that when soil is placed over sodic spoil with a low hydraulic conductivity 10 to 15 cm of upward Na migration can be expected (*46*). A northern Great Plains study found 7 to 14 cm of upward Na migration in test plots (*7*). In addition, this study found that 50 to 70 cm of soil over most spoil types is sufficient to maximize production of cool-season perennial grasses. A burial depth of at least 70 cm should, therefore, be sufficient to contain any upward migration of B and to provide adequate soil depth for revegetation in most situations.

Treatment to remove B. As mentioned, B can be removed from the growth medium by leaching with water. The quantity of water necessary to leach B to a particular depth varies a great deal. Much depends upon the original B concentration, the desired final B concentration, and the physical and chemical traits of the soil or overburden. Generally, B removal from soil requires about three times as much water as that required to remove most Na salts (*80*). The leaching process in one case was hastened by applying about 3 t/ha of concentrated sulfuric acid; consequently, the amount of water needed to leach a given amount of B was reduced about 80 percent compared to water leaching alone (*59*). But leaching is seldom feasible in surface mining situations. In addition, some soils can regenerate the soluble B lost through leaching. Also, any leaching operation must evaluate where

Table 5. Boron tolerances of some plant species used in revegetation.

Species	Yield Reduction or Injury	B Content in Growth Medium (ppm)	Method of B Analysis	Reference
Alkali sacaton	90% maximum yield	7.2	1N NH_4Ac	*76*
Fourwing saltbush	90% maximum yield	7.2	1N NH_4AC	*76*
Lewis flax	90% maximum yield	0.7	1N NH_4AC	*76*
Tall wheatgrass	90% maximum yield	10.2	1N NH_4AC	*76*
Western wheatgrass	90% maximum yield	8.5	1N NH_4AC	*76*
Tall wheatgrass	Normal	35	Saturated-paste extract	*67*
Giant bermuda	Normal	5	Saturated-paste extract	*60*
Great basin wildrye	Normal	5	Saturated-paste extract	*60*
Tall fescue	Normal	5	Saturated-paste extract	*60*
Tall wheatgrass	Normal	5	Saturated-paste extract	*60*
Fourwing saltbush	Normal	5	Saturated-paste extract	*60*
Quailbush saltbush	Normal	5	Saturated-paste extract	*60*
Desert saltbush	Normal	5	Saturated-paste extract	*60*
Gardner saltbush	Normal	60	Solution culture	*48*
Fourwing saltbush	Normal	60	Solution culture	*48*
Greasewood	Normal	20	Solution culture	*48*
Thickspike wheatgrass	Normal	15	Solution culture	*48*
Beardless wheatgrass	Normal	15	Solution culture	*48*
Indian ricegrass	Normal	10	Solution culture	*48*
Rubber Rabbitbrush	Normal	7.5	Solution culture	*48*
Big sagebrush	Normal	<5	Solution culture	*48*
Silky lupine	Normal	<5	Solution culture	*48*
Tall fescue	Normal	>9	Hot-water-soluble	*57*
Barley	Slight injury	5	Soluble B applied to soil	*20*
Kentucky bluegrass	Slight to moderate injury	1-25	Soluble B applied to soil	*20*
Oats	Mild injury	5	Soluble B applied to soil	*20*
Alfalfa	Moderate injury	15-25	Soluble B applied to soil	*20*
Sweet clover	Slight injury	10	Soluble B applied to soil	*20*
Lupine	Severe injury	5	Soluble B applied to soil	*20*
Vetch	Slight injury	5	Soluble B applied to soil	*20*
Tall wheatgrass (Jose)	50% root yield reduction	30	Water-saturated extract	*64*
Tall wheatgrass (Jose)	50% shoot yield reduction	66	Water-saturated extract	64
Basin wildrye (Magnar)	50% root yield reduction	22	Water-saturated extract	64
Basin wildrye (Magnar)	50% shoot yield reduction	37	Water-saturated extract	64

the leachate will travel and the consequences of this high-B leachate for the receiving system.

Species selection. Use of B-tolerant species to revegetate materials containing elevated levels of B is promising. However, limited information exists about the B tolerance of plants typically used in revegetation; a summary of the available information is presented in table 5. Lacking specific information, B tolerance can be estimated roughly by assuming an analogous relationship with salt tolerance (*80*). However exceptions to this apparent relationship exist, such as alfalfa, which is salt-sensitive and B-tolerant (personal communication with B. Schafter, Montana State University, Bozeman). Recent advances in genetic engineering are increasing the potential for breeding B-tolerant plant varieties. Until more information is available, however, selection of B-tolerant species will be dominated by trial-and-error experimentation.

As mentioned, B accumulates in leaf tips and margins. This apparent survival mechanism limits the extent of plant injury and allows plant B to be lost through leaf removal or abscission (*52*). Therefore, certain management practices following revegetation, such as forage removal by mowing and baling, may effectively reduce B toxicity.

There appears to be little concern about the secondary transfer of B from plants to animals. Cases of B toxicity from consumption of forage high in B are rare. The B in forage is rapidly and almost completely excreted (*79*).

REFERENCES

1. Adriano, D. C. 1986. *Trace elements in the terrestrial environment.* Springler-Verlag, New York, New York. pp. 73-105.
2. Anderson, B. M. 1982. *Effects of processed oil shale on the element content of* Atriplex cancescens. Open-file Report 82-1051. U.S. Geological Survey, Washington, D.C.
3. Aubert. H., and M. Pinta. 1977. *Trace elements in soils.* Elsevier Scientific Publishing Co., New York, New York. 395 pp.
4. Baird, G. B., and J. E. Dawson. 1955. *Determination of that portion of soil boron available to plants by modified soxhlet-extraction procedure.* Soil Science Society of American Proceedings 19: 219-222.
5. Baker, A. S., and W. P. Mortensen. 1966. *Residual effect of single boron applications on western Washington soils.* Soil Science 102: 163-179.
6. Barth, R. C. 1976. *An evaluation of the soils of tract C-a and adjacent areas. Rio Blanco Oil Shale Project.* Colorado School of Mines Research Institute, Golden. 129 pp.
7. Barth, R. C. 1984. *Soil-depth requirements to reestablish perennial grasses on surface-mined areas in the Northern Great Plains.* Mineral Energy Resources 27: 1-20.
8. Bauer, A., G. W. Gee, and J. E. Filley. 1976. *Physical, chemical, and biological aspects of strip-mined lands in western North Dakota.* Old West Regional Commission, Denver, Colorado. 605 pp.

9. Becic, J. N. 1983. *Viewpoint: Soil boron guidelines for reclaimed western soils.* Journal of Range Management 36: 673-674.
10. Berger, K. C., and E. Truog. 1940. *Boron deficiencies as revealed by plant and soil tests.* Journal of the American Society of Agronomy 32: 297-301.
11. Biggar, J. W., and M. Fireman. 1960. *Boron adsorption and release by soils.* Soil Science Society of America Proceedings 24: 115-120.
12. Bingham, F. T. 1982. *Boron.* In A. L. Page [editor] *Methods of Soil Analysis. Part 2. Chemical and Microbial Properties.* Agronomy Monograph No. 9. American Society of Agronomy, Madison, Wisconsin. pp. 431-447.
13. Bingham, F. T., and A. L. Page. 1971. *Specific character of boron adsorption by an amorphous soil.* Soil Science Society of America Proceedings 35: 892-893.
14. Bowen, H.J.M. 1966. *Trace elements in biochemistry.* Academic Press, New York, New York. 240 pp.
15. Bradford, G. R. 1965. *Boron.* In H. D. Chapman [editor] *Diagnostic Criteria for Plants and Soils.* Quality Printing Co., Inc., Abilene, Texas.
16. Buckman, H. O., and N. C. Brady. 1969. *The nature and properties of soil.* MacMillan Publishing Co., New York, New York. 639 pp.
17. Connor, J. J., J. R. Keith, and B. M. Anderson. 1976. *Trace-metal variation in soils and sagebrush in the Powder River Basin, Wyoming, and Montana* U.S. Geological Survey Journal of Research (4): 49-59.
18. Cox, F. R., and E. J. Kamprath. 1972. *Micronutrient soil tests.* In J. J. Mortvedt, P. M. Giordano, and W. L. Lindsay [editors] *Micronutrients in Agriculture.* Soil Science Society of America, Madison, Wisconsin. pp. 289-317.
19. Crock, J. G., and R. C. Severson. 1980. *Four reference soil and rock samples for measuring element availability in western energy regions.* Circular 841. U.S. Geological Survey, Washington, D.C. 16 pp.
20. Eaton, F. M. 1944. *Deficiency, toxicity, and accumulation of boron in plants.* Journal of Agriculture Research 69: 237-277.
21. Ebens, R. J., and H. T. Schacklette. 1982. *Geochemistry of some rocks, mine spoils, stream sediments, soils, and plants and water in the Western Energy Region of the conterminous United States.* Professional Paper 1237. U.S. Geological Survey, Washington, D.C.
22. Elseewi, A. A., and A. E. Elmalky. 1979. *Boron distribution in soils and waters of Egypt.* Soil Science Society of America Journal 43: 297-300.
23. Erd, R. C. 1980. *The minerals of boron.* In R. Thompson [editor] *Mellor's Comprehensive Treatise on Inorganic and Theoretical Chemistry. Supplement to volume V.* Longman, New York, New York. pp. 7-71.
24. Erdman, J. A., and R. J. Ebens. 1979. *Element content of crested wheatgrass grown on reclaimed coal spoils and on soils nearby.* Journal of Range Management 32: 159-161.
25. Fleming, G. A. 1980. *Essential micronutrients I: Boron and molybdenum.* In B. E. Davies [editor] *Applied Soil Trace Elements.* John Wiley & Sons, New York, New York. pp. 155-197.
26. Gauch, H. G., and W. M. Dugger, Jr. 1953. *The role of boron in the translation of sucrose.* Plant Physiology 28: 457-467.
27. Gestring, W. D., and P. N. Soltanpour. 1981. *Boron analysis in soil extracts and plant tissue by plasma emission spectroscopy.* Communications in Soil Science and Plant Analysis 12: 733-742.
28. Gestring, W. D., and P. N. Soltanpour. 1981. *Evaluation of wet and dry digestion methods for boron determination in plant samples.* Communications in Soil Science and Plant Analysis 12: 743-753.
29. Glaubig, B. A., and F. T. Bingham. 1985. *Boron toxicity characteristics of four northern California endemic tree species.* Journal of Environmental Quality 14: 72-77.

30. Gough, L. P., and R. C. Severson. 1981. *Biogeochemical variability of plants at native and altered sites, San Juan Basin, New Mexico.* Professional Paper 1134-D. U.S. Geological Survey, Washington, D.C.
31. Gough, L. P., and R. C. Severson. 1983. *Rehabilitation materials from surface-coal mines in western U.S.A II. Biogeochemistry of wheatgrass, alfalfa, and fourwing saltbush.* Reclamation and Revegetation Research 2: 103-122.
32. Gupta, S. K., and J.W.B. Stewart 1975. *The extraction and determination of plant-available boron in soils.* Schweizerische Landwirtschaftliche Forschung 14: 153-169.
33. Gupta, U. C. 1979. *Boron nutrition of crops.* Advances in Agronomy 31: 273-307.
34. Hatcher, J. T. 1960. *West digestion of plant material gives low boron values.* Analytical Chemistry 32: 726.
35. Jackson, J. F., and K.S.R. Chapman. 1975. *The role of boron in plants.* In D.J.D. Nicholas and A. R. Egan [editors] *Trace Elements in Soil-Plant-Animal Systems.* Academic Press, New York, New York. pp. 213-225.
36. Jones, J. B. 1972. *Plant tissue analysis for micronutrients.* In J. J. Mortvedt, P. M. Glordano, and W. L. Lindsay [editors] *Micronutrients in Agriculture.* Soil Science Society of America, Madison, Wisconsin. pp. 319-346.
37. Keren, R., and F. T. Bingham. 1985. *Boron in water, soils, and plants.* Advances in Soil Science 1: 230-276.
38. Keren, R., F. T. Bingham, and J. D. Rhoades. 1985. *Plant uptake of boron as affected by boron distribution between liquid and solid phases in soils.* Soil Science Society of America Journal 48: 297-302.
39. Keren, R., and R. G. Gast. 1981. *Effects of wetting and drying and of exchangeable cations on boron adsorption and release by montmorillonite.* Soil Science Society of America Journal 45: 478-482.
40. Kilkelly, M. K., and W. L. Lindsay. 1982. *Selected trace elements in plants grown on retorted oil shale.* Journal of Environmental Quality 11: 422-427.
41. Klusman, R. W., and J. Candito. 1978. *Trace element reconnaissance.* In *Trace Elements in Oil Shale.* DOE Contract EX-76-5-02-4017. University of Colorado, Boulder.
42. Krauskopf, K. B. 1972. *Geochemistry of micronutrients.* In J. L. Mortvedt, P. M. Glordano, and W. L. Lindsay [editors] *Micronutrients in Agriculture.* Soil Science Society of America, Madison, Wisconsin. pp. 7-40.
43. Lucas, R. E., and B. D. Knezek. 1972. *Climatic and soil conditions promoting micronutrient deficiencies in plants.* In J. J. Mortvedt, P. M. Glordano, and W. L. Lindsay [editors] *Micronutrients in Agriculture.* Soil Science Society of America, Madison, Wisconsin. pp. 265-288.
44. Mattigod, S. V. 1983. *A method of estimating the standard free energy of formation of borate minerals.* Soil Science Society of America Journal 47: 654-655.
45. McClung, A. C., and J. E. Dawson. 1951. *Some studies on the behavior of soil boron under cropping.* Soil Science Society of America Proceedings 15: 268-272.
46. Merrill, S. D., E. J. Doering, J. F. Power, and F. M. Sandoval. 1983. *Sodium movement in soil/mine spoil profiles: Diffusion and convection.* Soil Science 136: 308-316.
47. Mortvedt, J. J., P. M. Giodano, and W. L. Lindsay. 1972. *Micronutrients in agriculture.* Soil Science Society of America, Madison, Wisconsin. 666 pp.
48. Munshower, F. F. 1984. *Boron tolerance of some western range species.* Montana State University Research Report. Black Butte Coal Company, Rock Springs, Wyoming. 144 pp.
49. Norrish, K. 1975. *Geochemistry and mineralogy of trace elements.* In D.J.D. Nicholas and A. R. Egan [editors] *Trace Elements in Soil-Plant-Animal Systems.*

Academic Press, Inc., New York, New York. pp. 55-81.

50. Oertli, J. J. 1962. *Loss of boron from plants through guttation.* Soil Science 94(24): 214-219.
51. Oertli, J. J., and H. C. Kohl. 1961. *Some considerations about the tolerance of various plant species to excessive supplies of boron.* Soil Science 92. pp. 243-247.
52. Oertli, J. J., O. R. Lunt, and V. B. Younger. 1961. *Boron toxicity in several turf grass species.* Agronomy Journal 53: 262-265.
53. Olsen, S. R. 1972. *Micronutrient interactions.* In J. J. Mortvedt, P. H. Giordano, and W. L. Lindsay [editors] *Micronutrients in Agriculture.* Soil Science Society of America, Madison, Wisconsin. pp. 243-264.
54. Parker, D. R., and E. H. Gardner 1982. *Factors affecting the mobility and plant availability of boron in some western Oregon soils.* Soil Science Society of America Journal 46: 573-578.
55. Peryea, F. J., F. T. Bingham, and J. D. Rhoades. 1985. *Regeneration of soluble boron by reclaimed high-boron soils.* Soil Science Society of America Journal 49: 313-316.
56. Peryea, F. J., F. T. Bingham, and J. D. Rhoades. 1985. *Mechanisms for boron regeneration.* Soil Science Society of America Journal 49: 840-843.
57. Peterson, L. A., and R. C. Newman. 1976. *Influence of soil pH on the availability of added boron.* Soil Science Society of America Journal 40: 280-282.
58. Philipson, T. 1953. *Boron in plant and soil with special regard to Swedish agriculture.* Atca Agriculturae Scandinavica 3: 121-242.
59. Prather, R. J. 1977. *Sulfuric acid as an amendment for reclaiming soil high in boron.* Soil Science Society of America Journal 41: 1,098-1,101.
60. Pratt, P. F., E. C. Nord, and F. L. Bair. 1971. *Early growth tolerances of grasses, shrubs, and trees to boron in tunnel spoil.* Research Note PSW-232. Forest Service, U.S. Department of Agriculture, Washington, D.C.
61. Reinsenauder, H. M., L. M. Walsh, and R. G. Hoeft. 1973. *Testing soils for sulfur, boron, molybdenum, and chlorine.* In L. M. Walsh and J. D. Beaton [editors] *Soil Testing and Plant Analysis.* Soil Science Society of America, Madison, Wisconsin. pp. 273-300.
62. Rhoades, J. D., R. D. Ingvalson, and J. T. Hatcher. 1970. *Laboratory determinations of leachable soil boron.* Soil Science Society of America Proceedings 34: 871-875
63. Ringrose, C. D., R. W. Klusman, and W. E. Dean. 1976. *Soil chemistry in the Piceance Creek Basin.* In *Geochemical Survey of the Western Energy Regions.* Open-File Report 76-729. U.S. Geological Survey, Washington, D.C. pp. 101-111.
64. Roundy, B. A. 1985. *Germination and seedling growth of tall wheatgrass and basin wildrye in relation to boron.* Journal of Range Management 38: 270-272.
65. Sahota, N. S., and D. R. Bhumbla. 1969. *Effect of leaching saline-alkali soils with and without gypsum on the soluble salts, boron, and exchangeable calcium and sodium of the soils.* Indian Journal of Agricultural Science 40: 763-771.
66. Sandoval, F. M., and J. F. Power. 1977. *Laboratory methods recommended for chemical analysis of mined-land spoils and overburden in western United States.* Handbook No. 525. U.S. Department of Agriculture, Washington, D.C.
67. Schuman, G. E., 1969. *Boron tolerance of tall wheatgrass.* Agronomy Journal 61: 445-447.
68. Schwab, A. P., W. L. Lindsay, and P. J. Smith. 1983. *Elemental contents of plants growing on soil-covered retorted shale.* Journal of Environmental Quality 12: 301-304.
69. Severson, R. C., and L. P. Gough. 1981. *Geochemical variability of natural soils and reclaimed mine-spoil soils in the San Juan Basin, New Mexico.* Professional

Paper 1134-C. U.S. Geological Survey, Washington, D.C.
70. Severson, R. C., and L. P. Gough. 1983. *Boron in mine soils and rehabilitation plant species at selected surface coal mines in western United States.* Journal of Environmental Quality 12: 142-146.
71. Severson, R. C., and L. P. Gough. 1983. *Rehabilitation materials from surface-coal mines in western U.S.A. I. Chemical characteristics of spoil and replaced cover soil.* Reclamation and Revegetation Research 2: 83-102.
72. Severson, R. C., and R. R. Tidball. 1979. *Spatial variation in total element concentration in soil within the Northern Great Plains Coal Region.* Professional Paper 1134-A. U.S. Geological Survey, Washington, D.C.
73. Shacklette, H. T., J. C. Hamilton, J. G. Boerngen, and J. M. Bowles. 1971. *Elemental composition of surface materials in the conterminous United States.* Professional Paper 574-D. U.S. Geological Survey, Washington, D.C. 8 pp.
74. Sims, J. R., and F. T. Bingham. 1968. *Retention of boron by layer silicates, sesquioxides, and soil materials: II. Sesquioxides.* Soil Science Society of America Proceedings 32: 364-369.
75. Smidt, R. E., and J. S. Whitton. 1975. *Note on boron toxicity in a stand of radiata pine in Hawkes Bay.* New Zealand Journal of Science 18: 104-113.
76. Smith, P. J., and W. L. Lindsay. 1982. *Boron toxicity in range plants grown on retorted oil shale.* Department of Agronomy, Colorado State University, Fort Collins.
77. Stanley, M. A., G. E. Schuman, F. Rauzi, and L. I. Painter. 1982. *Quality and element content of forages grown on three reclaimed mine sites in Wyoming and Montana.* Reclamation and Revegetation Research 1: 311-326.
78. Stone, E. L., and G. Baird. 1956. *Boron level and boron toxicity in red and white pine.* Journal of Forestry 54: 11-12.
79. Underwood, E. J. 1971. *Trace elements in human and animal nutrition.* Academic Press, New York, New York. 543 pp.
80. U.S. Salinity Laboratory Staff. 1954. *Diagnosis and improvement of saline and alkali soils.* Agriculture Handbook 60. U.S. Department of Agriculture, Washington, D.C. 160 pp.
81. Vimmerstedt, J. P., and T. N. Glover. 1984. *Boron toxicity to sycamore of mine soil mixed with sewage sludge containing glass fibers.* Soil Science Society of America Journal 48: 389-393.

Chapter 8

Nitrogen

J. D. Reeder and Burns Sabey

A deficiency of plant-available nitrogen can limit the successful revegetation and long-term stability of land disturbed by surface mining. Deficiencies arise because disturbance drastically alters the flow of N through the soil-plant-microbial ecosystem, and decades or centuries are required for some disturbed land to accumulate N naturally and evolve toward a steady state (*74*).

Most mined land in the western United States is returned to rangeland for grazing and browsing by domestic livestock and/or wildlife (*6*). A major reclamation goal has been to achieve stabilization of the disturbed site with a desirable plant community that requires little or no long-term fertilizer inputs to maintain a level productivity comparable to that existing before disturbance (*114*). To realize this goal of a self-maintaining, desirable plant community, the reestablishment of an active biological N cycle is necessary (*21, 85*). Another reclamation concern involving N is for the potential for environmental degredation due to N losses from disturbed land by leaching, runoff, or soil erosion.

Nitrogen in the environment

Geochemical distribution of N. Ninety-seven percent of the total mass of N in the earth and surrounding atmosphere exists in the igneous rocks of the earth's mantle. Most of the remainder (2.3 percent) is found in the atmosphere. The mass of N in sedimentary rocks represents about 0.3 percent of the total, while the combined mass of N in coal and terrestrial soils, in the oceans and sea-bottom organic compounds, and in living organisms represents only 0.01 percent of the total mass of N in the earth and surrounding atmosphere (*97*). Most N found in the atmosphere and in the hydrosphere (all oceans and freshwaters) is in inorganic form, while the bulk of the N in the biosphere (all living organisms) is in organic form. Molecular N_2 comprises 78 percent of the gases in the atmosphere; other N compounds found in small quantities in the atmosphere are ammonium (NH_4^+), nitrate (NO_3^-), nitrous oxide (N_2O), and organic N associated with dust particles. In the hydrosphere 95 percent of the N occurs as molecular N_2; the remaining N compounds include NH_4^+, NO_2^-, NO_3^- and dissolved and particulate organic matter. The N in living organisms exists mainly in organic forms, such as proteins, amino acids, amino sugars, nucleic acids, and chitin (*97*).

Nitrogen is a common component of the lithosphere (soils, sediments, silicate minerals, fossils, and all rocks). The N associated with igneous rocks and with sediments and sedimentary rocks is bound either in organic form or as NH_4^+ held within the lattice structures of silicate minerals (*91, 92, 94*). The organic N found in sedimentary rocks, petroleum, and coal originated from the plant and animal residues comprising the humus fraction of prehistoric lake and sea bottom sediments (*93, 106*). Soil N is mostly organically bound; usually less than 2 percent of the total N content of soils is in the plant-available mineral forms of NH_4^+ and NO_3^- at any one time (*13*). Some soils contain significant amounts of fixed (nonexchangeable) NH_4^+, but the proportion of soil N in this form is rarely greater than 8 percent in surface soils, although it can exceed 40 percent in subsoils. About half of the N in most surface soils is in the form of bound amino acids, hexosamines, and purine and pyrimidine derivatives; the chemical nature of the remaining half remains obscure (*17*). The N content of mineral soils varies widely among different climatic zones: 0.03 to 0.16 percent N in arctic soils; 0.02 to 1.06 percent N in cool, temperate soils; 0.03 to 0.30 percent N in subtropical soils; and 0.24 to 1.61 percent N in tropical soils (*96*). Some peat soils contain more than 2.5 percent N (*17*).

Nitrogen cycles among the four spheres of the earth, flowing from the atmosphere and lithosphere into the biosphere and hydrosphere, then recycling back into and between the atmosphere and lithosphere. The rate of N flow between spheres can be quite rapid, or so slow as to be measured in centuries or millenia. Only some of these N flows directly affect plant growth within a given ecosystem. An understanding of how N cycles through a stable, undisturbed ecosystem is needed to understand the effects of surface mining on the ability of a reclaimed site to supply the N needed to sustain plant growth.

The N cycle in a grassland ecosystem. Annual N transfers between soil and vegetation far exceed other global N transfers; 95 percent of the N flow in the global terrestrial system occurs in the soil-plant-microbial system, while only 5 percent of the total N flow involves exchanges to and from the atmosphere and the hydrosphere (*77*). Figure 1 shows the components and interconnecting processes of the N cycle within a generalized grassland ecosystem. Table 1 presents N distribution in representative grassland ecosystems of Canada and the western United States (*113, 114*). The following sections describe processes by which N is lost from or added to these grassland soils, the N requirements of plants, and the sources of N for plant

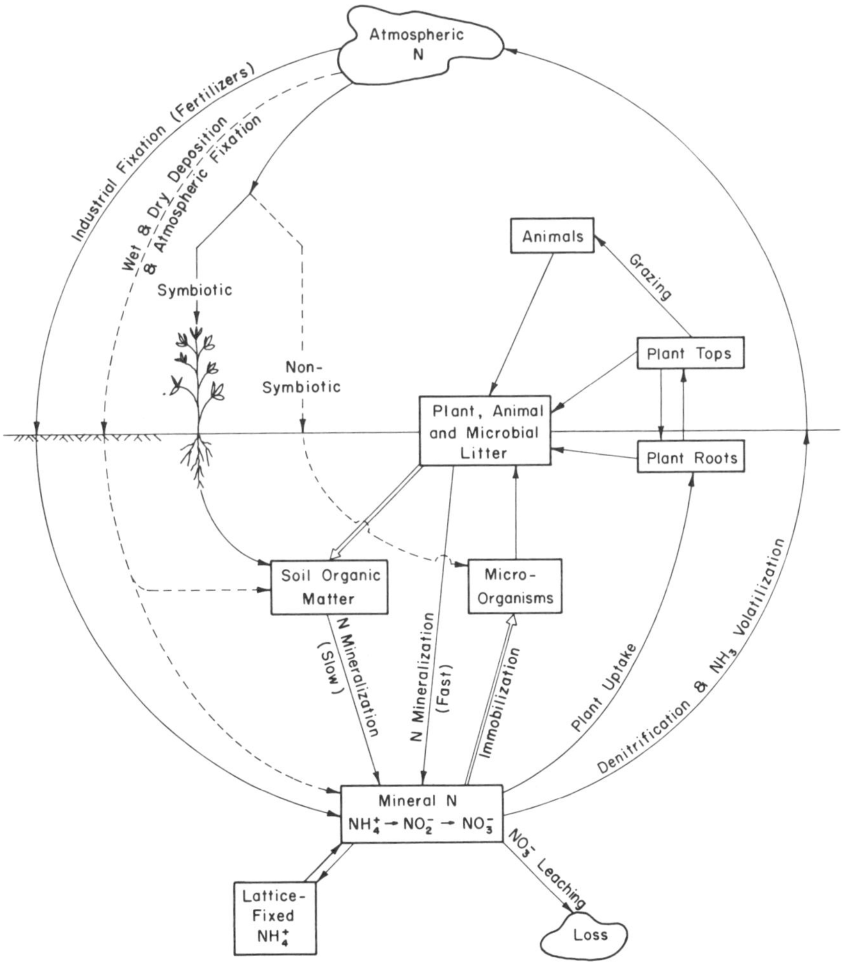

Figure 1. Components and interconnecting processes of the N cycle within a generalized grassland ecosystem.

Table 1. Inputs, losses, and nitrogen distribution in grassland ecosystems of the western United States and Canada (*113, 114*).

Ecosystem Type	Location	Inputs*	Losses	Net Gain	N in Biomass: Standing Live	Litter	Roots†	Microbes	Large Animals	N in Soil: Organic Matter	Mineral‡ N
		kg/ha/yr			kg/ha						
Great Basin shrub-steppe	Southeastern Washington	2	<0.5	2	30	30	140	50§	0-3	4,170	21
Desert grassland	Southern New Mexico	3	<0.5	3	20	20	20	20§	0-3	810	4
Shortgrass prairie	Northeastern Colorado	6	<0.5	6	30	90	240‖	50	0-3	5,500	28
Mixed-grass prairie (I)	Southwestern South Dakota	9	1	8	30	70	150‖	130§	0-3	8,400	42
Mixed-grass prairie (II)	Southwestern Saskatchewan	4	<0.5	4	20	70	140	130	0-3	5,400	27

*Principal inputs are assumed to be from the atmosphere as wet and dry deposition.
†Twenty percent of the N in roots is assumed to be living tissue.
‡Nitrogen in mineral form estimated as 0.5 percent of N in soil organic matter.
§Estimate (guess).
‖Includes crowns.

growth in nondisturbed grassland ecosystems.

Losses of N from western soils. The pathways of possible N loss from an undisturbed grassland ecosystem include nitrate leaching, denitrification, ammonia volatilization, erosion, and grazing by livestock. However, total annual N loss from all of these processes is usually low, ≤1 kg/ha/yr (Table 1). Because water rarely occurs in sufficient quantity to saturate the soil profile and because NO_3^- is usually found at very low concentrations in grassland soils, N losses by leaching or by biological denitrification are generally very small. Chemo-denitrification is limited largely to urine patches in grasslands and is not considered a major pathway of N loss. Nitrogen loss from grasslands as volatilized NH_3 from senescing plant parts, animal excreta, and decomposing litter is thought to be the major source of N loss in natural grasslands where the environment is not favorable for leaching or denitrification (*112*). Losses of N from wind or water erosion of soil are usually very low because of the fully developed plant cover on most grasslands, but sizeable losses may occur during intense storms on sites where plant cover has been reduced by fire, drought, or overgrazing (*113*). The amount of N removed in cattle tissue depends on grazing intensity. However, animals assimilate only a small fraction of the N they consume; 80 to 85 percent of the N ingested by cattle is returned to the soil, mostly in urine and feces (*113*).

Additions of N to western soils. Inputs of N from the atmosphere, in the forms of NH_4^+, NO_2^-, NO_3^- and organic N associated with dust particles, range from 2 to 9 kg/ha for the semiarid and arid regions of the western United States (Table 1) and probably represent the largest natural source of N to these areas (*114*). Annual additions of N from atmospheric fixation, for example, the formation of NO_3^- from atmospheric N_2 photochemically or by lightning discharges, is considered negligible (*88, 97*). Annual inputs of N by N_2 fixation are low because legumes are not a dominant component of the vegetation in most grasslands and usually represent only 1 to 5 percent of the primary production (*113*). Inputs of N by nonsymbiotic N_2 fixation are thought to be low, ranging from 1 to 2 kg/ha/yr (*64*). These levels of annual N additions to the soil from the atmosphere are insufficient to meet the annual growth requirement of a late-successional or climax-level plant community (*24, 114*).

Nitrogen may also be added to western rangeland by the application of nitrogenous fertilizers. The response of native rangeland to fertilizer N appears to vary widely with location and depends on such factors as climate, soil type, plant species, and management. Although fertilization can increase the quality and/or quantity of forage, the benefits are usually limited to the first few years following application (*26, 73*). Moisture is often the limiting growth factor on western rangeland, and native species may not respond sufficiently to added fertilizer N to make the practice economically sound (*53*). Increased production or improved forage quality as a response to added fertilizer N is affected by the supply of available soil water and by growing season precipitation supply and distribution (*6*). Growing conditions the first few months after fertilizer application appear to have a major effect upon the fate of added fertilizer N (*69*). In the West fertilization has been most useful on mountain meadows or in the Pacific Coast region where precipitation is high during relatively mild winters (*53*).

Nitrogen requirements of plants. Because of the high N requirement for growth and reproduction, far more N than any other nutrient is removed from the soil by growing plants (*54*). Nitrogen is an essential component of chlorophyll, amino acids, plant growth enzymes, and ATP (adenosine triphosphate), which is the carrier of free energy needed for macromolecule synthesis and active transport of ions. Nitrogen also stimulates root development and activity and supports the uptake of other nutrients, such as P (*62*). Under conditions of N deficiency, leaves are small and pale yellowish-green in the early stages of growth, and plants have thin stems and a sparse, spindly appearance (*13*). Deficiency symptoms are first noticed and more pronounced on older leaves because N is translocated from older tissues to young, actively growing regions (*35*).

Concentrations of N in healthy leaves of rangeland plants generally vary from 1.3 to 1.7 percent N in grass species, such as needlegrass, brome, and wheatgrass; 1.2 to 3.4 percent N in shrubs, such as sagebrush, blackbrush, and shadscale; and 4.5 to 5.0 percent N in introduced legumes, such as alfalfa (*20, 62*). The N concentrations of live roots and of root and herbage litter are usually lower than in green herbage. However, for a given hectare of land, more N is usually associated with the roots and litter than with the green herbage (Table 1). This is because, on an area basis, the mass of roots and litter is usually larger than the mass of green herbage. The distribution of N in grassland ecosystems generally ranges from 20 to 30 kg/ha N in the aboveground, live herbage; 20 to 90 kg/ha N in aboveground plant litter; 20 to 240 kg/ha N in living and dead roots; and 50 to 130 kg/ha N in microbial biomass (Table 1).

Sources of N for plants. Although a few plant species can meet their N needs by the symbiotic fixation of atmospheric N_2, most plants obtain needed N by absorbing NH_4^+ and NO_3^- from the soil. However, the amounts of plant-available NH_4^+ or NO_3^- in the soil at any one time are usually limiting (*77*). In most grassland soils mineral N represents less than 0.5 percent of the total N of the system (*113*). Therefore, plants need a continually renewed source of mineral N to sustain growth, and this N is supplied in natural grasslands by the rapid mineralization of organic N compounds found in plant, microbial, and animal litter and by the slow mineralization of soil organic matter (Figure 1). These organic forms of N in the soil are mineralized by a large variety of microorganisms that use nitrogenous organic substances as an energy source and transform N from the organic state into NH_4^+ (*46*). The NH_4^+ is oxidized (nitrified) by bacteria that derive energy from the oxidation of NH_4^+ to NO_2^-, or NO_2^- to NO_3^-, and use CO_2 as a source of C (*81*).

Annual plant requirements for N in stable grassland ecosystems are not met entirely by uptake of mineral N from the soil because stable systems tend to conserve essential nutrients (*60*). About one-third of the annual N requirement of a grassland community is met by internal translocation of N within living plants, whereby N from one season is stored overwinter in belowground plant parts and then moved to new growth during the next growing season (*22*). However, the principal source of N for plant growth in stable grassland ecosystems is the rapid mineralization of labile organic N in plant and microbial litter (*114*). Although the soil organic matter constitutes the largest reservoir of N in a grassland ecosystem (Table 1), the amount of N mineralized annually from this reservoir is small and insufficient to meet the N requirements of the plant community. This is because most of the N in the soil organic matter is held in organic compounds that are highly resistant to decomposition.

The amount of N found in climax or late-successional grassland ecosystems, for example, those in table 1, is the culmination of centuries or millennia of ecosystem development. During the long period of ecosystem development, a sequence of plant communities replaces one another as the soil develops and accumulates organic matter (*14, 95*). Late-successional or climax grassland species are generally not capable of fixing N_2, but rather must rely on the ability of the soil-litter system to supply N via mineralization of organic N in plant and microbial litter and from the soil organic matter. These plant species cannot dominate the site until pathways of fast-cycling N are well established and no longer limiting. The N in a climax or late-successional grassland ecosystem is assumed to be in steady-state equilibrium, meaning that the amount of N incorporated into living tissues of plants and microorganisms roughly equals that lost by death or exudation (*114*). Thus, the annual N requirements of climax or late-successional grassland communities are met largely by the cycling of N already present in the ecosystem (*24*).

The N cycle in a disturbed grassland ecosystem. Surface mining frequently results in a deficiency of plant-available N because the mining process disrupts the N cycle within the ecosystem. Nitrogen deficiency seldom limits seedling establishment on reclaimed mined land (*27, 114*). Moreover, in some cases plant productivity on reclaimed, N-deficient mined land is comparable to production on native, undisturbed sites, although the N content of the mined land forage is generally lower (*90, 115*). However, N deficiency can affect the long-term stability of a site by limiting plant productivity, such that N-deficient sites will generally not develop the vegetation needed to prevent erosion (*114*). Plant communities fail to persist on severely N-deficient sites, despite initial seedling establishment and fertilization, because of a lack of a continuing N supply from a mineralizing cycle (*15, 72*). Thus, the deficiency of plant-available N is a challenging, long-term management problem on some western mined land.

The mining process disrupts internal N cycling within a stable ecosystem by altering the forms and amounts of N found in the various components of the ecosystem and by disrupting the microbial subsystem responsible for N mineralization. The entire living plant biomass, containing perhaps 24 to 78 kg/ha N (as estimated from Table 1), is destroyed and must be reestablished. The microbial biomass, the hub of the organic portions of the N cycle and containing 20 to 130 kg/ha N (Table 1), is partly or entirely destroyed and may be slow to reestablish. Much of the original, aboveground litter and root litter components (36 to 282 kg/ha N as estimated from Table 1) are mixed within the topsoil and thus eventually are returned to the disturbed site, and additional residues are frequently surface applied as mulch. However, the total N content and the proportion of readily mineralizable N in this replaced litter component are often different from the original undisturbed components. Stockpiling, especially for long periods of time, can alter the form and content of N in the topsoil. If topsoil is not properly removed and then replaced, the large reservoir of N in the soil organic matter can be diluted with substrata.

Reclamation programs generally include regrading the disturbed site, replacing topsoil, and seeding the site with plant species that are selected to stabilize the site and achieve productivity comparable to that before disturbance. Most of the plant species that are reseeded rely on the soil for a large part of their N requirements, but the original soil-litter subsystem, which supplied rapidly mineralizable N to plants, has been altered and is usually seriously depleted by the mining process. For a number of years after reseeding, the developing ecosystem will tend to accumulate biomass and nutrients as it stabilizes. During this time, the plant and microbial biomasses and the litter component act as sinks for N. These N sinks withdraw N from the fast-cycling pools, thereby creating a deficiency of N available for new plant growth (*114*). Additions of fertilizer N may, therefore, be needed to sustain plant growth until the system becomes self-maintaining.

Although various amounts of time are required for the different components of a disturbed system to accumulate biomass and nutrient levels comparable to those before disturbance, diverse plant communities can develop in some disturbed lands that are lower in N than native ecosystems. In some instances productivity can exceed the potential of undisturbed soils (*80*). The principal requirements for a functioning system are a developed pathway of fast cycling N and an input of N that is equal to or greater than the amount of N channeled into the soil organic matter (*114*). Research has also suggested that some minimum level of organic matter is needed to sustain N cycling, and this minimum level is much less than the amounts found in well-established ecosystems (*15, 31, 115*).

Topsoil N. Federal and state laws require that topsoil be removed before mining and replaced over the graded overburden after a site has been mined. Use of topsoil for reclamation provides a good rooting medium, improves infiltration, reduces runoff, encourages faster reestablishment of nutrient cycles, and can increase species diversity of vegetational cover (*82*). Topsoil can alleviate N deficiency in revegetated disturbed land by providing plant and microbial residues as sources of mineralizable N and by serving as an inoculum of the microorganisms that mineralize N (*114*). By returning mineralizable N to the system, replacing topsoil should also shorten the time needed until the system becomes self-sustaining. However, sufficient depth and quality of topsoil to restore completely the N cycle are not always available. If topsoil is not replaced, or if inadequate quality or depth of topsoil (or other surface material) is replaced, the system will not likely be able to supply vegetation with adequate N levels. Biologically resistant N compounds will tend to accumulate and form soil organic matter, thus withdrawing N from the fast cycling portion of the system. Decades or centuries may be required for the soil organic matter component of the ecosystem to accrue N and reach a new equilibrium. During this time, the soil organic matter may act more as a sink than as a source of N.

To optimize the value of topsoil, it is important that no significant losses in chemical, physical, or biological properties occur. Whenever feasible, direct hauling and immediate respreading of topsoil on regraded sites is advisable. However, stockpiling may be necessary because it is not always possible to strip topsoil selectively for immediate use. Stockpiling topsoil can adversely affect its chemical and biological properties. The degree of damage depends upon the methods of stripping and stockpiling the soil, management of the topsoil stockpile, and length of storage time (*105*). Stored topsoil has little biological resemblance to undisturbed surface soil (*39*) and storage reportedly causes (a) loss of soil aggregation; (b) a loss in organic N with a subsequent increase in $NO_3^- - N$, which can leach down to deeper sections of the stockpile; (c) a decline in the similarity of fungi between stockpiled soil and undisturbed soil as storage time increases; (d) a decline in inoculum potential of vesicular-arbuscular mycorrhizal fungi; and (e) a decline in microbial biomass C in the surface soil of the stockpile (*105*). Compaction during topsoil storage may lead to poor aeration and conditions suitable for denitrification (*58*).

Management practices are needed that reduce the adverse effects of stockpiling on the topsoil. Minimizing the time that topsoil is stockpiled before use will minimize adverse effects on the microbial population. Reducing the size of topsoil stockpiles and seeding them provides vegetative cover that protects the soil from erosion and helps to stabilize the microbial population. Although the biological component may be significantly altered because of topsoil segregation and storage, the microbial population can begin to reestablish fairly rapidly in topsoiled, revegetated mined lands.

Under some conditions microbial populations of topsoiled, revegetated mined land can equal those found in undisturbed soil after the first growing season (*39*). Management practices, such as mulching and fertilizing, greatly benefit the reestablishment of the microbial population (*39*).

Geologic N. The geological materials that become surface or subsurface plant growth media on mined land may in some cases be sources of mineral or mineralizable N. The N content of geologic materials is highly variable, ranging from 11 $\mu g/g$ total N in some igneous rocks to more than 4,000 $\mu g/g$ N in some shales. Nitrogen in geologic materials is usually in organic form or as NH_4^+ fixed within the lattice structures of silicate minerals (*94*). However, some shales contain significant quantities of exchangeable NH_4^+ that could support vegetation for several growing seasons (*68*). The organic N content of many geologic materials can be comparable to the N content of topsoil. However, the mineralization potential of geologic organic N is usually much lower than the mineralization potential of soil organic N, meaning a significantly smaller proportion of organic geologic N becomes available to plants over time. Low mineralization potentials of geologic N are not surprising. Exposed geologic N may be physically less susceptible to microbial mineralization than the organic N of soils because the larger particle size of some geologic materials exposes significantly less surface area than soil particles (*114*). Moreover, the biochemicals found in geologic materials are those that have been preserved selectively through their ability to combine with metals and fine-grained minerals or that are resistant to microbial attack (*93*).

Because the organic N of geologic materials is not readily mineralizable, geologic materials exposed as surface or subsurface plant growth media will usually not supply much of the total N demand of plants. In general, vegetation growing on geologic spoils takes up less N and produces less biomass than vegetation growing in soil (*74, 76*). However, at present it is unknown whether long-term weathering of exposed geologic materials will result in the release of indigenous fixed NH_4^+ or whether the degraded silicate minerals will become a sink for NH_4^+ added as fertilizer or mineralized from indigenous organic N (*114*).

Sources of supplemental N for plants. Until pathways of fast-cycling N are developed and the ecosystem is able to support the N requirements of the plant community, supplemental forms of N will be needed. Without added N, the plant biomass will be stressed for N, and plant productivity will decline as the system tends toward some new equilibrium between N demand and supply (*114*). A new equilibrium may result in a plant community with inadequate biomass to control soil erosion or support grazing; it may also contain undesirable plant species that are able to invade and dominate the site because they can tolerate low nutrient conditions better than the desired species. In some instances where the depth and quality of topsoil is inadequate, supplemental forms of N may be needed for a long time until a pool of mineralizable N is built up. Supplemental forms of N may include N fertilizers and N from N-fixing legumes (*70*).

Nitrogen-fixing plant species included in the seeding mixture used to revegetate a disturbed site will help relieve N deficiency. The use of legumes has been considered by some researchers to be the most efficient way of accumulating N in mined land because significantly more mineralizable and hydrolyzable N has been found in mined land receiving supplemental N from legumes rather than from inorganic fertilizer (*63*). Legumes, such as sweet clover, are estimated to fix from 100 to 200 kg/ha/yr N; alfalfa fixes as much as 130 to 600 kg/ha/yr N (*97, 102*). However, the use of introduced legume species may not always be allowed if state reclamation laws require that only native species be used. Moreover, introduced species may not thrive in western disturbed lands that are too saline or sodic or that receive inadequate precipitation. On disturbed sites where introduced legume species are adapted to the climate and plant growth media conditions, the legumes may tend to reduce species diversity by dominating an N-deficient site for a number of years until the ecosystem has accrued enough N to support climax grass species (*10*). The use of introduced legumes on mined land also may require special management alternatives to prevent bloat of livestock. However, many native legumes are now commercially available, and use of these species could reduce some of the problems associated with introduced species. Additions of N from nonleguminous native species probably will be minor. Certain nonlegume species, such as sagebrush (*Artemisia ludoviciana*), prickly pear (*Optuntia fragilis*), and rabbitbrush (*Chrysothamnus viscidiflorus*), have been found to fix N_2, but the amount of N fixed by these species in the field may be very small (*36, 37*).

Although in some instances part of the supplemental N needed to sustain plant growth on N-deficient sites is supplied by N-fixing legumes, much of the supplemental N needed in revegetated disturbed land is usually supplied by fertilizer N. A number of studies on mined land in the western states have evaluated the effects of N fertilization rate, timing, and method of application on seedling establishment and herbage quality and yield (*2, 3, 34, 43, 55, 56, 71, 86*). In general, on mined land where toxic elements or salinity are not problems, the effect of fertilizer N on plant growth depends upon available moisture and upon the quality and depth of topsoil. Water is the controlling factor in plant growth and yield and thus affects not only the amount of N needed to correct an existing deficiency but also the response of plants to a given fertilizer application rate (*6*). Where adequate mineral and mineralizable N are present in the topsoil, N fertilization does not generally increase yields (*7*). On the other hand, N fertilization of mined land with inadequate topsoil usually results in increased yields where moisture is not limiting (*1, 10*). Yields can significantly increase with increasing rate of N applied (*43*). However, application rates of N up to 121 kg/ha have been found to promote the establishment of introduced grasses at the expense of the less competitive native grasses (*34*), and can thus be detrimental to the achievement of a diverse stand of native vegetation. Application of fertilizer N at the time of planting on disturbed land that has been topsoiled is not usually beneficial to seedling establishment and may tend to increase weed growth and subsequent competition for moisture and nutrients (*27, 55*). Therefore, application of fertilizer N in the fall after the first growing season or before the start of the second growing season has been recommended (*27*).

Because multiple applications of fertilizer are more expensive than a single application, reclamationists may consider a single application of nutrients in quantities to meet crop needs for more than one season. This type of fertilizer application may be feasible where leaching is not a potential hazard. Power and Alessi (*67*) found no negative effect on cumulative grass yields when N rates up to 540 kg/ha were applied all in one year, or one-third applied in each of 3 years, or one-sixth applied in each of 6 years. However, applications of high rates of fertilizer N at one time usually result in less efficient use than annual applications at rates needed to meet crop requirements. The reduction in efficiency is attributed to time-related chemical fixation and biological immobilization processes and to displacement leaching (*6*). Moreover, application of high rates of fertilizer N to a recently revegetated disturbed area is discouraged, especially if the site has not been topsoiled or if topsoil is shallow or low in nutrients because the disturbed area will not yet have full components of microbial and plant mass that can immobilize large amounts of added N and, therefore, prevent it from being lost by leaching or volatilization.

Initial growth response of herbaceous species to N fertilizer on N-deficient sites is usually good, but the effect is usually short lived, with a decline in productivity occurring several years after application (*71*). A single application of fertilizer N to establish vegetation may, therefore, be inadequate to sustain the vegetation until adequate N cycling has been reestablished. Periodic applications of fertilizer N may be necessary until N cycling is reestablished and the site becomes self-sustaining. The number of applications of fertilizer N that may be needed theoretically depends upon the rate of organic matter buildup in the soil or spoil and how the N cycle begins to evolve. These factors are a function of the quality and depth of topsoil replaced, the climate and its effect on the establishment of vegetative and microbial communities, and the N requirements of the plant species selected for revegetation. However, the need for long-term fertilizer N applications and the ultimate long-term stability of revegetated mined land is not well understood and requires further study.

Potential toxicity hazards. Losses of N by erosion and runoff or by leaching from disturbed land may in some cases negatively affect the environment or human and animal health. The N associated with waterborne sediments or windborne particles lost from mined land by erosion or runoff may eventually be deposited in streams, lakes, or estuaries and cause increases in biological productivity in these surface waters. As the level of N input increases, over-enrichment or eutrophication may result, which produces undesirable changes, such as proliferation of algae, depletion of dissolved oxygen, and a decline in water clarity (*48*).

Nitrates lost by leaching from disturbed lands may eventually reach and degrade groundwater resources. Water with greater than 10 mg/liter of NO_3-N is not considered safe for human consumption (*104*). Nitrate can be reduced to nitrite (NO_2^-) in the intestines of some animals and in the human infant during the first few months of life. The nitrite is then rapidly absorbed into the blood and oxidizes the iron of hemoglobin, forming methemoglobin, which cannot function in oxygen transport (*48*). The effects of nitrate/nitrite toxicity in livestock are generally similar to those observed for humans. But much larger doses of nitrate/nitrite are required and the response varies depending upon animal body weight, the rate of water consumption, and the rate and type of food consumed by the animal (*48*).

Nitrogen lost from disturbed land by erosion and runoff primarily includes organic N in the organic matter of replaced topsoil and mineral N mineralized from the plant growth media and added as fertilizer. The magnitude of N loss from disturbed land has not been adequately studied. However, the amount of N lost by erosion and runoff from disturbed land probably varies greatly and should depend upon the same factors that control losses via erosion from cropland: the amount of vegetative groundcover, infiltration rates, seasonal storm patterns, soil-plant management practices, and topography (*51*). If the disturbed site has been topsoiled, most of the N lost by erosion will be as soil organic matter associated with waterborne or windborne sediments. Runoff losses of soluble N (ammonium and nitrate) should be generally low except when high rates of N fertilizer are surface applied just before high rainfall events (*51*).

Sources of N lost from disturbed land by leaching include N mineralized from the plant growth media or added as fertilizer, unexploded ammonium nitrate/fuel oil left in the overburden after blasting operations, and in some cases exchangeable NH_4^+ and NO_3^- from certain geologic strata exposed by mining.

The amount of mineralized N or fertilizer N lost by leaching from disturbed land could be higher than from native grassland because the mining process often increases infiltration rates and because for a number of years the disturbed ecosystem does not have fully developed plant and microbial biomasses that immobilize mineral N and prevent it from being lost from the rooting zone by leaching (*74*). Residual ammonium nitrate/fuel oil may be only a minor source of N in most cases (*19*), but mineral N (NH_4^+ and NO_3^-) naturally occurring in geologic materials could be a major source of nitrate contamination of groundwater.

Significant quantities of exchangeable NH_4^+ have been reported in Paleocene and Paleozoic shales (*68, 94*). High levels of NO_3^- have been reported below the water table in weathered Canadian glacial tills (>100 mg/liter NO_3-N) (*42*); in various marine sediments in Utah, Colorado, Wyoming, and California (up to 1,691 mg/liter NO_3-N in water extracts) (*98*); and in the overburden at a surface coal mine in Wyoming (up to 400 mg/kg mineral N) (*19*). Geologic NO_3-N levels of 1,012 μg/g have been reported in shales and mudstones and 197 μg/g in sandstones and conglomerates of the Diablo Range in California (*100*).

Federal regulations require that natural toxic and acid-forming materials be handled during mining to prevent degredation of surface water and groundwater. Because the high levels of naturally occurring mineral N found in some geologic materials could degrade post-mining quality, some states require that the mineral N content of overburden materials be analyzed prior to mining. These state laws further specify that overburden which exceeds a specified concentration of NO_3-N (e.g., 50 μg/g in Wyoming) must be specially handled and placed above the predicted groundwater table. Exchangeable NH_4^+ present in geologic materials exposed during surface mining could be susceptible to leaching losses if conditions in the geologic material are conducive to rapid nitrification. Power and associates (*68*) reported rapid nitrification of NH_4^+ present in shales when the shales were exposed to atmospheric conditions. Other studies have suggested that bacterial nitrification of naturally occurring NH_4^+ in geologic materials has occurred after exposure of the geologic materials by surface mining (*19*) or in the geologic past (*42*). However, Reeder and Berg (*75*) reported no nitrification activity in an aerobically incubated Cretaceous shale, even when samples were inoculated with an active microbial population. These studies indicate that nitrification of indigenous NH_4^+ in geologic materials depends upon the suitability of the chemical and physical properties of the material for nitrifying bacteria; thus the potential losses of indigenous NH_4^+ in geologic materials by leaching cannot be automatically assumed.

Nitrogen losses from disturbed land can be minimized by several reclamation practices. Pre-mining analysis of deep core samples of the overburden for NH_4^+ and NO_3^- content and for nitrification potential may indicate the degree to which nitrate leaching could occur. Locating and properly handling overburden materials that contain high levels of mineral N will help minimize the degradation of groundwater quality. However, research is needed to develop better spoil disposal and management practices. On regraded, topsoiled sites, the correct timing and rate of fertilizer N application should help minimize levels of potentially leachable mineral N. The most efficient use of fertilizer N by plants would result if fertilizer N is applied just prior to the start of the seasonal rapid-growth phase (*6*). Applying fertilizer N in adequate but not excessive levels to meet plant requirements will reduce the amount of mineral N remaining in the profile after plant and microbial needs are met.

Sampling and testing procedures

Soil testing of mined land is used prior to disturbance and after regrading to detect toxicities, unsuitable materials, and deficiencies and to predict fertilizer treatments for correcting identified deficiencies. However, analytical detection methods and extraction solutions vary among states, and differences in interpretations of the analytical data vary in terms of the nutrient concentration levels that constitute specific degrees of toxicity, sufficiency, or deficiency (*6*). A uniform set of sampling and laboratory procedures for evaluating plant growth capabilities and limitations of mined land spoils in the western states would make reclamation specialists from government and industry better able to define and directly compare land restoration problems and results (*78*).

Determination of a site's N status involves a number of steps: the collection of representative samples; handling, preparation, and storage of these samples; analysis of the samples for N content; and interpretation of the results.

Representative sample collection. Obtaining a soil sample that is representative of the area is one of the most important aspects of soil testing (*102*). However, there is considerable opportunity for error in trying to analyze the nutrient status of an area. For example, of the 2.2 million kg of soil in the surface 15 cm of a 1-ha site, only a few kilograms may be sampled and composited, and from that composite only a few grams are actually analyzed for nutrient content. Moreover, obtaining a representative surface sample of mixed and graded spoil is difficult because of the heterogeneity due to the random and nonuniform mixing of materials. Nonetheless, it is vital to be able to characterize a site, both before and after mining, with some degree of accuracy in order to develop a reclamation plan that will ensure stabilization of the site and protect the surrounding environment from potential environmental degradation.

Most standard statistical texts outline the theory of adequately sampling a population, and the use of statistics for developing sampling plans for surface soils or materials has been advocated by a number of scientists (*25, 45, 65*). However, statistical methods require some test data to assess variability within the sample population and determine the number of samples necessary to overcome this variability. Therefore, this approach may not be appropriate for those taking samples for on-going reclamation projects (*11*). Moreover, a statistical determination of the number of samples needed may be academic when reclamation laws specify sampling frequency.

Guidelines for sampling disturbed land have stressed separate sampling of soil materials that are visibly different in color, texture, moisture content, surface elevation, erosion, or volunteer vegetation (*5, 9, 11, 32, 87*). It is difficult, however, to sample a reconstructed profile so that the samples accurately represent the distribution of mineral N within the profile. Nitrate and ammonium distributions in graded mine spoil or in respread topsoil materials are generally more variable and heterogeneous than in undisturbed soil or overburden profiles. Thus, the number of sample cores recommended for undisturbed rangelands may be inadequate to represent accurately the more heterogeneous distribution of mineral N in a disturbed-land profile. The NO_3^- concentration within a profile is constantly changing because of the mobility of NO_3^- and because of the dynamic nature of the N mineralization and immobilization processes. Therefore, the number of samples needed to assess adequately the available N status may be greater than that needed for P or K (*6*). Underestimating the NO_3^- content of a regraded site may result in overfertilization and subsequent leaching of excess NO_3^- into the groundwater. In contrast, if the NO_3^- content of the site is overestimated, fertilizer N may be applied in levels inadequate to sustain plant growth.

The cost of taking a large number of samples and separately analyzing each sample can become prohibitive. Sampling a disturbed site to determine N distribution within the profile and thus infer the site's ability to supply N for plant growth becomes a compromise between representative sampling and economic feasibility.

Processing and preservation of samples for N analysis. Soil and overburden samples must be properly handled, prepared, and stored prior to analysis to avoid significant changes in the mineral N contents by chemical and microbial processes. Mineral N can accumulate in or be lost from samples between sampling and analysis. For example, storage of samples in a warm, moist condition prior to analysis may stimulate microorganisms to mineralize organic N and result in the accumulation of NH_4^+ and/or NO_3^- in the samples, whereas storage of wet samples may result in denitrification and loss of NO_3^- from samples. Attempts to preserve the integrity of samples usually involve drying, freezing, or the addition of a reagent lethal to the microbial population, but these processes themselves may cause increases or losses of mineral N. A completely satisfactory method of handling soil samples destined for N analysis has not yet been found because each method in some way modifies the N content of samples.

Sandoval and Power (*78*) recommend using polyethylene bags for transporting overburden and spoil samples from the field because canvas bags or ordinary paper bags may absorb soluble salts from wet samples. They also suggest that moist or wet samples should be immediately frozen or spread to dry on a waterproof material, ground to 2-mm size, then stored in closed, water-resistant containers until analyzed. However, many spoil materials contain large amounts (up to 90 percent) of coarse fragments (>2mm), and the handling and accounting for these coarse fragments when using soil tests poses a problem. One processing procedure is to sieve out the coarse fragments, determine the mass in relation to the entire sample, and later correct the results to include the coarse fragments as inert material (*11*). An alternative procedure is to grind the entire sample, usually to <2 mm, but some procedures grind the entire sample to <0.25 mm (*87*). If the coarse fragments include shales that weather rapidly, these fragments may contribute to the N fertility of the site, but will not be accounted for with the first processing procedure. In comparison, the second procedure may overestimate the N fertility of the site because the grinding procedure increases the surface area of the material (*11*). This increase in surface area may represent hundreds of years of natural weathering and thus misrepresent the current status of the sampled site.

Traditionally, it was thought that the best approach to sample processing was to air dry (or oven dry at 55 °C) soil samples as quickly as possible to inhibit microbial and chemical activity (*66*). However, considerable change in NO_3^- and NH_4^+ content can occur during drying (*4, 12, 28, 29*). Generally, greater changes occur in NH_4^+ content than in NO_3^-. Several workers have shown that upon drying, or upon storage of samples after drying, there is a marked increase in the content of exchangeable NH_4^+ (*29, 41, 61, 101, 107*). This increase has been observed most often in samples stored in open bags (nonairtight), but it has also been observed in airtight storage containers (*29*). The changes may be due to mineralization of organic N or sorption of NH_3 from the atmosphere. Soil samples may also lose N as volatilized NH_3 during the drying process, thus further complicating interpretation of the changes in mineral N content of dried samples.

In an attempt to avoid the problems of drying samples, compounds toxic to organisms, such as toluene, phenyl mercuric acetate, and chloroform, have been added to samples to prevent microbial activity that alters NH_4^+ and NO_3^- content of the samples during the delay between obtaining the sample and analysis. However, these toxic compounds have often been found to be ineffective and can affect the results of N analysis (*29, 33, 52, 66*). A further attempt to maintain the integrity of soil sample NH_4^+ and NO_3^- contents has been to use deep-freeze storage (*40*). However, Allen and Grimshaw (*4*) observed considerable changes in exchangeable NH_4^+ content and smaller changes in NO_3^- content in soil samples stored at −15 °C.

Although there is not a completely satisfactory method of handling soil samples, two methods appear to be used most often for stabilizing the N content of samples prior to analysis. These two methods are deep freezing or drying at room temperatures (*59, 83, 99, 108*). Keeney and Nelson (*49*) suggest three precautions for drying and storage: dry samples at low temperatures (i.e., room temperature); dry samples in a NH_3-free atmosphere;

and store dried samples in airtight containers, such as sealed plastic bags, plastic vials, or glass jars. If NO_2^- concentration is to be determined, that analysis should be conducted on soil samples as soon as possible after the samples have been collected and before the samples are dried (*49*).

Samples of overburden and spoils can be collected and prepared for laboratory analysis by a variety of methods. However, no matter which procedure is selected, it is critical that the same procedure be used consistently for all samples. Final selection of a procedure should reflect the ability to process all samples in the same way and within the same amount of time, given available equipment and staff.

Laboratory tests and applicability of test results to field evaluation. Agronmists traditionally have tried to correlate plant growth and response to nutrient levels in the soil, and many laboratory tests have been developed for estimating available N for plant growth on agricultural land (*57*). The most common soil tests used by states for making N fertilizer recommendations include measurements of NO_3-N in the soil profile, NO_3-N or NH_4-N+NO_3-N in the surface layer, organic matter content of the surface layer, or mineralizable N by aerobic incubation (*47*). Because these soil tests measure only a part of the total N content of the soil, the values are of little use in themselves and cannot be used in predicting N needs of crops until they are calibrated against N rate experiments in the field (*102*). However, numerous soil, climate, and management variables affect plant yield and, hence, the N needs of the crop (*47*). Therefore, most soil tests for N have only fair to poor correlations with field results, and only limited success has been experienced in developing these N soil tests on agricultural soils used for agronomic crops.

Field research has not been conducted to develop soil N tests for disturbed land or to correlate measured N levels in spoils or overburdens with the growth response of native plant species. Therefore, the application of traditional soil fertility tests, which have been developed largely for intensively managed annual crops growing on agricultural soils, may not be appropriate on sites where little or no topsoil is replaced and the plant growth medium is comprised of overburden materials (*6*). However, soil N test interpretations that are applicable to undisturbed dryland may be applicable to disturbed areas where spoils are covered with topsoil. This approach has been used by the Colorado Mined Land Reclamation Division, which has recommended that the "Guide to Fertilizer Recommendations in Colorado" (*89*) be consulted for fertilizer requirements on reclaimed mined land (*55*).

The most common soil N tests used to analyze the N content of mined land soils, overburdens, and spoils are the analyses (a) for inorganic N (NH_4^+, NO_2^-, and NO_3^-), (b) for easily oxidized organic matter (the Wakley Black method) as a measure of N availability, and (c) for Kjeldahl N (organic N + a portion of the inorganic N) (*78, 87*). Measurement of inorganic N is an accepted procedure on certain intensively fertilized crops where substantial carryover of N may occur (*11*). However, the ability of plant growth media to provide N for plant growth on mined land cannot be easily assessed from mineral N data alone because mined land is not usually as intensively fertilized as agronomic cropland (*109*). The mineral N status of a soil, spoil, or overburden continuously fluctuates and represents the net result of all the interacting factors of N production and consumption at that point in time. The determination of organic matter by the Wakley-Black method on spoils or overburden materials containing carbonaceous shales is not considered very meaningful if plant growth correlations have not been made because the determination may include other carbonaceous materials (e.g., coal) besides organic matter. Significant increases that have been reported in the (Wakley-Black) organic matter content of some spoils over a period of years were originally thought to be due to organic matter buildup from contemporary vegetation growth and decomposition, but may in fact be due to increasing amounts of geologic carbon being detected by the Wakley-Black procedure as the geologic materials weather and oxidize (*11*). Kjeldahl N measurements have shown little promise in predicting the amount of available N produced in soils, even though there is a general relationship between the total N content of a soil and the amount of available N released during a single growing season (*79*). Kjeldahl N analysis on revegetated spoil can be misinterpreted if the N naturally present in geologic materials is overlooked and it is assumed that the N accumulation has all occurred after disturbance and revegetation (*11*). However, total N measurements, when used in conjunction with a laboratory index of N availability, may be useful in evaluating the ability of overburdens and spoils to provide N for plant growth.

Biological laboratory indexes of soil N availability have been highly correlated with N uptake by plants grown under both greenhouse and field conditions and provide a valid relative measure of the soil's ability to release N for growth (*47*). However, these methods are not usually used for routine determinations of the N-supplying power of soils because the procedures are more complex and time-consuming than the procedures for determining the total N or inorganic N contents of soils. Although the procedures are not routinely used for soils, they could be useful in estimating the plant-available N potentials of western mined land. High correlations have been found between the inorganic N contents of laboratory-incubated samples of coal spoils and topsoil and both N uptake and dry matter yield of greenhouse-grown barley (*76*). However, to date field research has not been conducted on mined land to evaluate the relationship between the N value of the analyzed sample and plant N uptake or plant yield with increasing increments of N fertilizer.

A combination of the following soil N tests on the surface and subsurface plant growth media of disturbed land may help to estimate the N-supplying capability of the system:

1. NH_4^+ and NO_3^- (*49*): Mineral N content is determined by extracting soil or spoil samples with 2<u>M</u> KCl (10 ml extractant per g of soil), and analyzing the extract by alkaline steam distillation and titration, or by a colorimetric procedure. Colorimetric determination can be made either manually with a spectometer, or automatically with a continuous-flow system, such as Technichon or Lachat.[1] Colorimetric chemistries commonly used on automated continuous-flow systems include the salicylate hypochlorite and alkaline phenol methods for NH_4^+ and the copperized cadmium reduction column method for NO_2^- and NO_3^-.

2. Kjeldahl N (*18*): The standard acid digest of the Kjeldahl procedure recovers organic forms of N but does not quantatively recover NO_2^- or NO_3^- in the soil. Sample pretreatments with salicylic acid or permanganate-reduced iron, which recover NO_2^- and NO_3^- in the Kjeldahl digest, are probably not necessary for analysis of spoils and topsoils. The N content of the acid digested samples can be determined either by alkaline steam distillation and titration or colorimetrically on an automatic continuous-flow system with chemistry similar to that described above for NH_4^+.

3. Laboratory index of N availability: Biological laboratory indexes involve a short-term incubation of soil samples under anaerobic (*47*) or aerobic conditions (*16*). To date, only aerobic incubation techniques have been used on mined land spoils and topsoils (*21, 44, 50, 75*). Until further research is conducted to test alternative laboratory indexes on mined land soil and spoil materials, an aerobic incubation technique, such as that described by Bremner (*16*) or Reeder and Berg (*75*), is recommended.

The time of the year that spoils and topsoils are sampled and analyzed for mineral N content is important to the interpretation of the analytical results. Determination of the mineral N content of mined land plant growth media in the spring, before the plants are actively growing, will provide information about the potential supply of N for that season's plant growth. Low levels of mineral N may indicate that fertilization is needed to sustain plant growth. High levels may indicate that the system can meet plant requirements without fertilization. High levels of NO_3^- in the profile in the autumn may suggest a possible groundwater contamination hazard if the potential for leaching is high. Measurement of the Kjeldahl N content of soils or spoils is less sensitive to the time of the year that samples are taken. The Kjeldahl N analysis measures the size of the N pool within a spoil or soil material and thus provides a rough idea of the amount of N that potentially may be mineralized and become available to plants. A laboratory index of N availability is a better measurement of the long-term potential availability

[1]Trade names are included only for the benefit of the reader and do not constitute any preferential endorsement over similar products on the market.

of the organic N pool. Comparing the N mineralization potentials of the mined land plant growth media and the native, undisturbed soil will provide information concerning the N cycling status of the mined land system and its ability to sustain a desirable plant community.

Every existing or new soil N test used for mined land spoils requires research to determine the field calibration for the particular plant species being seeded and for each particular type of spoil and/or topsoil acting as surface or subsurface plant growth media. Because this needed field research has not been conducted, the N values derived from the above three soil tests, used with regional fertilizer recommendations for undisturbed dryland, may provide the best possible fertilizer recommendations currently available for disturbed land.

Scientists have tried to construct reasonable scenarios for ecosystem development that will help estimate the levels of fertilizer N needed and the number of years N applications will be needed to sustain the desired vegetative biomass until the mined land system becomes self-maintaining. One approach has been to compare the N contents and distributions within undisturbed grassland ecosystems and mined land ecosystems (*114* and unpublished data from Reeder). Estimates of the amount of N in the plant, microbial, and soil organic matter components of stable grassland ecosystems (e.g., those in table 1) have been taken from Clark and Paul (*23*), Paul (*64*), Woodmansee (*111*), and French and colleagues (*38*). Other sources of information include Dahlman and Kucera (*30*), Ursek and associates (*103*), Clark (*22*), and Sims and colleagues (*84*) for decomposition rates of above- and below-ground litter and Beeson (*8*) for data concerning the mineral composition of numerous plant species, including some introduced grasses. Estimates of the amount of N in the various components of the mined land ecosystem are best made from direct measurements because little of this type of information has been published and because the N distribution within a mined land ecosystem is site-specific, depending upon such things as climate, the depth and quality of topsoil replaced, the plant species reseeded, and management practices. The amount of time needed for a mined land ecosystem to stabilize can be estimated by determining the difference in N contents between the stable ecosystem and the revegetated mined land ecosystem, then dividing this amount of N by the annual N inputs into the revegetated system (e.g., atmospheric inputs as in table 1, inputs by N-fixing legumes, and inputs as fertilizer). Estimates of the amount of time needed for the mined land ecosystem to stabilize can be shortened or lengthened depending upon the levels of fertilizer N applied. The research approach described here is limited by the amount and applicability of data available to estimate the N levels and distributions in both native and disturbed ecosystems. It also is limited by the current understanding of how ecosystems develop on disturbed lands. More basic research is needed to better describe N distribution and dynamics in stable self-maintaining ecosystems, and more long-term field research is needed to evaluate the development of stable ecosystems on disturbed lands under various conditions of climate, topsoil quality, and management.

Conclusions and recommendations

The deficiency of plant-available N can be a limiting factor in the successful revegetation and long-term stability of land disturbed by surface mining. Deficiencies arise because disturbance drastically alters the flow of N through the soil-plant-microbial ecosystem, and long periods of time are required for some disturbed land to accumulate N naturally and evolve toward a new steady state. To realize the goal of a self-maintaining desirable plant community on disturbed land, the reestablishment of an active biological N cycle is necessary, and this requires the reestablishment of a viable, balanced microorganism subsystem.

Reestablishing an active biological N cycle can be achieved by a number of management techniques. Replacing topsoil that has not been stockpiled for a long time should shorten the time needed until the system becomes self-sustaining because topsoil returns mineralizable N to the system and serves as an inoculum for microorganisms. However, sufficient depth and quality of topsoil to restore completely the N cycle are not always available, and supplemental forms of N may be needed. Nitrogen deficiency can be relieved by the use of N-fixing plant species and plant species that are less sensitive to N supply or are highly capable of internally recycling much of their N. However, applications of fertilizer N may be required until pathways of fast-cycling N are developed and the ecosystem is able to support the N requirements of the desired plant biomass. A single application of fertilizer N to establish vegetation may be inadequate to sustain the vegetation until N cycling has been reestablished. The number of applications of fertilizer N that may be needed theoretically depends upon the quality and depth of the plant growth media, upon the management practices used, upon the climate and its effect on th establishment of vegetative and microbial communities, and upon the N requirements of the plant species selected for revegetation. However, the need for long-term fertilizer applications and the ultimate long-term stability of revegetated mined land are not well understood. Long-term field studies are needed to evaluate the development of stable ecosystems on disturbed land under various conditions of climate, topsoil quality, and management. Evaluation of seasonal transformations and flows of indigenous and fertilizer N through the soil-plant-microbial ecosystem on disturbed land should result in a better understanding of how disturbed ecosystems reestablish a functional N-cycling ecosystem and of the length of time needed to achieve a self-sustaining system. This type of information is critical in evaluating the long-term success of various reclamation and management programs on land disturbed by mining.

Geologic materials exposed as surface or subsurface plant growth media will usually not supply much of the N requirements of plants because the organic N of geologic materials is usually not readily mineralizable. Research is needed to study the effects of long-term weathering on exposed geologic materials and determine if weathering will result in the release of indigenous, fixed NH_4^+ or if the degraded silicate minerals will become a sink for NH_4^+ added as fertilizer or mineralized from indigenous organic N.

Research is needed to better understand the sources and magnitude of N losses from mined land. Losses of N by erosion and runoff or by leaching from disturbed land may in some cases negatively affect the environment or human and animal health. Nitrogen lost from mined land may enrich surface waters and result in eutrophication, which produces undesirable changes, such as proliferation of algae, depletion of dissolved oxygen, and a reduction in water clarity. Nitrates lost by leaching may eventually reach the groundwater and contaminate it to levels considered unsafe for human or livestock consumption. Significant quantities of naturally occurring mineral N in some geologic materials could be a major source of groundwater contamination. Exchangeable NH_4^+ present in geologic materials exposed during surface mining could be susceptible to leaching losses if conditions in the geologic material are conducive to rapid nitrification. However, conditions are not always conducive to nitrification; some studies have reported rapid nitrification rates of NH_4^+ present in exposed geologic materials, while other studies have reported no nitrification activity in exposed shales. Research is needed to develop spoil disposal and management practices that minimize contamination of groundwater by nitrates leached from mined land.

Determination of the pre- and post-mining N status of a disturbed site involves the collection of representative samples from the site, the correct handling and storage of samples, analysis of the samples for N content, and, finally, interpretation of the results. Research is needed to evaluate the number of samples necessary to characterize the N-supplying capability of a site. Statistical procedures can be helpful in trying to determine the number of samples needed to represent adequately a disturbed site, although this approach may be precluded by reclamation laws that specify sampling frequency. The samples collected from a site should all be processed in the same way and within the same amount of time. The recommended processing procedure is to (a) store collected samples in closed polyethylene bags in a cooler until they can be returned to the laboratory (preferably on the same day as sampled), (b) spread samples to dry at room temperature in an NH_3-free atmosphere, (c) store dried samples in airtight containers, and (d) analyze samples as soon as possible. A completely satisfactory method of handling soil samples destined for analysis has not yet been found. Therefore, further research is needed to improve sample processing techniques so that the integrity of collected samples is preserved and subsequent N analysis truly represents the N status of the field. Research is needed to evaluate the efficacy of soil N tests for assessing fertility needs of native and introduced species on disturbed land. Until this research is conducted, a combination of three soil N tests (mineral N, Kjeldahl N, and N mineralization potential) on the surface and subsurface plant growth media, used with regional fertilizer recommendations for undisturbed dryland, may provide the best possible fertilizer recommendations for disturbed land.

REFERENCES

1. Aldon, Earl F. 1978. *Reclamation of coal-mined land in the Southwest.* J. Soil and Water Conservation 33: 75-79.
2. Aldon, E. F., and H. W. Springfield. 1973. *Revegetating coal mine spoils in New Mexico: A laboratory study.* Research Note RM-245. Rocky Mountain Forest and Range Experiment Station, Forest Service, Fort Collins, Colorado.
3. Aldon, E. F., H. W. Springfield, and George Garacia. 1975. *Can soil amendments aid revegetation of New Mexico coal mine spoils?* Research Note RM-292. Rocky Mountain Forest and Range Experiment Station, Forest Service, Fort Collins, Colorado.
4. Allen, S. E., and H. M. Grimshaw. 1962. *Effect of storage temperature on the extractable nutrient ions in soils.* Journal of the Science of Food and Agriculture 13: 525-529.
5. Barnhisel, R. I. 1975. *Sampling surface-mined coal spoils.* AGR-41. Department of Agronomy, University of Kentucky, Lexington. 4 pp.
6. Bauer, Armand, William A. Berg and Walter L. Gould. 1978. *Correction of nutrient deficiencies and toxicities in strip-mined lands in semiarid and arid regions.* In F. W. Schaller and P. Sutton [editors] *Reclamation of Drastically Disturbed Lands.* American Society of Agronomy, Madison, Wisconsin. pp. 451-466.
7. Bauer, A. P., P. Nyren, G. Riechman, G. Gee, and J. Gilley. 1978. *Fertilization of wheat, corn and a grass-legume mixture grown on reclaimed spoil banks.* Research Report 67. North Dakota Agricultural Experiment Station, Fargo.
8. Beeson, K. C. 1941. *The mineral composition of crops with particular reference to the soil in which they were grown.* Miscellaneous Publication No. 369. U.S. Department of Agriculture, Washington, D.C. 164 pp.
9. Berg, W. A. 1969. *Determining pH of strip-mine spoils.* Research Note NE-98. Northeast Forest Experiment Station, Forest Service, Upper Darby, Pennsylvania. 7 pp.
10. Berg, W. A. 1975. *Revegetation of land disturbed by surface mining in Colorado.* In M. K. Wali [editor] *Practices and Problems of Land Reclamation in Western North America.* University of North Dakota Press, Grand Forks. pp. 79-89.
11. Berg, W. A. 1978. *Limitations in the use of soil tests on drastically disturbed lands.* In F. W. Schaller and P. Sutton [editors] *Reclamation of Drastically Disturbed Lands.* American Society of Agronomy, Madison, Wisconsin. pp. 653-664.
12. Birch, H. G. 1958. *The effect of soil drying on humus decomposition and nitrogen availability.* Plant Soil 10: 9-31.
13. Black, C. A. 1968. *Nitrogen.* In C. A. Black [editor] *Soil-plant Relationships.* John Wiley & Sons, Inc., New York, New York. pp. 405-557.
14. Bradshaw, A. D. 1983. *The reconstruction of ecosystems.* Journal of Applied Ecology 20: 1-17.

15. Bradshaw A. D., R. H. Marrs, R. D. Roberts, and R. A. Skeffington. 1982. *The creation of nitrogen cycles in derelict land.* Philosophical Transactions of the Royal Society of London, Series B 296: 557-561.
16. Bremner, J. A. 1965. *Nitrogen availability indexes.* In C. A. Black [editor] *Methods of Soil Analysis.* American Society of Agronomy, Madison, Wisconsin. pp. 1,324-1,345.
17. Bremner, J. A. 1965. *Organic nitrogen in soils.* In W. V. Bartholomew and F. E. Clark [editors] *Soil Nitrogen.* American Society of Agronomy, Madison, Wisconsin. pp. 93-149.
18. Bremner, J. A., and C. S. Mulvaney. 1982. *Nitrogen—total.* In A. L. Page, R. H. Miller, and D. R. Keeney [editors] *Methods of Soil Analysis* (Second Edition). American Society of Agronomy, Madison, Wisconsin. pp. 595-624.
19. Carter Mining. 1985. *Predicted post-mining ground-water quality and existing backfill ground-water quality at the Caballo mine.* Report submitted to the Wyoming Department of Environmental Quality, Land Quality Division, Cheyenne. 204 pp.
20. Charley, J. L. 1977. *Mineral cycling in rangeland ecosystems.* In R. E. Sosebee [editor] *Rangeland Plant Physiology.* Society for Range Management, Denver, Colorado pp. 215-256.
21. Chichester, F. W., and S. J. Smith. 1983. *Biological cycling of* ^{15}*N-labeled fertilizer nitrogen in lignite minesoil materials.* Soil Science Society of America Journal 47: 676-682.
22. Clark, F. E. 1977. *Internal cycling of* ^{15}*nitrogen in shortgrass prairie.* Ecology 58: 1,322-1,333.
23. Clark, F. E., and E. A. Paul. 1970. *The microflora of grasslands.* Advances in Agronomy 23: 375-435.
24. Clark, F. E., C. V. Cole, and R. A. Bowman. 1980. *Nutrient cycling.* In A. J. Breymeyer and G. M. VanDyne [editors] *Grasslands, Systems Analysis and Man.* Cambridge University Press, Cambridge, England. pp. 659-712.
25. Cline, M. G. 1944. *Principles of soil sampling.* Soil Science 58(4): 275-288.
26. Cook, C. W., and G. Q. Bateman. 1966. *Hay and pasture seedings for the intermountain area.* In H. D. Hughes, M. E. Heath, and D. S. Metcalfe [editors] *Forages, the Science of Grassland Agriculture.* Iowa State University Press, Ames. pp. 459-467.
27. Cook, C. Wayne, Robert M. Hyde, and Phillip L. Sims. 1974. *Revegetation guidelines for surface mined areas.* Science Series No. 16. Range Science Department, Colorado State University, Fort Collins.
28. Cooke, G. W., and R. K. Cunningham. 1958. *Soil nitrogen. III. Mineralizable nitrogen determination by an incubation technique.* Journal of the Science of Food and Agriculture 9: 324-330.
29. Cunningham, R. K. 1962. *Mineral nitrogen in tropical forest soils.* Journal of Agriculture Science 59: 257-262.
30. Dahlman, R. E., and C. L. Kucera. 1965. *Root productivity and turnover in native prairie.* Ecology 46: 84-90.
31. Dancer, W. S., J. F. Handley, and A. D. Bradshaw. 1977. *Nitrogen accumulation in kaolen mining wastes in Cornwall: I. Natural communities.* Plant Soil 48: 153-167.
32. Davis, G., chairman. 1971. *A guide for revegetating bituminous stripmine spoils in Pennsylvania.* Northeastern Forest Experiment Station, U.S. Forest Service, Upper Darby, Pennsylvania 46 pp.
33. Davies, E. B., M. R. Coup, F. B. Thompson, and R. P. Hansen. 1940. *Studies on nitrate and ammonium in soils under permanent pasture. I. The stabilization*

of nitrate in soil samples. New Zealand Journal of Science and Technology 21a: 348-351.

34. DePuit, E. J., and J. G. Coenenberg. 1979. *Responses of revegetated coal strip mine spoils to variable fertilization rates, longevity of fertilization program and season of seeding.* Research Report 150. Montana Agricultural Experiment Station, Bozeman.
35. Epstein, Emanual. 1972. *Mineral nutrition of plants: principles and perspectives.* John Wiley and Sons, Inc., New York, New York.
36. Farnsworth, R. B. 1975. *Physiological and biochemical studies of nitrogen fixation in the nodules of sagebrush (*Artemisia ludoviciana*) and other desert nitrogen-fixing systems.* US/IBP Desert Biome Research Memo 75-136. Utah State University, Logan.
37. Farnsworth, R. B., and M. W. Hammond. 1968. *Root nodules and isolation of endophyte on* Artemesia ludoviciana. Proceedings, Utah Academy of Science, Arts, and Letters 45: 182-188.
38. French, N. R., R. K. Steinhorst, and D. M. Swift. 1979. *Grassland biomass trophic pyramids.* In N. R. French [editor] *Perspectives in Grassland Ecology.* Ecology Studies 32: 59-88.
39. Fresquez, P. R., Earl F. Aldon, and W. C. Lindemann. 1986. *Microbial reestablishment and the diversity of fungal genera in reclaimed coal mine spoils and soil.* Reclamation and Revegetation Research 4: 245-258.
40. Gasser, J.K.R. 1958. *Use of deep freezing on the preservation and preparation of fresh soil samples.* Nature 181: 1,334-1,335.
41. Gasser, J.K.R. 1961. *Effects of air drying and air drying storage on the mineralizable nitrogen of soils.* Journal of the Science of Food and Agriculture 12: 778-784.
42. Hendry, M. J., R.G.L. McCready, and W. D. Gould. 1984. *Distribution, source and evolution of nitrate in a glacial till of southern Alberta, Canada.* Canadian Journal of Hydrology 70: 177-198.
43. Hons, F. M., L. R. Hossner, and E. L. Whiteley. 1979. *Yield and rooting activity of forage grasses on a surface-mined soil of Texas.* Agronomy Journal 71: 113-116.
44. Hons, F. M., and L. R. Hossner. 1980. *Soil nitrogen relationships in spoil material generated by the surface mining of lignite coal.* Soil Science 129: 222-228.
45. Jacobs, W. C., and A. Klute. 1956. *Sampling soils for physical and chemical properties.* Soil Science Society of America Proceedings 10: 257-259.
46. Jansson, S. L., and J. Persson. 1982. *Mineralization and immobilization of soil nitrogen.* In F. J. Stevenson [editor] *Nitrogen in Agricultural Soils.* Agronomy 22: 229-252.
47. Keeney, D. R. 1982. *Nitrogen-availability indices.* In A. L. Page [editor] *Methods of Soil Analysis.* (Part 2, Second Edition). American Society of Agronomy, Madison, Wisconsin. pp. 711-733.
48. Keeney, D. R. 1982. *Nitrogen management for maximum efficiency and minimum pollution.* In F. J. Stevenson [editor]. *Nitrogen in Agricultural Soils.* American Society of Agronomy, Madison, Wisconsin. pp. 605-649.
49. Keeney, D. R., and D. W. Nelson. 1982. *Nitrogen-inorganic forms.* In A. L. Page [editor] *Methods of Soil Analysis* (Part 2, Second Edition). American Society of Agronomy, Madison, Wisconsin. pp. 643-698.
50. Legg, J. O., F. W. Chichester, G. Stanford, and W. H. DeMar. 1971. *Incorporation of ^{15}N-tagged mineral nitrogen into stable forms of soil organic nitrogen.* Soil Science Society of America Proceedings 35: 273-276.
51. Legg, J. O., and J. J. Meisinger. 1982. *Soil nitrogen budgets.* In F. J. Stevenson [editor] *Nitrogen in Agricultural Soils.* American Society of Agronomy, Madison, Wisconsin. pp. 503-566.

52. Lewis, D. G. 1961. *Determination of inorganic nitrogen in soil.* Journal of the Science of Food and Agriculture 12: 735-742.
53. Love, R. M. 1966. *Range pastures and their improvement.* In H. D. Hughes, M. E. Heath, and D. S. Metcalfe [editors] *Forages, the Science of Grassland Agriculture.* Iowa State University Press, Ames. pp. 589-599.
54. Marrs, R. H., and A. D. Bradshaw. 1982. *Nitrogen accumulation, cycling, and the reclamation of China clay wastes.* Journal of Environmental Management 15: 139-157.
55. McGinnies, William J., and Kent A. Crofts. 1986. *Effects of N and P fertilizer placement on establishment of seeded species on redistributed mine topsoil.* Journal of Range Management 39: 118-121.
56. McGinnies, W. J., and P. J. Nicholas. 1980. *Effects of topsoil thickness and nitrogen fertilizer on the revegetation of coal mine spoils.* Journal of Environmental Quality 9: 681-685.
57. Melsted, S. W., and T. R. Peck. 1977. *The Mitscherlich-Bray growth function.* In T. R. Peck, J. T. Cope, Jr., and D. A. Whitney [editors] *Soil Testing: Correlating and Interpreting the Analytical Results.* Special Publication No. 29. American Society of Agronomy, Madison, Wisconsin. pp. 1-18.
58. Miller, R. M. 1977. *Below-ground ecosystems project.* In *Land Reclamation Program, Annual Report. 1977.* Argonne National Laboratory, Argonne, Illinois. pp. 95-120.
59. Nelson, D. W., and J. M. Bremner. 1972. *Preservation of soil samples for inorganic nitrogen analysis.* Agronomy Journal 64: 196-199.
60. Odum, H. T. 1968. *Work circuits and systems stress.* In H. E. Young [editor] *Symposium on Primary Production and Mineral Cycling in Natural Ecosystems.* University of Maine Press, Orono. pp. 81-138.
61. Olson, R. A., M. W. Meyer, W. E. Lamke, A. D. Woltemath, and R. E. Weiss. 1960. *Nitrate production rate as a soil test for estimating fertilizer nitrogen requirements of cereal crops.* Transactions, Seventh International Congress of Soil Science 2 463-470.
62. Olson, R. A., and L. T. Kurtz. 1982. *Crop nitrogen requirements, utilization, and fertilization.* In F. J. Stevenson [editor] *Nitrogen in Agricultural Soils.* American Society of Agronomy, Madison, Wisconsin. pp. 567-604.
63. Palmer, J. P., and M. J. Chadwick. 1985. *Factors affecting the accumulation of nitrogen in colliery spoil.* Journal of Applied Ecology 22: 249-257.
64. Paul, E. A. 1976. *Nitrogen cycling in terrestrial ecosystems.* Environmental Biogeochemical 1: 225-242.
65. Peterson, R. G., and L. D. Calvin. 1965. *Sampling.* In C. A. Black [editor] *Methods of Soil Analysis.* American Society of Agronomy, Madison, Wisconsin. pp. 54-72.
66. Piper, C. S. 1944. *Soil and plant analysis.* Inter-Science Publishers, Inc., New York, New York.
67. Power, J. F., and J. Alessi. 1971. *Nitrogen fertilization of semi-arid grasslands: Plant growth and soil mineral N levels.* Agronomy Journal 63: 277-280.
68. Power, J. F., J. J. Bond, F. M. Sandoval, and W. O. Willis. 1974. *Nitrification in Paleocene shale.* Science 183: 1,077-1,079.
69. Power, J. F., and J. O. Legg. 1984. *Nitrogen-15 recovery for five years after application of ammonium nitrate to crested wheatgrass.* Soil Science Society of America Journal 48: 322-326.
70. Power, J. F., R. E. Ries, and F. M. Sandoval. 1978. *Reclamation of coal-mined land in the northern Great Plains.* Journal of Soil and Water Conservation 33: 69-74.
71. Rauzi, Frank, and Gerald E. Schuman. 1982. *Fertilization of four grasses estab-*

lished on topsoiled and nontopsoiled mined land. Agricultural Research Results, Western Series No. 24. Agricultural Research Service, U.S. Department of Agriculture, Washington, D.C. 45 pp.

72. Rauzi, F., and R. L. Tresler. 1978. *A preliminary report on herbage yields, stand evaluation, soils, and chemical content of selected grasses and a legume grown on topsoil, White River and Wind River geologic materials.* Research Journal 124. Wyoming Agricultural Experiment Station, Laramie.
73. Rauzi, Frank, Robert L. Land, and L. I. Painter. 1968. *Effects of nitrogen fertilization on native rangeland.* Journal of Range Management 21: 287-291.
74. Reeder, J. D. 1985. *Fate of Nitrogen-15-labeled fertilizer nitrogen in revegetated Cretaceous coal spoils.* Journal of Environmental Quality 14: 126-131.
75. Reeder, J. D., and W. A. Berg. 1977. *Nitrogen mineralization and nitrification in Cretaceous shale and coal mine spoils.* Soil Science Society of America Journal 41: 922-947.
76. Reeder, J. D., and W. A. Berg. 1977. *Plant uptake of indigenous and fertilizer nitrogen from a Cretaceous shale and coal mine spoils.* Soil Science Society of America Journal 41: 919-921.
77. Rosswall, T. 1976. *The internal nitrogen cycle between microorganisms, vegetation and soil.* In B. H. Svensson and R. Soderlund [editors] *Nitrogen, Phosphorus and Sulphur—Global Cycles.* Ecology Bulletin 22: 157-167.
78. Sandoval, F. M., and J. F. Power. 1977. *Laboratory methods recommended for chemical analyses of mined-land spoils and overburden in western United States.* Agriculture Handbook No. 525. U.S. Department of Agriculture, Washington, D.C.
79. Scarsbrook, C. E. 1965. *Nitrogen availability.* In V. W. Bartholomew and F. E. Clark [editors] *Soil Nitrogen.* American Society of Agronomy, Madison, Wisconsin. pp. 481-502.
80. Schafer, William. 1984. *Minesoil restoration and maturity: A guide for managing minesoil development.* In F. F. Munshower and S. E. Fisher, Jr. [editors] *Third Biennial Symposium on Surface Coal Mine Reclamation on the Great Plains.* Cooperative Extension Service, Bozeman, Montana. pp. 172-185.
81. Schmidt, E. L. 1982. *Nitrification in soil.* In F. J. Stevenson [editor] *Nitrogen in Agricultural Soils.* American Society of Agronomy, Madison, Wisconsin. pp. 253-288.
82. Schuman, Gerald E., and James F. Power. 1981. *Topsoil management on mined lands.* Journal of Soil and Water Conservation 36: 77-78.
83. Selmer-Olsen, A. R., A. Quen, R. Baerug, and I. Lyngstad. 1971. *Pretreatment and storage of soil samples prior to mineral nitrogen determination.* Acta. Agriculture Scandinavia 21: 57-63.
84. Sims, P. L., J. S. Singh, and W. K. Laurenroth. 1978. *The structure and function of ten western North American Grasslands. I. Abiotic and vegetional characteristics.* Journal of Ecology 66: 251-285.
85. Skeffington, R. A., and A. D. Bradshaw. 1981. *Nitrogen accumulation in kaolen wastes in Cornwall. IV. Sward activity and the development of a nitrogen cycle.* Plant Soil 62: 439-451.
86. Smith, J. A., E. J. DePuit, and G. E. Schuman. 1986. *Wood residue and fertilizer amendment on Bentonite mine spoils: II. Plant Species Responses.* Journal of Environmental Quality 15: 427-435.
87. Sobek, Andrew A., William A. Schuller, John R. Freeman, and Richard M. Smith. 1978. *Field and laboratory methods applicable to overburdens and minesoils.* EPA-600/2-78-054. Environmental Protection Agency, Cincinnati, Ohio.
88. Soderlund, R., and B. H. Svensson. 1976. *The global nitrogen cycle.* In B. H. Svensson and R. Soderlund [editors] *Nitrogen, Phosphorus and Sulphur—Glob-*

al Cycles. SCOPE Report 7. Ecology Bulletin 22: 23-73.

89. Soltanpour, P. N. 1978. *Guide to fertilizer recommendations in Colorado.* Cooperative Extension Service, Colorado State University, Fort Collins.
90. Stanley, M. A., G. E. Schuman, F. Rauzi, and L. I. Painter. 1982. *Quality and element content of forages grown on three reclaimed mine sites in Wyoming and Montana.* Reclamation and Revegetation Research 1: 311-326.
91. Stevenson, F. J. 1959. *On the presence of fixed ammonium in rocks.* Science 130: 221-222.
92. Stevenson, F. J. 1960. *Chemical state of the N in sedimentary and igneous rocks.* Geology Society of America Bulletin 71: 1985.
93. Stevenson, F. J. 1960. *Some aspects of the distribution of biochemicals in geologic environments.* Geochemical Cosmochim. Acta 19: 261-271.
94. Stevenson, F. J. 1962. *Chemical state of the nitrogen in rocks.* Geochemical Cosmochim. Acta 26: 797.
95. Stevenson, F. J. 1965. *Origin and distribution of nitrogen in soil.* In W. V. Bartholomew and F. E. Clark [editors] *Soil Nitrogen.* American Society of Agronomy, Madison, Wisconsin pp. 1-42.
96. Stevenson, F. J. 1982. *Organic forms of soil nitrogen.* In F. J. Stevenson [editors] *Nitrogen in Agricultural Soils.* American Society of Agronomy, Madison, Wisconsin. pp. 67-122.
97. Stevenson, F. J. 1982. *Origin and distribution of nitrogen in soil.* In F. J. Stevenson [editor] *Nitrogen in Agricultural Soils.* American Society of Agronomy, Madison, Wisconsin. pp. 1-42.
98. Stewart, R., and W. Peterson. 1917. *Further studies of the nitric nitrogen content of the country rocks.* Bulletin 150. Utah Agricultural College Experiment Station, Logan. 20 pp.
99. Storrier, R. R. 1966. *The pretreatment and storage of soil samples for nitrogen analyses.* Journal of the Australian Institute of Agriculture Science 32: 106-113.
100. Sullivan, Patrick J., Garrison Sposito, S. M. Strathouse, and C. L. Hansen. 1979. *Geologic nitrogen and the occurrence of high nitrate soils in the western San Joaquin Valley, California.* Hilgardia 47(2): 15-49.
101. Timmons, D. R., V. J. Kilmer, and V. F. Stahly. 1961. *A possible source of serious error in the determination of nitrate production in soils.* Soil Science Society of America Proceedings 25: 328.
102. Tisdale, S. L., and W. L. Nelson. 1975. *Soil fertility evaluation.* In *Soil Fertility and Fertilizers.* Macmillan Publishing Company, Inc., New York, New York. pp. 439-490.
103. Uresk, D. W., P. L. Sims, and D. A. Jameson. 1975. *Dynamics of blue grama within a shortgrass ecosystem.* Journal of Range Management 28: 205-208.
104. U.S. Environmental Protection Agency. 1975. *Water programs: National interior primary drinking water regulations.* Federal Register 40(248).
105. Visser, S., J. Fujikawa, C. L. Griffiths, and D. Parkinson. 1984. *Effect of topsoil storage on microbial activity, primary production and decomposition potential.* Plant and Soil 82: 41-50.
106. Waksman, S. A. 1933. *On the distribution of organic matter in the sea bottom and the chemical nature and origin of marine humus.* Soil Science 36: 125-147.
107. Walsh, L. M., and J. T. Murdock. 1960. *Native fixed ammonium and fixation of applied ammonium on several Wisconsin soils.* Soil Science 89: 183-193.
108. Westfall, D. G., M. A. Henson, and E. P. Evans. 1978. *The effect of soil sample handling between collection and drying on nitrate concentration.* Communication in Soil Science Plant Analyses 9: 169-185.
109. Williams, P. H. 1975. *Investigations into the nitrogen cycle in collier spoil.* In M. J. Chadwick and G. T. Goodman [editors] *The Ecology of Resource Degra-*

dation and Renewal. Blackwell Scientific Publication, Oxford, England. pp. 259-274.

110. Wilson, H. A., and G. Stewart. 1955. *Ammonification and nitrification in a strip mine spoil.* Bulletin 379T. West Virginia University Agricultural Experiment Station, Blacksburg. 160 pp.
111. Woodmansee, R. G. 1978. *Additions and losses of nitrogen in grassland ecosystems.* BioScience 28: 448-453.
112. Woodmansee, R. G. 1979. *Factors influencing input and output of nitrogen in grasslands.* In N. R. French [editor] *Perspectives in Grassland Ecology.* Springer-Verlag, New York, New York. pp. 117-134.
113. Woodmansee, R. G., I. Vallis, and J. J. Mott. 1981. *Grassland nitrogen.* In F. E. Clark and T. Rosswall [editors] *Terrestrial Nitrogen Cycles.* Ecology Bulletin (Stockholm) 33: 443-462.
114. Woodmansee, R. G., J. D. Reeder, and W. A. Berg. 1978. *Nitrogen in drastically disturbed lands.* In C. T. Youngberg [editor] *Forest and Soil Land Use.* Department of Forest and Wood Sciences, Colorado State University, Fort Collins. pp. 376-392.
115. Woods, L. E., and G. E. Schuman. 1986. *Influence of soil organic matter concentrations on carbon and nitrogen activity.* Soil Science Society of America Journal 50: 1,242-1,245.

Chapter 9

Soil organic matter

Peter L. Smith, Edward F. Redente, and Everett Hooper

Soil organic matter has long been considered "an elixir of plant life" (*4*). Throughout the history of agriculture, dark soils (organic matter-rich) have been equated with relatively high vegetation production. With the onset of mining and, more specifically, the revegetation of mined land, reclamation specialists have again recognized the importance of soil organic matter.

Compared with other soil property data, soil organic matter data are of minor to moderate significance in planning mined land reclamation. Organic matter is most important with regard to its nutritive and physical effects on topsoil.[1] If pre-mining assessments of revegetation potential are performed, soil organic matter data can be of minor to moderate importance in projecting the success of revegetation efforts. With respect to mining-related impact assessments of the soil resource, organic matter[2] plays a minor role in affecting erosion hazard calculations.

[1]Topsoil is soil material that is appropriate for use as a plant growth medium. It may include none, part, or all of the A, E, B, and C soil horizons. Subsoil is defined as soil material of any horizon except the A horizon.

[2]Organic matter is that portion of the soil that includes microflora and microfauna in the soil (living and dead) and residual decomposition products of plant and animal tissue (*4*) or, stated more simply by Jenny (*36*), any carbon assembly (exclusive of carbonates), large or small, dead or alive, inside soil space. Additionally, it is defined somewhat by the analytical methods used to determine its quantity within the soil and includes only those organic materials that accompany soil particles through a 2 mm sieve. For purposes of this chapter, the definition excludes coal fragments or carbonaceous shale associated with bedrock materials.

Regulatory agency perceptions

Various regulatory agencies in the western United States have guidelines to assist land users in rating the quality of soil and overburden for reclamation. Other agencies have published information that does not rate soil materials for use as topsoil but that applies to the management of topsoil. A review of these guidelines and information facilitates understanding the perceived importance that regulatory agencies place on the role of organic matter in reclamation.

Organic matter and associated suitability rating levels were listed in tables produced by Utah, North Dakota, and the Forest Service. The Utah Department of Natural Resources recently published draft guidelines specifically for the state (*85*). According to these guidelines, topsoil material rated as fair or good should have more than 2 percent organic matter. Soil with less than 2 percent organic matter is rated poor or unsuitable. An optimum level of 2 percent or more organic matter is also recommended for topsoil in North Dakota by the North Dakota State University Agricultural Experiment Station (*60*). According to the Forest Service (*84*), good topsoil has 1.5 percent or more organic matter; fair topsoil has 1.5 to 0.5 percent organic matter; and poor topsoil has less than 0.5 percent.

The remaining state and federal agencies that provide organic matter recommendations, but do not include rating levels, include the North Dakota Public Service Commission, Soil Conservation Service (*76*), Montana Department of State Lands (*53*), and Wyoming Department of Environmental Quality (*89*). The North Dakota Public Service Commission recommends that soil horizon(s) (usually the A and upper B) having higher organic matter levels and other acceptable characteristics be used as "first-lift" topsoil rather than subsoil horizons qualifying as topsoil but with lower organic matter levels. SCS indicates that topsoil having a sandy clay loam, clay loam, or silty clay loam texture should be rated good rather than fair if more than 3 percent organic matter occurs therein. In discussing important qualities of reclaimed prime farmland, the Office of Surface Mining does not mention organic matter levels. The Wyoming Department of Environmental Quality lists organic matter as one of the parameters to be evaluated in determining the fertility of the replaced plant growth media. Finally, the Montana Department of State Lands mentions that organic matter percentages are needed to calculate site-specific K factors for use in the universal soil loss equation and, much like the North Dakota Public Service Commission, to aid in determining "first-lift" (stripping) depths.

Formation, nature, and analysis of organic matter

Chapters in several books discuss the formation and nature of soil organic matter (*2, 4, 6, 9, 34, 42, 71, 80*). Given this detailed literature, only highlights are given here.

Formation and decomposition of organic matter. Accumulation and biological formation of organic matter is important to the development of most mineral soils. Organic matter accumulates rapidly at first, then decreases with time until a constant or equilibrium level is reached (*34*).

The decay of plant and animal matter is one phase of an unending cycle in which the components return to the same state again and again (*34*). In the soil organic matter cycle, C is transformed by decomposition from plant and animal tissue to microbial tissue and humus, then to CO_2. Eventually, CO_2 is reabsorbed by growing plants and, through photosynthesis, converted again into plant tissue (*34*). Figure 1 illustrates the relationship between the processes involved in the formation and decomposition of soil organic matter.

Soil organic matter is primarily humus. Humus is formed during the microbial decomposition or alteration of plant and animal residues (organic) and products synthesized by soil organisms. Constituents are acidic substances of microbiological, polyphenolic, and lignin origin, with a high

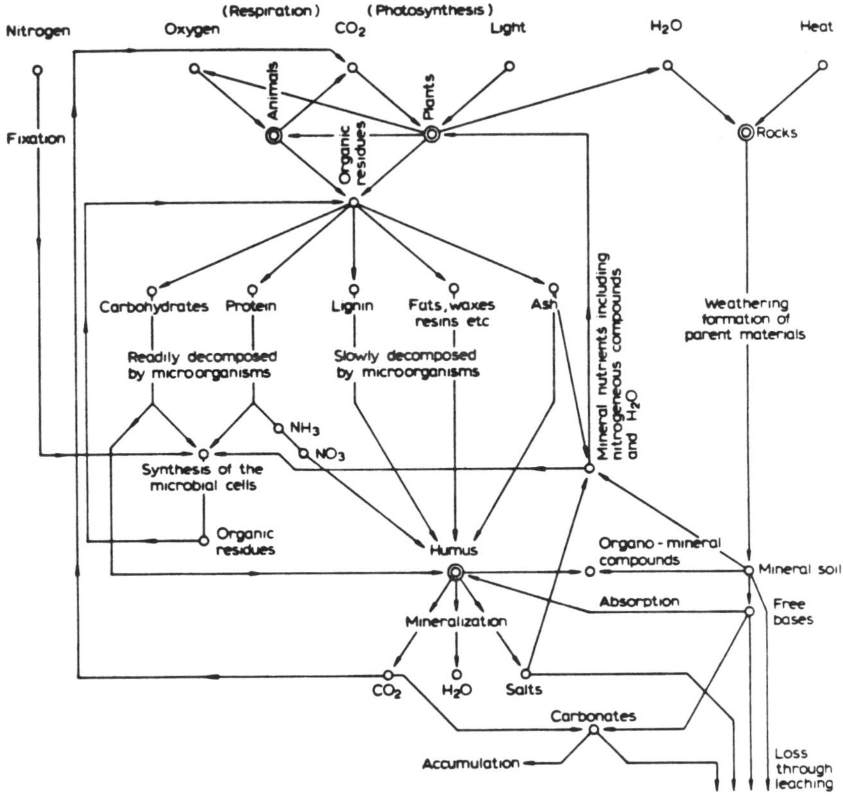

Figure 1. Formation and decomposition of soil organic matter (4).

molecular weight. Generally, 15 to 30 percent of the C in humus remains after 1 year as resistant microbial cells or new humus (*37*).

Organic residues of plants deposited on or in the soil vary widely in their susceptibility to and rate of decomposition by microorganisms under aerobic conditions (*2*). Simple carbohydrates and proteins decompose rapidly. Lignin fats, tannins, and waxes are more resistant to biological decay (*4*). Water-soluble residues are the quickest to be digested by soil microorganisms. In addition, succulent substances decompose more rapidly than the same material in a drier condition. Cellulose and hemicellulose are converted readily to soluble forms and used quickly by soil organisms (*69*).

Organic residue from animals contains higher levels of protein and fat than plant tissue. Animal tissue also does not contain lignified tissues as does plant tissue. Much animal tissue is used by other animals, reaching the soil through excretion.

Soil organisms affect organic residue decomposition rates. Fresh plant tissue, deposited on or in the soil , creates favorable conditions for rapid microbial population increases. Heterotrophic microflora (organic compound digesters) and, to a lesser extent, microfauna use organic substances for energy supply. Microfauna, especially earthworms (*Lumbricus* spp.), speed up decomposition by directly digesting organic debris and/or soil and leaving trails of castings (*81*). Earthworms, however, generally do not occur in arid and semiarid soils, except in agricultural and riparian areas where they have been introduced (*24*).

As microorganisms and their activity increase, there is a corresponding increase in CO_2 emission, which is a direct measure of organic matter decomposition. Assuming no organic residues are added, the CO_2 emitted decreases with time, but the rate of decomposition will be equal to or greater than rates during periods when fresh residues are added (*69*). Therefore, the amount of organic C evolved as CO_2 directly relates to the soil's C content.

Temperature, moisture, pH, and O level affect the size and composition of organism communities. Warm temperatures and moist soil conditions are usually optimal for organism activity (*2*). But decomposition also occurs in the northern Rockies and in cold deserts during the winter when soils are moist (*29*). Anaerobic conditions also create conditions that totally change and generally reduce organism activity.

Inhibitory substances present in, or deposited on, soils affect decomposition rates. For example, excessive salt concentrations in saline or sodic soils or herbicides can cause a change in the composition and size of organism communities, thereby affecting the rate and possibly the end products of decay. Other substances, such as Na, facilitate dispersion of organic matter on the surface of the soil.

Nature of organic matter. Included in the organic residues contributed by plant and animal tissues are carbohydrates (cellulose, hemicellulose, sugars), protein, alcohols, lignins, fats (waxes and resins), tannin, glucosides, alkaloids, and pigments (*4*). Humus is produced as these residues decompose.

Physical and chemical properties. The physical nature of soil organic matter freshly added to the soil depends largely upon the type of plant or animal residue. For example, forest ecosystems contribute different plant and animal litter than do prairie ecosystems (*4*). During decomposition, the material is altered physically and will, many times, be unlike the original material.

Organic residues constitute a minor percentage of soil organic matter, usually less than 35 percent. Sources of these organic residues include such materials as plant roots, animal manure, dead animals, microorganisms, and aboveground plant remains. Of these materials, plant tissue contributes more to soil organic matter than animal tissue (*4*).

Humus, the primary constituent of soil organic matter, is amorphous, dark brown to black, and nearly insoluble in water, but soluble in dilute alkali (NaOH or KOH) solutions. Humus contains about 30 percent each of N-rich proteins, the slow-to-decompose lignins, and complex sugars. The complex sugars are a large percentage of the varied organic substances that cement soil aggregates together. Humus has about 60 percent C, 5 percent N, some O, and lesser amounts of S, P, and other elements. Its cation exchange capacity, on a dry weight basis, is many times greater than that of clay colloids (*28*). The particle density of humus is nearly half that of clay or quartz minerals, ranging from 1.3 to 1.5 g/cm^3.

Humic acid (proteins and lignins), the major constituent of humus, is that fraction of humus that is soluble in alkali solution and precipitated by acid (pH 1.0 to 2.0). It is dark brown to black in color.

Of the several chemical elements found in organic matter, C and N are the most important. While the N level in humus is about 5 percent (*4*), C levels range from 48 to 58 percent (*70*). Hence, the organic C content of various soil horizons is a useful indicator of the extent of organic matter accumulation.

Biological properties. A productive soil includes a myriad of microorganisms and small animals that represent a source of C, energy, and nutrients for plants and other microorganisms. Such factors as the food material available, plant species, amounts and kinds of organic amendments applied, soil texture, pH, moisture, aeration, and amounts of salts influence the numbers, kinds, and activities of these organisms. They live by the rule of the survival of the fittest. That is, those organisms active in any particular microenvironment are best adapted to the conditions present or represent the most successful competitors (*28*). These organisms include bacteria, actinomycetes, fungi, algae protozoa, nematodes, insects, and small mammals.

The chief role of soil fauna seems to be mixing soil and organic matter, translocating microorganisms, fragmenting plant material (*4*), and releasing such plant nutrients as C, N, and S (*47*). Earthworms in arid and semiarid agricultural and riparian soils frequently comprise half a soil's biomass. They can digest 49 to 861 t/ha of soil per year. Beetle grubs and other soil invertebrates that inhabit most arid and semiarid soils in the West do not completely replace the function of earthworms (*24*).

Bacteria grow best in moist, but aerobic, warm, and slightly alkaline to moderately acid soils. Actinomycetes grow better in drier soils with a neutral pH and in warmer climates (*2*). Optimum soil conditions for fungi range from very low to very high pH and moist and warm soils. Yeasts, however, do well in anaerobic conditions of stockpiled topsoil (E. Allen, Utah State Univerity, Logan, personal communication).

Vesicular-arbuscular mycorrhiza are soil-borne fungi that invade the roots of vascular plants and establish a symbiotic relationship. The mycorrhiza hyphae, extending from plant roots, increase the surface area for absorption of nutrients and water. In the symbiosis the mycorrhiza fungus provides nutrients to the plant, while the plant provides favorable growth mileu and carbohydrates for metabolism by the fungi (*21*).

Surface mining can disrupt mycorrhiza populations in soil because plant communities are temporarily destroyed and soils are either stored for long periods of time or mixed or diluted with spoil and subsoil material (*44*). Such disruptions usually lead to lower propagule numbers and infection levels in mined land relative to adjacent, undisturbed areas (*3, 23, 78*). Repopulation by mycorrhiza in such areas can be very slow because the fungi do not readily adapt to chemical and physical changes in soils caused by disturbance, especially when topsoil is diluted with chemically undesirable material (*54*).

Types of applied organic matter. Soil organic matter derives from plant and animal tissue. Under natural conditions, plants supply large amounts of residue annually. These plant materials are decomposed by soil organisms and become part of the underlying horizons, either by leaching or physical incorporation (*9*).

Machinery operations usually mix crop remains with the soil. If incorporated into the soil, crop remains facilitate greater water infiltration and erosion control. Also, decomposition rates are greater when crop remains are mixed with soil rather than left on the soil surface. Root remains are superior to top remains, such as stubble. Roots are better sources of C, with the exception of leguminous residues, and as roots decompose they build better soil structure (*4*).

Manure, which supplies numerous micronutrients, is an excellent source of organic matter. It helps maintain good soil tilth, improve water infiltration, increase CO_2 release, and increase organic matter content in some soils (*82*). Manure contains 50 to 80 percent water and 20 to 50 percent volatile solids. Thus, an application of 10 kg/ha would supply 1 to 4 kg of organic matter. Aged manure and compost decompose slowly because they are usually well rotted already. Aged manure can also serve as a growth stimulant, influencing the growth and development of plants and, in turn, yield (*8*). But applications of green manure extraction have retarded seedling germination and development (*73*).

Peat and mineral fertilizers can be combined effectively to improve soil conditions and plant growth. (Peat is undecomposed or slightly decomposed organic matter accumulated under conditions of excessive moisture.) Adding peat to a soil can improve water-holding capacity, increase aeration and water infiltration in fine-textured soils, provide a material capable of absorbing and retaining plant nutrients in an available form, improve root penetration, and increase a soil's buffering capacity (*46*).

Sewage sludge contains about 50 percent organic matter and can, therefore, be used to build and maintain fertile soil from spoil or soil (*83, 84*). Municipal sewage sludge can be applied in liquid or solid form in sufficient quantity to rebuild disturbed soil with one application (*33*). There are concerns about using sewage sludge as a soil amendment. Metallic elements in the sludge can be toxic to plants and eventually enter the food chain. Also, the applied sludge can increase the soluble salt concentration sufficiently to retard plant growth (*83*). Many sewage sludge amendment studies and application guidelines exist (*33*).

Wood wastes can be effective sources of organic matter. In theory, adding wood wastes to soil may have a depressing effect on plant growth because N may be immobilized. But data indicating immobilization are scarce. It may be necessary to add N with wood wastes to provide additional N for plant growth. Addition of wood wastes to soil has many positive effects on the physical properties of soil, such as improved friability with decomposition, which improves infiltration and subsequent water storage (*72*).

Methods of analyzing organic matter. Because the total weight of C within soil organic matter (passing a 2-mm mesh sieve) is 48 to 58 percent, organic C often is used to estimate organic matter. Organic matter is estimated by multiplying organic C times 1.724 (*58*). Broadbent suggested this factor should be 1.9 and 2.5 for surface horizons and subsoils, respectively (*11*). Other researchers recommended that the appropriate factor be determined experimentally for each soil (*58*). These same researchers also suggested that reporting only laboratory data on organic C is more appropriate than estimating organic matter unless the correction factor is presented with organic matter values. But Allison (*4*) concluded that using a correction factor to estimate organic matter is not objectionable "if extreme accuracy is not required."

There are three methods for determining organic C (*58*): (1) analysis of total C and inorganic C and subtraction of inorganic C from total C, (2) analysis of total C after destruction of inorganic C, and (3) reduction of $Cr_2O_7^{2-}$ by organic C compounds and subsequent determination of unreduced $Cr_2O_7^{2-}$ by oxidation-reduction titration with Fe or by colormetric techniques. Each method has inherent problems (Table 1). But the first method may yield the most accurate results. However, method three is most amenable to routine analysis of soils (L. Sommers, Colorado State University, Ft. Collins, personal communication).

The dichromate method is a modification of the technique described by Mebius (*50*). Heating of the chromic acid-soil mixture oxidizes organic C to CO_2. The amount of reduced $Cr_2O_7^{2-}$ then can be quantified. Mebius and Nelson and Sommers recommended using N-phenylanthranilic acid during titration (quantification of reduced $Cr_2O_7^{2-}$) because it produces a sharp, clean endpoint (*50, 58*). Nelson and Sommers (*57*) recommended heating the chromic acid-soil mixture for 30 minutes at 150 °C in Folin-Wu nonprotein N tubes placed in an aluminum block on a hot plate. Quantitative results provided by this technique represent the best combination of digestion reagents, heating procedure, and titration reagents of the modern dichromate determination methods. The technique uses soil-comprising organic material that has been dried and passed through a 2-mm sieve. This technique assumes that C in soil organic matter has an average valence of zero and an equivalent weight of 3 gm per equivalent when reacted with $Cr_2O_7^{2-}$.

Several researchers have recommended loss-on-ignition as a method of direct estimation of organic matter (*35, 58*). This method gives quantitative oxidation of organic matter. But inorganic constituents of the soil, largely hydrated aluminosilicates, lose structural water and carbonate minerals lose CO_2 upon heating at high temperatures. If an ignition temperature of about 430 °C is used (*27*) for 24 hours, the cumbersome pretreatment of soil with HC1 or HF to eliminate carbonates and subsequent correctional calculations are unnecessary. Although direct estimation of organic matter content using the loss-on-ignition technique at reduced ignition temperature (430 °C) may be a promising method to determine organic matter content, Nelson and Sommers (*58*) believe it cannot be recommended at this time because of insufficient evaluation of its usefulness and applicability.

Humus is the single largest constituent of soil organic matter. Analytical methods for determining humus content involve extracting humus substances from soil using dilute NaOH (O.1N or O.5N) in a N atmosphere. After extraction, analytical methods can be used to study humus: nondegradative and degradative (*70*). These methods are lengthy and not applicable to reclamation planning.

In summary, because of the difficulty in directly estimating or calculating organic matter present in soil, a more appropriate procedure is to determine and express organic C content as a measure of organic matter. Organic C can be accurately and precisely measured by the modified Mebius tech-

Table 1. Comparison of methodologies used for determination of organic C in soil (*58*).

Method	Principle	Advantages	Disadvantages
Difference between total C and inorganic C.	Total C and inorganic C are determined on separate samples: Organic C = total C − inorganic C.	Useful if total C and inorganic C are routinely determined.	Two separate analyses are required. Total C determination requires special equipment. Organic C calculated by difference has some inherent error.
Determined as total C after removal of inorganic C.	Total C is determined in soil sample after removal of inorganic C with an acid pretreatment: Organic C = total C.	Accurate if dolomite is absent from soil.	Not all dolomite in soil may be removed by acid treatment. Specialized equipment needed.
Dichromate oxidation without external heat.	$Cr_2O_7^{2-}$, oxidizes organic C to CO_2 in acid medium. Amounts of $Cr_2O_7^{2-}$ reduced is quantitatively related to organic C present. Not all organic C in samples is oxidized when external heat is omitted, and a correction factor is required.	Very rapid and simple. No specific equipment required.	Incomplete oxidation of organic C necessiitates use of correction factors, which often results in erroneous values. Chloride, Fe, and MnO_4 interfere with method. It assumes soil organic C has an average valence of 0.
Dichromate oxidation with external heat.*	This is the same as the dichromate method above except that all organic C in the sample is oxidized, and no correction factor is required.	Rapid and simple. Complete oxidation of organic C occurs.	Chloride, Fe^{2+}, and MnO_2 interfere with method. Some specialized equipment is needed. It assumes soil organic C has an average valence of 0.

*Recommended procedure for determination of organic C.

nique, whereas organic matter content can only be estimated. Therefore, measuring organic C concentration is preferable to determining soil organic matter content because the latter is not, at this point in the development of soil science, an appropriate or accurately measurable soil constituent (*58*).

Effects of organic matter on plant growth and growth media

Organic substances play a direct role in the formation of a fertile soil because they provide plant nutrients that become available during mineralization. Organic matter also has a fundamental effect on the physical properties of soil, such as water-holding capacity, structure, and heat regime. It also influences such physiochemical properties as cation exchange and buffering capacities. These properties influence nutrient uptake, water and nutrient availability, and the deleterious effect of soil acidity and alkalinity (*42*).

Effects on plant growth. The direct effects of organic matter on plant growth and development are well documented. Humus has positive effects on plant physiological and biological processes, promoting development of green tissue and uptake of nutrients (*42*).

Several studies confirm that plants take in CO_2, mainly CO_2 liberated from the soil pool, where considerable CO_2 is formed as a result of organic matter decomposition. Decomposed organic fertilizers are also a source of CO_2. Crop yields can be increased by CO_2 (*8*).

Organic matter levels in disturbed areas correlate strongly with plant growth after disturbance. The greater the predisturbance level of organic matter, the better the reestablishment of vegetation. In a California study (*66*), a desert area was seeded to barley and, in some areas, fertilized with N. Yields on soils beneath old shrub areas with moderately high levels of organic matter and without N fertilizer were three times greater than yields on old, bare areas with low soil organic matter levels receiving fertilizer. In a study of reforested areas, growth of cedars (*Thuja* spp.) and junipers (*Juniperus* spp.) was best on soils with a moderately thick to thick organic matter-rich layer (*25*).

In more severely disturbed areas organic matter applications appear to benefit plant growth temporarily. In sorghum and corn fields of Texas and Montana, where land leveling had removed 30 to 90 cm of the upper soil profile, additions of N, P, and/or manure produced a plant response equal to or slightly less than on nearby undisturbed areas (*64, 75*). In Texas during drier-and warmer-than-normal seasons, yields were considerably lower, probably due to "intensified drought and heat damage related to higher bulk density and lower porosity" (*75*). Infiltration declined because of the lower porosity and higher bulk density. This resulted in the failure to build a normal topsoil over 31 years, which emphasizes the value of the virgin A horizon.

Such research underscores an important principle: applied fertilizers and organic matter can stimulate plant growth in severely disturbed areas but, in themselves, do not create a self-sustaining, productive A horizon upon application. Rather, the organic matter usually accelerates nutrient cycling and plant growth, which augments desirable long-term soil development and increased fertility. Similarly, humus appears to exert a temporary beneficial effect on plant growth (*69*). Although Schnitzer's review pertained to vegetable crops, similar results would probably occur with typical mined-land revegetation species.

Only in a few instances has the application of organic matter resulted in an undesirable plant growth response. In one test high rates of manure application (400 t/ha) reduced the size and yield of corn because of increased salinity levels (*67*). Other studies involving a high manure application (7,400 t/ha) have shown similar yield declines (*48, 56, 62*). Even at an application rate of 112 t/ha, plant stress was evident early in the growing season (*49*). Salt build-up and NH_3 toxicity are the two most often cited causes of reduced germination and/or plant growth. In areas where 2.5 t/ha of manure were applied and sprayed with water-soluble asphalt emulsion, depressed growth and development of 12 transplant species were evident because of prolonged, less-than-optimum soil temperatures (*12*).

N and other minerals are important elements in soil organic matter from an economic viewpoint (*4*). Plants use N and other nutrients under conditions in which substances are cycled by biological and geological processes. These elements are scattered throughout the atmosphere, hydrosphere, and earth's crust. They accumulate in autotrophic—inorganic compound or element digestors—and heterotrophic soil organisms. These elements are subsequently liberated during the life and after the death of the organism (*42*).

Because N makes up more than 78 percent of the gaseous weight of the atmosphere, the main reservoir pool is the atmosphere, even though most N cycling takes place in the soil (*86*). Atmospheric N is unavailable to plants directly. Other mechanisms are needed to convert N to a form that plants can use. This process involves mineralization of organic matter by soil microorganisms.

Most N in surface soils is organically combined (*10*), bound largely to organic matter and mineral matter with only a few kg/ha existing in available mineral forms as nitrates and exchangeable ammonium at any one time (*79*).

Total N content of soils ranges from less than 0.02 percent in subsoils to more than 2.5 percent in A horizons rich in organic matter (*10*). Determinations of the C/N ratio in natural soils have shown that the amount of organic N relates to the amount of organic C. Generally, throughout mining areas in the West, C/N ratios range from 7:1 to 10:1. A C/N ratio of 17:1 was found in one 40-year-old mine soil, much higher than the 10:1 ratio in adjacent natural soil (*68*). The ratio of organic C to N is a useful indication of the degree of organic matter decomposition in soils; it narrows with increasing decomposition (*14*). As soil reaches a relatively static C/N ratio, the soil organic matter level is relatively constant.

Regardless of soil conditions, the direct effects of organic matter are the same: generally, plants respond favorably because nutrients, especially N, are provided through decomposition and mineralization (*43*). Campbell stated, "Crop yields are often directly proportional to the N released from organic matter, the other nutrients are also important but N is required in much larger amounts and is more likely to be deficient" (*15*).

Effects on soil. Soil organic matter directly or indirectly influences the physical environment (*4*). As the percentage of organic matter decreases in a given soil, the bulk density increases, with an associated reduction in porosity. The combination of increased bulk density and reduced aeration restricts root growth, impairs normal root absorption, and inhibits microbiological activity.

Soil structure greatly influences fertility (*42*). Organic matter is one of the most important factors in soil structure development. Soils with good structure provide the best conditions for supplying water and nutrients to plants (*42*). The best water and nutrient regimes occur in soils with granular structure in which the aggregate size is 1-2 mm. Such soil structure is formed most readily under a mixture of perennial grasses and legumes. The prerequisites for the formation of good soil structure include (1) consolidation of the soil particles under pressure from highly branched root systems of grasses, (2) the formation of humin and humic acid, (3) penetration into the soil crumbs of the highly dispersed solutions of humin and humic acids, and (4) conversion of these acids into a cohesive soil aggregate by means of Ca supplied by the legume roots (*42*).

Also critical is the fact that conversion of humus substances into a soil aggregate is irreversible under natural processes (*42*). After plowing or disking, formation of humus substances that bind soil crumbs is slow and requires a specific environmental condition (*42*). Similarly, increased stability of aggregates usually accompanies an increase in organic matter (*15*). Thus, to maintain good soil structure there must be an absence of continued plowing and a source for newly formed humus substances on a regular basis.

Incorporation of manure into soil influences microorganism activity. Organic material itself, without biological transformation, has little effect on soil structure. Microorganisms, without organic materials as a source of energy, are ineffective in producing soil aggregation. Fungi and actinomycetes produce mycelia and have metabolic processes that synthesize complex organic molecules. Decomposition products remain in the soil. The sum of these effects is the production of stable soil aggregates. This stability, therefore, can result from the mechanical binding action of the cells and filaments of the organisms; cementation effects of the products of microbial synthesis; and the stabilizing action of the decomposition products.

Organic matter usually improves infiltration, reduces evaporation, improves drainage in fine-textured soils, and encourages more extensive and deeper root systems. Collectively, these actions should improve water availability and the water use efficiency of plants (*4*). These effects vary with plant species, soil type, and climate. If fertilizers are added to the soil, the effects of organic matter may increase (*15*). Except for clayey soils, organic matter in general increases plant-available water-holding capacity.

Organic matter increases infiltration by helping to hold water on the soil's surface long enough for it to enter the soil and by improving the physical condition of the soil. Organic matter also reduces crust formation, which increases the rate of infiltration. Similarly, if organic matter is mixed into a soil, eventually improving soil aggregation and soil structure, percolation increases. Contrary to this generally favorable principle of organic matter increasing infiltration, some forms of organic matter reduce infiltration because of the presence of hydrophobic (nonwettable) organic substances.

One of the most important chemical properties of soil is its ability to retain and exchange positively charged ions on colloidal surfaces. The primary controlling factors of cation exchange are the amount and type of clay and the quantity of organic matter in the soil (*34*). The organic matter in most mineral soils accounts for about 30 to 65 percent of the total cation exchange capacity (*15*). In sandy and organic soils, more than 50 percent of the cation exchange capacity is due to the organic component of the soil.

A buffer solution is one that resists changes in pH upon the addition of acid or base. Soils behave like buffers. Most of this ability is a result of their colloidal properties associated with the humus and clay minerals, which hold H, Al, and other cations (*15*). Resistance to pH change thus is low in soils with lower organic matter levels but higher in soils with higher organic matter levels.

There is a wide range of chemically stable organic anions capable of binding metal cations into weakly ionized forms (*34*). These substances—chelating agents—occur naturally in soils. They are associated with the solid organic fraction as well as soluble organic materials in the soil solution. Organic matter can hold metallic ions both by cation exchange and by chelation (*15*). In soils having metallic ion concentrations that are toxic to plants, chelation may hinder adverse effects through the immobilization of the metallic ion.

Soil organic matter and revegetation planning

Because virgin organic matter levels vary from region to region and within a particular mine site, there should be no single set of suitability levels for organic matter to rate topsoil as good, fair, poor, or unsuitable. The same is true of surface soil horizons used for agricultural purposes (*15*).

Rather, suitability levels should vary with location and equilibrium level of the organic matter in the surface soil horizon under the proposed post-mining land use. These levels should be such that they supply nutrients to the plant, help ameliorate undersirable soil or site conditions affecting plant growth, and facilitate a self-sustaining soil system.

Mined land in the western United States has been reclaimed in recent years largely to one land use: grazing or browsing by livestock and/or wildlife. In a few locations mined land has been returned to forest land or cropland. Recognizing that cultivation of rangeland reduces organic matter levels by as much as 50 percent (*4*), a reduction in organic matter of up to 25 percent from baseline levels in the post-mining soil is a good target for proposed post-mining cropland areas. Proposed rangeland areas should be 100 percent of baseline levels.

Many researchers believe that the optimum level of organic matter for a mined area is already expressed in the adjacent undisturbed area. But one important consideration is the fact that disturbed systems are in the process of arriving at a balance. As such, adding organic matter may expedite that balance or delay it, depending on the particular set of soil-forming factors in the area.

Relationships between baseline conditions and processes of soil organic matter formation. In the surface horizons of undisturbed land, equilibrium organic matter levels generally vary with temperature, rainfall, vegetation cover, soil texture, physiographic position, and land use management (*4*). Equilibrium organic matter levels in surface horizons of native soils in mining areas of the western United States range from 0.5 to 8 percent (*9, 16, 17, 18, 19, 20, 31, 77*). In nonmountainous areas of the West organic matter levels range from 0.5 to 3.4 percent and 1 to 6 percent in southern and northern zones, respectively. Organic matter levels in mountainous areas range from 2 to 12.0 percent or higher. Coarser textured soils in the southeastern United States have a stable fraction level of 1 percent (*26*). Finer textured soils of the more northern states have a stable fraction level of 3 to 5 percent.

At least three organic matter level studies have been done on reclaimed land in the West. At all three sites topsoil salvaging or replacement techniques were not used. Organic C levels in a 50-year-old mine soil near Colstrip, Montana, were lower throughout the profile than in adjacent natural soil profiles, except in the upper 5 cm where mine soil organic C levels were higher (*68*). On 28- to 40-year-old mine soils in Saskatchewan, Canada, humus levels were similar to natural soils (*5*). In a North Dakota mine soil, organic C levels were lower than those in natural soils (*87*). In Montana, calculated accumulation rates were about 45.4 gm/yr in a 50-year-old mine soil and about 135 gm/yr in a 10-year-old mine soil that was subjected to heavy fertilization and seeding with vigorous introduced perennial grasses and alfalfa (*68*). Accumulation rates of about 28 gm/yr have been calculated in Canada (*5*).

Generally, equilibrium organic matter levels increase with increasing precipitation and decrease with increasing temperature (*9*). Therefore, the organic matter content of soils in western mining areas should be highest in eastern Montana, followed in order by eastern Wyoming, eastern Colorado, and eastern New Mexico. Furthermore, with all other conditions being equal, organic matter levels generally decrease with decreasing clay and/or silt content. That is, organic matter content in clay is greater than that in loam, which is greater than that in sand. Finally, more concave or depositional physiographic positions have higher organic matter levels in the upper horizons because of the additional organic matter-enriched runoff received from upslope surfaces or increased plant production facilitated by additional moisture.

Equilibrium levels of organic matter also vary with the type of land management. Organic matter levels are commonly lower in cultivated areas adjacent to virgin soils having the same aspect and landscape position than organic matter levels in the virgin soil. In addition, organic matter levels in cultivated soils tend to remain lower because tillage breaks up organic residues and brings them into contact with soil organisms, "thereby increasing the rate of decomposition" (*9*).

In most mining regions of the western United States, topsoil use is becoming more common. In these situations low organic matter content will not hinder revegetation efforts as much as other undesirable soil conditions, such as low N or P levels, adverse (high) sodicity or salinity levels, a skeletal particle size class (greater than 35 percent by volume rock fragments), or a sand or clay texture. Even though organic matter levels in surface soil horizons at any given mine location will vary as much as 5 percent, the immediate adverse effect of low organic matter levels on revegetation is less than other undesirable soil conditions, such as excess Na. In the long term, however, organic matter is probably a controlling influence insofar as continued nutrient cycling relates to revegetation success.

Mitigating undesirable soil or site conditions with organic matter. Our assumption is that ample topsoil generally is available for use at most mining sites in the West. But "ample" topsoil varies from one site to the next. Moreover, some western mining operations are on very shallow, sodic/saline, or skeletal soils or on steeply sloping areas providing little or no topsoil. In such areas as the San Juan Basin in New Mexico or sodic/saline locations in southwestern Wyoming revegetation is being attempted on spoil or nontopsoil material. In some of these cases applied organic matter may be beneficial. Organic matter probably would increase the water-holding capacity of skeletal spoil, but no more than loamy soil material. In sodic/saline subsoil or spoil, which is often medium to fine textured, organic matter may enhance infiltration and percolation to facilitate leaching of Na cations and prevent the formation of a claypan caused by excessive Na.

In revegetated spoil areas without topsoil, organic matter can play an important role in nutrient supply. In spoil material minor elements usually are adequate for plant growth, but there may be N and P deficiencies. As mentioned, organic matter is a viable nutrient source enhancing revegetation efforts. Furthermore, applications of organic matter increase microbial growth, creating a more fertile environment for plants that can enhance reclamation of relatively infertile, biologically inactive spoil or subsoil material.

Determining the soil organic matter content needed to mitigate undesirable characteristics in plant growth media or site conditions varies with the undesirable characteristic or degree to which such characteristics are expressed. Following are the organic matter levels achievable through organic matter application that may mitigate adverse soil properties or site conditions.

Texture. Increased water-holding capacity is the dominant positive influence that organic matter has on sandy, skeletal, and loamy particle size classes. In one study the water-holding capacities of a 50 percent sand and 50 percent peat mix and 100 percent peat mix was 36 percent and 82 percent, respectively, greater than that of sand (*38*). In addition, research shows that organic matter with high C/N ratios reduced permeability in sand, while low C/N ratios caused reduced permeability for only short periods of time. In another study a 20 percent addition of peat raised the water-holding capacity of a clay loam and loamy fine sand about 30 and 51 percent, respectively (*13*). A thin layer of peat or organic matter on the surface may preclude surface crusting, thus facilitating better seedling emergence (*4*). Such a layer is also beneficial from an infiltration perspective (*74*). On the other hand, minimizing organic matter build-up in the surface layer benefits plants because it promotes increased N availability through enhanced nutrient cycling (*68*).

In summary, sandy, loamy, and skeletal particle size classes benefit most from the standpoint of water-holding capacity because of higher organic matter levels. Clayey textured soils benefit more from increased infiltration and percolation rates and reduced crusting.

Reaction. Organic matter indirectly and directly benefits very high or very low pH materials. In soil and/or spoil materials that have high pH levels (>8.4) organic matter applications may indirectly reduce the pH one unit or more by increasing infiltration and percolation. This assumes that adequate precipitation causes leaching and the proper types of organic matter are applied.

In an Ohio study mulch treatments in sandy loam spoils having a pH from 2.4 to 4.9 produced survival rates among Ponderosa pine (*Pinus ponderosa*), jack pine (*Pinus banksiana*), pitch pine (*Pinus rigida*), and white pine (*Pinus strobus*) that were 42 to 60 percent better than on control areas (*45*). The mulch no doubt exerted a beneficial influence on other spoil properties. But a pH of 2.4 is prohibitive for most revegetation species, including these four tree species. The mulch probably had a direct or indirect effect on pH as well.

On sandy soils with a pH of 5.0 to 6.6, one-time peat applications raised the pH about 0.3 unit after 18 months (*55*). Generally, in the Rocky Mountain region and elsewhere in the West the pH of a soil is lower—at or near neutral—where the organic matter level is higher. This is probably a function of an increased or higher cation exchange capacity, hence a lower base saturation (*55*) and higher hydrogen ion concentration.

Sodicity. Organic matter applied to sodic (Na adsorption ratio greater than or equal to 13) spoil used as a plant growth medium partially reduces the effects of Na. Applying organic matter likely will be beneficial if it is cultivated or mixed into the top 15 to 30 cm of soil. Because of the dispersing (repelling) effect Na has on soil particles, creating a low-permeability claypan, organic matter will facilitate greater percolation. Under irrigation or in higher rainfall areas, the Na effect may be ameliorated by replacement of the Na cation if suitable amendments, such as $CaSO_4$, are subsequently added.

Salinity. In reclaimed areas with potential salinity problems resulting from saline spoil organic matter mixed in the surface can be beneficial. In the desert Southwest a 1 percent increase in organic matter, from manure applications, reduced infiltration time 31 percent and increased percolation rates—saturated hydraulic conductivity (*51*). The manure's effect on water infiltration rates lasted only one year. The higher percolation rates provided for a more frequent and complete flushing of soluble salts in the upper layers of the spoil, thereby reducing salinity. The conclusion: proper residue management in the desert Southwest could increase organic matter and reduce salinity problems and flooding times (*51*).

Nutrient levels. Organic matter can play an important role in supplying nutrients when revegetating spoil areas without topsoil in the western United States. In spoil material minor elements are usually adequate for plant growth. But N and usually P can be deficient. As mentioned, organic matter is a nutrient source and a good vehicle for the retention of nutrients. But organic matter is an inefficient method of fertilization. A kilogram of cow manure contains about 5 g N, 2.5 g potash, and 2.5 g phosphate (*4*).

Organic amendments, such as manure or compost, are important sources of energy for soil microorganisms. They also intensify biological processes in the soil. As a result, soil nutrients are converted more rapidly into ions and compounds available to plants (*8*). Manure additions add a large quantity of organic matter to the soil and, via microbiological processes, add to the nutrient reserves of the soil and, temporarily to the C supply.

Undesirable organic matter levels. Increasing organic matter levels may aid revegetation efforts. Organic matter can be separated into two general fractions: stable and dynamic (*26*). The stable fraction has an equilibrium level that varies with geographic location. Little can be done to increase this stable fraction. Thus, temporary increases must be directed at changing the dynamic fraction, such as bringing organic matter to the site (exogenous) or growing cover crops.

The applied organic matter should have a C/N ratio as close to the target level of the soil as possible. Otherwise, N applications may be necessary to compensate for immobilized N. Even leguminous mulches have C/N ratios of about 18:1 or greater, much higher than average soil ratios of 10:1 (*4*). Some straw mulches and wood residues in the western United States have C/N ratios as high as 2,000:1 (*4*). Alternatively, areas amended with mulches or residues that have high C/N ratios can be left fallow, allowing the area to return to an "available N" condition. But such an alternative may not be feasible because controlling weeds that may invade the area may be more costly than N fertilization.

The amount of organic matter applied should be consistent with the intended effects on topsoil, seedling establishment, and continued plant growth. If the purpose is to provide nutrients for plant growth, rather sizeable amounts may be required, depending upon the type of organic matter applied. If the purpose is to improve physical condition of the topsoil, smaller amounts may suffice. Recommended amounts that provide some plant nutrients and improve physical conditions vary considerably, but generally start at 0.5 t/ha.

Timing and frequency of application should be consistent with the intended effect on plant growth medium. Generally, a one-time application improves physical condition and provides some nutrients. Even if organic matter levels in the topsoil are relatively high, a one-time application provides some benefits, particularly on soils that are low or high in clay content.

Adverse effects of applying organic matter in the form of straw mulch include the possible immobilization of N during decomposition (*7*), interception of light, and possibly a physical barrier to seedling development (*39*). Also, litter accumulation on minesoil may result in a formation of high C/N ratio and/or stagnation of plant succession at an early seral stage (*68*). But one study indicated that these potential problems can be minimized (*30*).

Mechanical means are the best methods to apply mulches, wood wastes, composts, sludge, and other organic matter. Material should be incorporated into the upper 15 cm of soil. Except in low precipitation (<25 cm/yr) areas, an effective method of mulch incorporation is planting an annual nurse crop the year prior to seeding. Mowing or disking the crop in the fall adds residues to the soil and creates a physical barrier to trap snow and reduce runoff in the spring.

There has been little research in the West regarding permanent increases in organic matter levels. Eastern research (*55, 61*) provides recommendations, but sound recommendations for organic residue application rates in the West are only speculative.

In recent years a new technique has been used successfully at surface coal mines in northwestern Colorado. The method temporarily increases organic matter content of the dynamic fraction before topsoil stripping and replacement activities. It involves "rotoclearing" (rototilling) of native vegetation, including trees (aspen up to 50 cm in diameter), shrubs, forbs, and grasses, into the top 40 cm of the soil profile just prior to topsoil stripping (personal communication, T. Johnson, Colorado Yampa Coal Company, Steamboat Springs, Colorado, and J. Monarch, Pittsburgh and Midway Coal Company, Denver, Colorado). Rotoclearing involves pulling a 425-horsepower, 3-m wide, 9,525-kg rototiller-like machine through vegetation with a bulldozer. This action incorporates significant but unquantified amounts of organic material into the topsoil. The material ranges from wood chips and small tree stems with high C/N ratios to well-pulverized, rapidly decomposable leguminous plants with low C/N ratios. If direct replacement of 30 cm of soil follows soon thereafter (less than 2 months), the effect upon first- and second-year revegetation has been beneficial. The effects of stockpiling and replacing this material is unknown because the technique has not been used long enough to observe revegetation efforts.

Erosion. A primary goal of revegetation planning is to minimize erosion. Reducing erosion is very important with respect to maintaining organic matter in the surface horizon. Erosion robs the surface horizon of organic matter. Usually, eroded material contains 1.5 to 8 times as much organic matter and N as undisturbed soil (*4*). Humus has a low particle density and, therefore, is more susceptible to detachment and transport by raindrop splash than to clay or quartz minerals. This accounts for the high concentration of organic matter in eroded materials (*41*).

Organic matter in the form of mulch or nurse crop residues should reduce the soil erodibility and crop management factors in the universal soil loss equation. Therefore, the higher the organic matter content in the topsoil or plant growth medium, the lower the erosion rate. USLE-calculated erosion rates for two similar soils having 0 and 4 percent organic matter under the same management and vegetation, on the same slope, and in the same location will differ by 50 percent; the lower organic matter soil will have the higher erosion rate (*1, 88*). This agrees with our earlier discussion about organic matter increasing the infiltration rate and reducing runoff.

Research regarding effects of mulching and crop residue on wind and water erosion have shown that relatively low application rates can significantly reduce water erosion. Near Zap, North Dakota, 4.5 t/ha of wheat straw reduced erosion 93 percent on loamy topsoil relative to control plots (*32*). In western North Dakota erosion rates on a sandy loam topsoil, with a 9 percent slope on which 4 t/ha wheat straw had been applied, were 0.16

t/ha, compared to 32 t/ha on bare soil (*59*). In northwestern Colorado no topsoil loss was observed on reclaimed, 30 percent slopes (loamy texture) with 3.36 t/ha of crimped-in straw (*63*).

In summary, organic matter applied and incorporated or in-situ reduces erosion. Erosion rates on bare topsoil with high organic matter levels are 25 to 75 percent lower than on bare soils with low organic matter levels. Applications of straw mulch in excess of about 3.0 t/ha probably will reduce erosion to barely measurable levels.

Climate. In the mining areas of the arid and semiarid West annual precipitation ranges from 15 to 46 cm. While a few mining sites in mountainous areas or in the Northwest receive 45 to 125 cm, plant growth at most western mine sites commonly is limited by water stress related to low precipitation. Also, precipitation often occurs as short-duration, high-intensity rainstorms or rapid snowmelt. Moderate to high infiltration and percolation rates in the upper part of the soil profile, which are influenced by organic matter levels and additions of mulch, are important for minimizing erosion and maximizing water storage and intake.

Again, higher organic matter levels increase infiltration rates and the water-holding capacities of soils with various textures. Also, vesicular arbuscular mycorrhiza populations tend to increase the effective root surface area for absorption of water and nutrients. Both of these mitigative characteristics of organic matter tend to maximize the use of precipitation in semiarid areas.

Effects of topsoil stockpiling on organic matter. Based on studies in North Dakota and Wyoming, long-term stockpiling is detrimental to topsoil in arid and semiarid mining environments (*21*). Four years after direct replacement, native vegetation production on topsoil that had not been stockpiled was five times greater than that on a similar area covered with topsoil that had been stockpiled for 2 years. The researchers attributed the difference to adverse changes in organic matter.

Several factors affect humus formation in soil, and stockpiling affects many of these factors. Assuming the stripped area originally supported a dynamic plant community, the following factors of organic matter formation are likely to be affected: soil organisms, moisture, temperature, O supply, and plant material supply.

The first and most obvious change occurring to buried topsoil is the loss of a large seasonal addition and smaller daily additions of dead plant tissue (future humus). After the initial decay of easily decomposable carbohydrates, amino acids, and proteins, more slowly decomposable lignins, tannins, fats, oils, waxes, and other substances continue to decay. Microbial communities in the soil directly or indirectly compete with one another for the slowly declining supply of organic residues when moisture, temperature regime, and O supply are decreasing. The composition of microbial communities changes as a result, and populations generally decline (*21, 40*).

Changes in organic matter levels relate to the duration of storage and location within the stockpiled topsoil. After 3½ months of storage and without establishing vegetation on a topsoil stockpile, Klein and associates (*40*) found the organic matter levels declined (1.0% to 0.2%) on the margins of the stockpile compared to the interior of the pile. The researchers concluded that the natural wetting and drying on the margins of the pile facilitated greater availability of organic matter to microbiological decomposition than on the interior of the stockpile. As cellulose within the plant tissue breaks down, water-soluble organic C presumably increases (*21*).

With respect to topsoil stockpiling, one group of researchers concluded that vesicular-arbuscular mycorrhiza losses occur from topsoil in the interior of the stockpile and that 3 years of storage reduced the infection potential of the soil (*65*). Similarly, Christensen and Allen (*22*) recovered relatively few vesicular-arbuscular mycorrhiza spores from stockpiles over 5 years old. In Wyoming one scientist found that the potential for vesicular-arbuscular mycorrhiza infection (growth) declined with increasing stockpile age (*52*). Thus, vesicular-arbuscular mycorrhiza are limited by a lack of plant roots in the interior of the stockpiles and apparently can endure only a certain interval of stockpile storage (*44*).

In summary, the microbial populations, organic C levels, and vesicular-arbuscular mycorrhiza growth in the margins of topsoil stockpiles decline in time with reduced O levels, reduced soil moisture, changes in organic matter supply, and loss of plants and roots. Upon relatively immediate decomposition of the readily decomposable organic substances, NH_3 is produced, followed by CO_2, organic residues, and humus. At this point, the organisms responsible for this process in the stockpiled topsoil should be less active and not deteriorate further (*4*).

Soil sampling for organic matter characterization. During field investigations, scientists use soil color as the principal indicator of organic matter content in soils with less than 12 percent organic matter. As the soil formation factors that influence organic matter characteristics change from one soil map unit to the next, the thickness and color of the A horizon changes accordingly. As a result, understanding the relationship between organic matter field characteristics (color and horizon thickness) and the factors that determine those characteristics is essential for characterizing baseline conditions.

A taxonomically based soil survey and sampling program should be used for field characterization of baseline organic matter conditions, specifically, the soil taxonomic system outlined by the Soil Survey Staff (*77*). The soil survey should be conducted by a soil classifier familiar with organic matter characteristics and soil-forming factors of the mine permit area. The sampling program should insure collection of soil from one typifying pedon (all genetic horizons) of each soil series mapped. If several map units of the same soil series exist within the permit area, because the organic matter content of each map unit varies by more than 33 percent, one typifying pedon of each map unit should be sampled.

The detail of the soil survey depends upon whether topsoil or overburden/interburden will be used as the plant growth medium during reclamation. Where topsoil salvage and replacement is proposed, an order-1 soil survey should be used to characterize baseline organic matter and other soil properties. Where use of overburden/interburden is proposed, an order-2 soil survey provides sufficient baseline characterization to perform topsoil substitution evaluations.

The primary intent of an order-1 soil survey is to identify topsoil stripping depths for each map unit, along with obvious topsoil stripping depth criteria, such as texture or salinity. Thus, organic matter content should be included as a basis for setting up additional phases (map units) of any soil series. Once in the field, if the soil scientist suspects that the proportionate amount of organic matter might vary more than 33 percent within any soil series (e.g., 2.5 percent organic matter versus 4.0 percent), different phases of the soil series should be established and mapped, provided the different phases are distinguishable on aerial photographs and topographic or native vegetation differences are observed in the field. For example, if in a cultivated field the soil scientist could delineate on aerial photography an eroded condition of the same soil series occurring in an adjacent, less eroded, rangeland setting, different phases of the same soil series should be mapped to distinguish the soil with a lower organic matter content from the soil with a higher organic matter content, assuming a minimum map unit size criterion is fulfilled.

Similarly, the thickness of the A horizon should be included as a basis for additional map unit phases, assuming no other limiting soil conditions occur in the A horizon. If the thickness of the A horizon within a particular series varies by more than 15 cm because of an identifiable management, topographic, or native vegetation difference, different phases of the soil series should be set up in the map unit legend. For example, a slightly more concave position with a thicker A horizon due to a greater amount of deposition should be identified as a "thick-A" phase. It should be delineated separately on aerial photography from adjacent, less concave areas because the phase difference is distinguishable based on topography and fulfillment of minimum size delineation criteria.

Summary and recommendations

Reestablishment of a functional soil system is critical to the successful reclamation of surface mined areas in the arid and semiarid West (*21*). An important, integral part of this functional ecosystem is the process of soil organic matter formation. Hence, reestablishing the dynamic processes of soil organic matter formation may be more important than having topsoil with the optimum organic matter level. The most important input to soil organic matter formation is vegetative detritus, which must be deposited on a regular basis. Vegetation and inputs of lesser importance stimulate microbial populations to predisturbance levels. Rejuvenation of microbial activity starts the active cycling of various organic compounds resulting in end products of decomposition that will, in turn, be used by plants and microorganisms. As the cycles continue, soil organic matter levels and fertility will increase and eventually reach an equilibrium.

What is the optimum level of soil organic matter in topsoil for mined land? Research has not yet identified a specific level for any one location; we believe it will even vary within a specific mine. We believe the optimum level should be the stable, baseline level. The stable level, however, varies from mine to mine and within each mine because the stable organic matter level is a function of several varying factors: texture; chemical characteristics; microclimate; site-specific vegetation it will support; and future slope, management, topography, and land use. Furthermore, soil organic matter may be less important than some of these factors, yet all fertile soils, including those replaced on reclaimed land, must have a minimal supply of organic matter and will have a minimal supply if revegetation is successful (*68*).

Several characteristics of soil organic matter can be quantified, but few are useful as they relate to indicating the value of the material as plant growth media. Soil organic matter in soil baseline reports should be quantified on the basis of total organic C by soil series or by soil map unit rather than total organic matter. In addition, spore counts should be performed to quantify vesicular-arbuscular mycorrhiza populations; these counts will provide planning target levels. Total organic C should be determined by the modified Mebius technique (*58*). Although direct estimates of organic matter content using the loss-on-ignition technique and a reduced ignition temperature (430 °C) has promise as a method for determining organic matter content, the method is not recommended because of a lack of complete evaluation of its usefulness and applicability.

The sampling program should insure collection of soil from one typifying pedon—all genetic horizons—of each soil series mapped. If several map units of the same soil series exist within the permit area and are suspected of having an A horizon thickness or organic matter content variability greater than 33 percent, at a minimum one typifying pedon of each map unit should be sampled.

Many species of plants that are desirable for disturbed land revegetation may be dependent on vesicular-arbuscular mycorrhiza fungi. Reduced inoculum levels in disturbed soil may result in vesicular-arbuscular mycorrhiza choosing against such plants, resulting in a contribution to the establishment of nonmycorrhizal plants. Experimental inoculation of revegetation species with vesicular-arbuscular mycorrhiza fungi has enhanced seedling establishment, stress tolerance, and growth of revegetation species. Thus, this inoculation may improve the plants' competitive ability and productivity during revegetation. Commercial application of vesicular-arbuscular mycorrhiza to reclaimed land, however, has not been demonstrated even though development of commercial application methods are in progress. Given the present state of progress in developing commercial application methods, current reclamation goals should emphasize methods of maintaining inoculum levels by minimizing topsoil storage periods or using direct replacement methods.

As a rule of thumb, an optimum organic matter level in the plant growth medium should be approximately equal to baseline conditions under a topography and land use similar to that proposed in the post-mining reclamation plan. The implicit principle of this recommendation is important. Maintenance of humus at a higher level than surrounding undisturbed areas is not only difficult but expensive because the rate at which C is lost from the soil increases rapidly as organic matter content increases. Likewise, humus levels lower than surrounding, undisturbed soils may indicate long-term inferiority.

Where topsoil is stripped and topsoil is in abundant supply for reclamation, a first-lift stripping of the uppermost horizon(s) having a greater percentage of organic C (the mollic epipedon) should be stockpiled separately from lower horizons in the profile that meet topsoil criteria. During overburden and soil handling phases of reclamation, the first-lift topsoil should be replaced last, creating maximum fertility in the minesoil profile.

As a second rule of thumb, where cultivation is the proposed post-mining land use, optimum organic C levels should be about 75 percent of uncultivated baseline levels or 100 percent of cultivated baseline conditions under similar topographic conditions. For example, if the pre-mining land use is rangeland and the proposed post-mining land use is cultivated cropland, optimum topsoil levels should be 75 percent of baseline values.

Baseline levels of organic matter provide an estimate of what these rule-of-thumb levels should be and how they may vary with location. Throughout the western United States, organic matter content ranges from 0.5 to 12 percent and higher. But in most mining areas of the West, levels are usually 1 to 5 percent. In the dry, arid environment of the Southwest virgin soil organic matter content is less than 1 percent, whereas organic matter levels in the northern Great Plains range from 2.5 to 6 percent. Mountainous areas in the central Rocky Mountains have organic matter levels of 2 to 12 percent or higher. These provide general target levels for organic matter in western mined land.

Applications of organic matter can improve site and soil quality. Organic matter should not be regarded only as a limiting criterion of soil material being evaluated for use as topsoil. The following organic matter application and management recommendations can mitigate adverse conditions that may exist in nontopsoiled plant growth media:

► If the major limitation of the plant growth media is a clay, loamy sand or sand texture, or a loamy or sandy skeletal particle size class, at least 2.2 t/ha of organic matter should be incorporated into the upper 8 cm of soil.

► If the major limitation of the plant growth medium is its sodic condition, up to 4.2 t/ha of organic matter could be incorporated into the upper 20 cm of soil to increase infiltration and permeability. This condition will create an environment where, under irrigation or in higher rainfall areas, application of Na replacement amendments, such as $CaSO_3$, will optimize Na cation displacement.

► If the slope of the reclaimed land exceeds 25 percent and annual precipitation exceeds 30 cm, or the area receives intense thunderstorms, 2.2 t/ha or more of organic matter (straw mulch) should be applied, all other soil conditions not limiting, to increase infiltration and minimize water erosion.

► Any area receiving applied organic matter with a C/N ratio exceeding 20:1 may need N fertilization.

► Soil areas to be stripped and then have the topsoil replaced directly should be given consideration for rotoclearing prior to stripping.

► With the possible exception of the northern Great Plains, cropland use of mined land is discouraged for several years after mining because of the organic matter loss and, hence, accelerated erosion that may occur under this form of management.

► Excessive applications ($\geq$200 t/ha) of manure or sewage sludge should be avoided in any type of plant growth media.

► Applications of organic matter will raise the cation exchange capacity of the soil.

► Application of organic matter to nontopsoiled or spoil plant growth medium via the use of a leguminous covercrop should appreciably increase the dynamic fraction and possibly accelerate processes of soil organic formation when moisture is adequate.

REFERENCES

1. Agricultural Research Service, U.S. Department of Agriculture. 1965. *Predicting rainfall-erosion losses from cropland east of the Rocky Mountains.* Agricultural Handbook 282. Washington, D.C.
2. Alexander, M. 1977. *Introduction to soil microbiology.* John Wiley and Sons, Inc., New York, New York. 467 pp.
3. Allen, E. B., and M. F. Allen. 1980. *Natural reestablishment of vesicular-arbuscular mycorrhizae following strip-mine reclamation in Wyoming.* Journal of Applied Ecology 17: 139-147.
4. Allison, F. E. 1973. *Soil organic matter and its role in crop production. Developments in soil science, 3.* Elsevier Scientific Publishing Company, Amsterdam, The Netherlands. 637 pp.

5. Anderson, D. W. 1977. *Early stages of soil formation on glacial till mine spoil in a semi-arid climate.* Geoderma 19: 11-19.
6. Baver, L. D. 1968. *The effect of organic matter and soil fertility.* North Holland Publishing Company, Amsterdam, The Netherlands, and John Wiley and Sons, Inc., New York, New York. 1,017 pp.
7. Berg, W. A. 1980. *Nitrogen and phosphorus fertilization and mined lands.* In *Adequate Reclamation of Mined Lands.* Soil Conservation Society of America and WRCC-21, Billings, Montana. pp. 20-1-20-8.
8. Bodrova, E. M., and Z. D. Ozolina. 1968. *Simultaneous application of organic and mineral fertilizers.* U.S. Department of Agriculture, National Science Foundation, and Israel Program for Scientific Translations, Washington, D.C. 100 pp.
9. Brady, N. C. 1974. *The nature and properties of soils.* MacMillan Publishing Company, New York, New York.
10. Bremner, J. M. 1965. *Organic nitrogen in soils.* In W. V. Bartholomew and Francis E. Clark [editors] *Soil Nitrogen.* ASA Monograph 10. American Society of Agronomy, Madison, Wisconsin. pp. 93-149.
11. Broadbent, F. E. 1953. *The soil organic fraction.* Advances in Agron. 5: 153-183.
12. Brown, R. W., and R. S. Johnson. 1978. *Rehabilitation of a high elevation mine disturbance.* In S. T. Kenny [editor] *Proceedings: High Altitude Revegetation Workshop No. 3.* Colorado Water Resources Research Institute Information Series 28. Colorado State University, Fort Collins. pp. 116-130.
13. Buestel, I. C., and H. G. Barnes. 1936. *The comparitive moisture-absorbing and moisture-retaining capacities of peat and soil mixtures.* Technical Bulletin 532, G. U.S. Department of Agriculture, Washington, D.C.
14. Buol, S. W., F. D. Hole, and R. J. McCracken. 1980. *Soil genesis and classification.* Iowa State University Press, Ames.
15. Campbell, C. A. 1978. *Soil organic carbon, nitrogen, and fertility.* In M. Schnitzer and S. U. Khan [editors] *Soil Organic Matter.* Elsevier Scientific Publishing Company, Amsterdam, The Netherlands. pp. 173-271.
16. Camp Dresser & McKee, Inc. 1981. *Soil baseline report, Lost Day Quarry.* Wheat Ridge, Colorado.
17. Camp Dresser & McKee, Inc. 1981. *Soil baseline studies for the proposed permit area adjacent to the Sweetwater Uranium Mine.* Wheat Ridge, Colorado.
18. Camp Dresser & McKee, Inc. 1983. *Soil baseline report, Pacific Shale Project.* Wheat Ridge, Colorado.
19. Camp Dresser & McKee, Inc. 1983. *Soil baseline studies, Summitville Project, Colorado.* Wheat Ridge, Colorado.
20. Camp Dresser & McKee, Inc. 1983. *Soil survey of the CX Ranch area, Wyoming.* Wheat Ridge, Colorado.
21. Carter, R. P., R. R. Hinchman, and S. D. Zellmer. 1981. *Land reclamation program, 1980 annual report.* Argonne National Laboratory, Argonne, Illinois.
22. Christensen, M., and M. F. Allen. 1979. *Effect of VA mycorrhizae on water stress tolerance and hormone balance in native western plant species.* Final Report to the Rocky Mountain Institute of Energy and Environment. Institute of Energy and Environment, Laramie, Wyoming.
23. Christensen, M., and S. E. Williams. 1977. *Occurrence of vesicular-arbuscular mycorrhizae on shrubs and grasses in strip-mine regions of Wyoming.* Abstract in Proceedings of the Third North American Conference on Mycorrhiza, Southeast Forest Experiment Station, Forest Service, U.S. Department of Agriculture, Athens, Georgia.
24. Crawford, J. 1981. *The biology of desert and invertebrates.* Springer Verlag,

New York, New York.
25. Dam, J.G.C. Van, and F. A. Wopereis. 1977. *Suitability of sandy soils in southern Nederlands for the cultivation of ornamental conifers.* Bedrijfsontwikkeling 8(4): 365-374.
26. Davey, C. B., and H. H. Krause. 1980. *Functions and maintenance of OM in forestry nursery soils.* North American Forestry Tree Nursery Soils Workshop Proceedings. State University of New York, Syracuse. pp. 130-165.
27. Davies, B. E. 1974. *Loss-on-ignition as an estimate of soil organic matter.* Soil Science Society of America Proceedings 38: 150-151.
28. Donahue, R. L., R. W. Miller, and J. C. Shickluma. 1977. *An introduction to soils and plant growth. Chapter 5.* Prentice-Hall, Inc., Englewood Cliffs, New Jersey.
29. Fahey, T. 1983. *Nutrient dynamics of aboveground detritus in lodgepole pine ecosystems, Southeast Wyoming.* Ecological Monographs 53: a51-72.
30. Fimbel, R. A. 1982. *Revegetating carbonaceous spoils from an underground coal mine in southwestern Colorado.* M.S. thesis. Colorado State University, Fort Collins.
31. Fuller, W. H., D. F. McAllister, and D. S. Metcalfe. 1960. *Agronomy in the Southwest United States.* Advances in Agronomy 12: 197-227.
32. Gilley, J. E., G. W. Gee, A. Bauer, W. O. Willis, and R. A. Young. 1977. *Runoffs and erosion characteristics of surface-mined sites in western North Dakota.* Transactions, American Society of Agricultural Engineers 20(4): 697-704.
33. Halderson, J. L., and D. R. Zenz. 1978. *Use of municipal sewage sludge in reclamation of soils.* In F. W. Schaller and P. Sutton [editors] *Reclamation of Drastically Disturbed lands.* American Society of Agronomy, Madison, Wisconsin.
34. Hausenbuiller, R. L. 1972. *Soil science—principles and practices.* W. C. Brown Co., Dubuque, Iowa. 504 pp.
35. Jackson, M. L. 1958. *Soil chemistry analysis.* Prentice Hall Inc., Englewood Cliffs, New Jersey.
36. Jenny, H. 1980. *The soil resource, origin and behavior.* Ecological Studies 37. Springer-Verlag, New York, New York.
37. Jenkinson, D. S. 1971. *Studies on decomposition of C14-labeled organic C in soil.* Soil Science 111: 64-70.
38. Juncker, P. H., and J. J. Madison. 1967. *Soil moisture characteristics of sand-peat mixes.* Soil Science Society of America Proceedings 31: 5-7.
39. Kay, B. L. 1978. *Mulches for erosion control and plant establishment on disturbed sites.* In S. T. Kenney [editor] *Proceedings, High Altitude Revegetation Workshop No. 3.* Colorado Water Resources Research Institute Information Series No. 28. Colorado State University, Fort Collins. pp. 182-204.
40. Klein, D. A. L. E. Hersman, and F. B. Reeves. 1979. *Storage effects on the microbiological characteristics of surface soils used in oil shale revegetation programs.* Colorado State University, Fort Collins.
41. Kohnke, H. 1968. *Soil physics.* McGraw-Hill Inc., New York, New York.
42. Kononova, M. M., T. Z. Nowakowski, and A.C.D. Newman. 1966. *Soil organic matter, its nature, its role in soil formation and in soil fertility.* Pergamon Press, New York, New York. 574 pp.
43. Lee, Y. S., and R. J. Bartlett. 1976. *Stimulation of plant growth by humic substances.* Soil Science Society of America Journal 40: 876-879.
44. Loree, N. A., and S. E. Williams. 1984. *Vesicular arbuscular mycorrhizae and severe land disturbance.* In S. Williams and M. Allen [editors] *VA Mycorrhizae and Proceedings, Reclamation of Arid and Semi Arid Lands.* Wyoming Agricul-

tural Experiment Station, Cheyenne.

45. Lowry, G. L. 1960. *Conifer establishment on cool spoils as influenced by certain site factors and organic additions at planting time.* Soil Science Society of America Proceedings 24: 316-318.
46. Lucas, R. E., D. E. Ricke, and R. S. Farnham. 1965. *Peats for soil improvement and soil mixes.* Agricultural Extension Bulletin 516. Michigan State University, East Lansing.
47. Martin, J. P., and D. D. Focht. 1977. *Biological properties of soils.* In L. F. Elliot and F. J. Stevenson [editors] *Soils for Management of Organic Wastes and Waste Waters.* Soil Science Society of America, Madison, Wisconsin.
48. Mathers, A. C., and B. A. Stewart. 1971. *Crop production and soil analyses as affected by applications of cattle feedlot waste.* In *Livestock Waste Management and Pollution Abatement.* American Society for Agricultural Engineers, St. Joseph, Michigan. pp. 229-231.
49. Mathers, A. C., and B. A. Stewart. 1974. *Corn silage yield and soil chemical properties as affected by cattle feedlot manure.* Journal of Environmental Quality 3: 143-147.
50. Mebius, L. J. 1960. *A rapid method for the determination of organic carbon in soil.* Analytica Chemica Acta 22: 120-124.
51. Meek, B., L. Graham, and T. Donovan. 1982. *Long-term effects of manure on soil nitrogen, phosphorus, sodium, organic matter, and water infiltration rate.* Soil Science Society of America Journal 46: 1,014-1,019.
52. Miller, R. M. 1979. *Some occurrences of vesicular-arbuscular mycorrhiza in natural and disturbed ecosystems of the Red Desert.* Canadian Journal of Botany 57: 619-623.
53. Montana Department of State Lands. 1983. *Reclamation division, draft soil and overburden guidelines.* Helena.
54. Mosse, B., A. Warner, and C. A. Clarke. 1982. *Plant growth responses to vesicular-arbuscular mycorrhizae. XIII. Spread of an introduced VA endophyte in the field and residual growth effects of inocculation in the second year.* The New Phytologist 90: 521-528.
55. Munson, K. R. 1982. *Decomposition, function, and maintenance of organic matter in a sandy nursery soil.* Ph.D. dissertation. University of Florida, Gainesville. 95 pp.
56. Murphy, L. S., W. L. Wallingford, W. L. Powers, and H. L. Manges. 1972. *Effects of solid beef feedlot wastes on soil conditions and plant growth.* In *Waste Management Research.* Proceedings, Cornell Agricultural Waste Management Conference, Cornell University, Ithaca, New York. pp. 449-464.
57. Nelson, D. W., and L. E. Sommers. 1975. *A rapid and accurate procedure for estimation of organic carbon in soil.* Proceedings, Indiana Academy Science 84: 456-562.
58. Nelson, D. W., and L. E. Sommers. 1982. *Total carbon, organic carbon, and organic matter.* In Page et al. [editors] *Methods of Soil Analysis, Part 2.* American Society of Agronomy, Madison, Wisconsin. pp. 539-580.
59. North Dakota Agricultural Experiment Station. 1977. *North Dakota progress report on research on reclamation of strip mined lands—update 1977.* Northern Great Plains Research Center, Agricultural Research Service, Mandan, North Dakota.
60. Omodt, H. W., F. W. Schraer, and D. D. Paterson. 1975. *The properties of important agricultural soils as criteria for mined land reclamation.* Bulletin 492. North Dakota Agricultural Experiment Station, Fargo.
61. Pritchett, W. L. 1979. *Properties and management of forest soils.* John Wiley and Sons, Inc., New York, New York. 500 pp.

62. Reddell, D. L., W. H. Johnson, P. J. Lyerly, and P. Hobgood. 1971. *Disposal of manure by deep plowing.* In *Livestock Waste Management and Pollution Abatement.* American Society for Agricultural Engineers, St. Joseph, Michigan. pp. 235-238.
63. Redente, E. F., and N. E. Hargis. 1980. *Revegetating disturbed areas to protect the environment.* Quarterly Progress Report. Trapper Mine, Utah International Inc., Craig, Utah.
64. Reuss, J. O., and R. E. Campbell. 1961. *Restoration productivity to leveled land.* Soil Science Society of America Proceedings 25: 302-304.
65. Rives, C. S., M. K. Bajwa, A. E. Liberta, and R. M. Miller. 1980. *Effects of topsoil storage during surface mining on the viability of VA Mycorrhizae.* Soil Science 129: 253-257.
66. Romney, E. M., A. Wallace, J. W. Cha, and J. D. Childress. 1976. *Different levels of soil organic matter in desert soil and nitrogen fertilizer on yields and mineral composition of barley grown in the soil.* Communications in Soil Science and Plant Analysis 7(1): 51-55.
67. Ruehr, T. A. 1976. *Effect of beef cattle manure on soil properties and plant growth.* Ph.D. dissertation. Colorado State University, Fort Collins. 150 pp.
68. Schafer, W. M., G. A. Nielsen, and W. D. Nettleton. 1980. *Minesoil genesis and morphology in a spoil chronosequence in Montana.* Soil Science Society of America Journal 44: 802-807.
69. Schnitzer M. 1978. *Humic substances: Chemistry and reactions.* In M. Schnitzer and S. U. Kahn [editors] *Soil Organic Matter.* Elsevier Scientific Publishing Company, Amsterdam, The Netherlands.
70. Schnitzer, M. 1982. *Organic matter characterization.* In Page et al. [editors] *Methods of Soil Analysis, Part 2.* American Society of Agronomy, Madison, Wisconsin. pp. 581-594.
71. Schnitzer, M., and S. U. Khan, editors. 1978. *Soil organic matter.* Elsevier Scientific Publishing Company, Amsterdam, The Netherlands.
72. Schuman, G. E., and T. A. Sedbrook. 1984. *Sawmill wood residue for reclaiming bentonite spoils.* Forest Products Journal 34: 65-68.
73. Schuman, G. E., and T. M. McCalla. 1976. *Effect of short-chain fatty acids extracted from beef cattle manure on germination and seedling development.* Applied and Environmental Microbiology 31(5): 655-660.
74. Slick, B. M., and W. R. Curtis. 1985. *A guide for the use of organic materials.* Northeastern Forest Experiment Station, Forest Service, Upper Darby, Pennsylvania.
75. Smith, R. M., R. C. Henderson, E. D. Cook, J. E. Adams, and D. O. Thompson. 1967. *Renewal of desurfaced Austin clay.* Soil Science 103(2): 126-130.
76. Soil Conservation Service, U.S. Department of Agriculture. 1974. *National soils handbook, Part II, Section 403.3(d).* Washington, D.C.
77. Soil Survey Staff, Soil Conservation Service, U.S. Department of Agriculture. 1975. *Soil taxonomy.* Agriculture Handbook No. 436. Washington, D.C.
78. Stahl, P. D., M. F. Allen, E. B. Allen, and M. Christensen. 1979. *Occurrence of vesicular-arbuscular mycorrhizal propagules in disturbed and undisturbed sites in Wyoming.* Abstract in Proceedings, Fourth North American Conference on Mycorrhiza. Colorado State University, Fort Collins.
79. Stevenson, F. J. 1965. *Origin and distribution of nitrogen in soil.* In W. V. Bartholomew and Francis E. Clark [editors] *Soil Nitrogen.* No. 10 in Series Agronomy. American Society of Agronomy, Madison, Wisconsin. pp. 1-42.
80. Stevenson, F. J. 1982. *Humus chemistry: Genesis, composition, reactions.* John Wiley and Sons, Inc., New York, New York.
81. Storer T. I., R. L. Usinger, and J. W. Nybakken. 1968. *Elements of zoology.*

McGraw-Hill, Inc., New York, New York.
82. Tisdale, S. L., and W. L. Nelson. 1975. *Soil fertility and fertilizers.* MacMillan Publishing Company, New York, New York. 694 pp.
83. Topper, K. F., and B. R. Sabey. 1986. *Sewage sludge as a coal mine spoil amendment for revegetation in Colorado.* Journal of Environmental Quality 15: 44-49.
84. U.S. Forest Service. 1979. *Users guide to soils, mining and reclamation in the West.* General Technical Report INT-68. Intermountain Forest and Range Experiment Station, Ogden, Utah.
85. Utah Department of Natural Resources. 1984. *Guidelines for preparation of a topsoil management plan.* Salt Lake City.
86. Watts, D. 1971. *Principles of biogeography.* McGraw-Hill Book Company, New York, New York. 402 pp.
87. Wali, M. K., and P. G. Freeman. 1973. *Ecology of some mined areas in North Dakota.* In M. K. Walli [editor] *Some Environmental Aspects of Strip Mining in North Dakota.* Education Series 5. North Dakota Geological Survey, Grand Forks.
88. Wischmeier, W. H., D. D. Smith, and R. E. Uhland. 1958. *Evaluation of factors in the soil loss equation.* Agricultural Engineers 39(8): 458-462, 474.
89. Wyoming Department of Environmental Quality. 1984. *Land quality division, guideline 1.* Cheyenne.

Chapter 10

Copper and molybdenum

Dennis R. Neuman, James L. Schrack, and Larry P. Gough

Copper (Cu) and molybdenum (Mo) are among the elements essential for proper plant and animal metabolism. An imbalance of these two elements in the overburden-plant cover-grazer ecosystem is of concern in western mined-land reclamation. The interaction of Cu and Mo, as well as sulfide and sulfate, in ruminants can result in the metabolic unavailability of Cu. This disease is called molybdenosis. Symptoms of the disease include severe diarrhea, weight loss, color change, and roughening of the animal's coat (*4*). The condition occurs in locations where forage is naturally high in Mo and S and low in Cu. Outbreaks of the disease have been reported in California, Colorado, Montana, Nevada, Oregon, and South Dakota.

Mining can induce molybdenosis in areas previously free of the disease. Evidence of this has been found at a lignite ashing plant (*8*), a uranium mine (*14*), a clay pit area (*17*), and a uranium-bearing lignite area (*74*).

The metabolic role of copper and molybdenum

Molybdenum, discovered about 1782, was not recognized as an essential element until H. Bortels showed earlier in this century that Mo was required for nitrogen fixation by the bacterium *Azotobacter chroococcum* (*7*). Thereafter, its essentiality was established for higher plants and animals. In plants Mo is necessary for the activity of two enzymes, nitrate reductase, which reduces nitrate to nitrite, and nitrogenase, which fixes N (*62*). In animals a Mo-containing metallo-protein, xanthine oxidase, is essential for the degradation of purines to uric acid (*12*). Other Mo-dependent enzymes in animal tissues include aldehyde oxidase and sulfite oxidase (*62*).

Copper is essential to the growth and development of almost all vegetation. Specific Cu proteins that have been isolated from plant tissues include cytochrome oxidases, blue oxidases, laccase, plastocyanin, and tyrosinase (*56*). Copper serves many biological roles in animals. The element is essential to animals, but it can also be toxic. At the cellular level, Cu proteins function as oxidative enzymes. As many as 20 mammalian Cu proteins have been isolated (*56*).

The metabolic interaction of Cu and Mo in animals was highlighted when it was found that Mo toxicosis in cattle could be controlled by additions of Cu to feeds (*20*). Later, it was shown that Cu toxicosis in sheep could be controlled by Mo supplementation (*13*). The concentration of Cu relative to Mo, Zn, Fe, and S is important in defining Cu deficiency (*51*). The ratio of Cu to Mo and the interactions of the other elements are as important as the actual Cu level in the diet (*56*).

Determining copper and molybdenum concentrations

Total levels in soils and overburden. In soils and overburden Cu and Mo are effectively dissolved using a multi-acid, low-temperature, low-pressure procedure (*9, 76*). The soil or overburden sample is digested in a mixture of concentrated HF, HNO_3, and $HClO_4$ in a platinum dish or teflon beaker on a hot plate. After evaporation the residue is dissolved in HCl and brought to a given volume. Similar digestion procedures, using $HClO_4$ (*66*), Na_2CO_3 fusion (*65*), and HF-$HClO_4$ (*49*), are available.

Atomic absorption spectroscopy or inductively coupled plasma emission spectroscopy is used to determine the Cu concentration in the digestant. Use of flame atomic absorption spectroscopy to measure total Mo at levels normally encountered in soils and overburden will not provide an accurate determination because of its relatively poor detection limit. Extraction and concentration of Mo from the digestant solution, followed by flame atomic absorption spectroscopy (*40*) or the use of graphite furnace atomic absorption spectroscopy or inductively coupled plasma, are necessary to achieve acceptable analytical detection limits and reproducibility.

Total levels in plant tissue. The two most common methods of extracting Cu and Mo from plant tissue are dry ashing and wet digestion. Total Cu content can be determined directly using flame atomic absorption spectroscopy after NHO_3-$HClO_4$ digestion of plant material (*53*). This method has a relative error of 2.5 percent at a level of 12 ± 1 ppm Cu. Recovery of added Cu to plant material was 97.7 percent. Graphite furnace atomic absorption spectroscopy techniques have been successful in determining low levels of Mo in plant tissue (*37, 39, 60, 85*). Preparation of Mo standards in an appropriate matrix is important (*60, 85*). In one test the recovery of Mo added to plant digestants ranged from 95 to 100 percent (*85*) and averaged 94.3

percent in a second study (*60*). The accuracy of this latter method was near 10 percent (relative standard deviation) at a tissue level of 0.50 ppm Mo.

Extraction procedures in soils and overburden. Plant uptake of Mo can be related to extractable soil Mo using ammonium oxalate-oxalic acid buffered to a pH of 3.3 (*28*). After extraction, the Mo concentration is measured by the thiocynate method (*66*) or by graphite furnace atomic absorption spectroscopy (*59*). This extraction method is the only method that has been extensively field calibrated with plant response. One researcher summarized responses of several crops to extractable Mo (*67*). But other researchers found little correlation between oxalate-extractable Mo and plant uptake in various soils (*41, 49*). The ability of plants to extract Mo from soils depends upon physical, chemical, and environmental factors. Thus, a variety of extractants have been used in attempts to measure available Mo. These include water (*30, 45*), hot-water-soluble (*63*), anion-exchange resin (*5*), and ammonium acetate and EDTA (*84*). One test involving a 12-hour extraction with ammonium carbonate at pH 9 produced good correlations between extractable Mo levels and Mo uptake by alfalfa (*79*).

Researchers also have used different extractants to measure the availability of soil Cu to vegetation. More extraction and calibration work has been done with HC1 than with any other acid (*29*). In the past 10 years chelating agents have become more important in determining Cu availability in soils. The advantage of chelators over acids and neutral salts is that the pH of the extractant can be maintained near the natural pH of the soil. The DTPA test was designed to extract Zn, Fe, Mn, and Cu and to detect soils deficient in these elements (*21*). The DTPA test is used extensively in the western United States for both agronomic soils and in overburden analysis. The ammonium bicarbonate-DTPA (AB-DTPA) soil test was developed for extracting metals from alkaline soils having a pH range of 7.0 to 9.2 (*72*). This test was highly correlated with the DTPA test for Cu but extracted approximately 0.3 ppm more Cu than did the DTPA extractant.

Some controversy exists over the DTPA test for determining metal concentrations in soils relative to those in vegetation. In one test, for example, the concentration of Cu in corn did not relate closely to the amount extracted by DTPA in several soils in Ontario (*33*). Elsewhere, Cu concentrations in legumes grown on mine soils and DTPA extractable Cu concentrations were unrelated (*61*). Other researchers were unable to correlate Cu levels in native vegetation to extractable soil levels using simple linear or multiple linear regression techniques (*22*).

Occurrence of copper and molybdenum

Native soils and mine soils. The forms of Cu and Mo in western soils and their availability for plant uptake and translocation have been studied extensively (*2, 10, 47, 68*). Depending upon local geology, hydrology, vegetative cover, and climate, Cu and Mo can occur in soils in various proportions of the following forms: water-soluble, adsorbed on soil colloids, held in the crystal lattice of minerals, and present in organic matter. Generally, the metals in soluble forms or adsorbed onto clays are most available for uptake by plants.

Mine soil conditions are probably dramatically different than those of native soils, and the relative importance of each of these sources of metals will not be the same. Regulations that seek to compare metal concentrations in native soils and mine soils are inappropriate because mine soils represent a youthful stage of soil development; rapid changes in their chemical equilibria can be expected. The sedimentary materials that were previously confined to a presumed reducing and biologically inactive zone are exposed by mining to an oxidizing environment that is biologically active.

Availability of Cu and Mo. The average total Cu concentration in agriculturally productive soils is near 20 ppm, with a range of 1 to 50 ppm. Higher Cu levels may occur in mineralized parent rock. Several factors influence the relationship between Cu in parent material and the derived spoil: pH, redox potential, weathering, drainage, and organic matter (*3*). Cu mobility is greatly influenced by pH. Acid conditions promote solubility of Cu. Alkaline conditions favor precipitation. The clay content and subsequent cation exchange capacity of a soil also play a significant role in the movement and retention of Cu. Uptake of Cu by plants is favored in soils with impeded drainage, while Cu uptake is inhibited in soils with high or-

Table 1. Total copper and molybdenum concentrations in spoil from selected western coal mines (*16, 24*).

Mine (number of samples)	Concentrations (ppm)			
	Cu		Mo	
	Geometric Mean	Observed Range	Geometric Mean	Observed Range
Montana				
Big Sky (10)	19	12-34	4.6	2.5-6.5
Savage (10)	20	7.6-29	3.2	<.8-9.9
New Mexico				
San Juan (6)	18	6.0-28	2.7	2.1-3.2
North Dakota				
Beulah North (10)	24	5-41	5.5	2.1-9.3
Kincaid (10)	24	8.8-47	4.9	1.9-11
Velva (10)	19	12-28	4.4	2.5-6.1
Wyoming				
Dave Johnston (10)	16	12-29	2.0	<.56-6.2
Hidden Valley (10)	18	9.6-26	3.8	1.1-7.8
Saskatchewan				
Utility (10)	17	11-23	4.1	2.2-7.6

Table 2. DTPA-extractable copper concentrations in cover soil and spoil from selected western coal mines (*69, 70*).

	DTPA-Extractable Cu Concentrations (ppm)			
	Cover Soil		Spoil	
Mine (number of samples)	Geometric Mean	Observed Range	Geometric Mean	Observed Range
Colorado				
Energy Fuels (10)	0.9	0.7-1.1	1.7	1.2-2.3
Seneca II (10)	1.4	.8-2.0	1.2	.2-2.1
New Mexico				
San Juan (6)	.7	.5-2.1	2.0	1.5-2.3
North Dakota				
Husky (10)	1.8	.8-2.8	1.2	.3-2.2
Beulah South (10)	1.2	.8-2.0	4.4	2.8-6.7
Velva (10)	1.5	1.1-2.1	2.0	1.5-2.4
Montana				
Absaloka (10)	.5	.3-0.9	.6	.3-1.0
Big Sky (10)	.4	.3-0.5	2.1	.7-5.8
Decker (10)	.4	.2-1.0	.5	.2-1.0
Wyoming				
Dave Johnston (5)	1.2	.7-2.3	1.6	.9-3.7
Jim Bridger (5)	.8	.6-1.4	1.2	.9-1.7
Seminoe II (5)	1.6	1.5-1.8	1.3	1.1-1.8

ganic matter. Total Cu content in a soil is not an accurate indicator of a deficiency or excess because of the numerous factors controlling the metal's availability.

Parent materials in the West that contain Mo in sufficient quantities to cause high levels in vegetation include granites, granitic conglomerates, and shales. Wet bottomlands, where molybdeniferous rocks are major contributors to the development of wet soils, are prime suspect areas for elevated Mo levels in plants, particularly legumes, growing on these sites (*42*).

Tables 1, 2, and 3 present examples of the total and extractable concentrations of Cu and Mo generally expected in mine soil materials in the western coal region. Extractable Cu concentrations were measured using the DTPA method (*48*); Mo extractable concentrations were measured using ammonium oxalate (*61*).

Copper and Mo in reclamation plant species. Trends are beginning to emerge that will help predict the behavior of Cu and Mo in plant tissues under various reclamation conditions. For example, although considerable variation occurs in the ability of different plant species to absorb Cu and Mo from soils, higher concentrations of both elements characteristically occur in roots than in above-ground plant parts (*68*). This may help explain why, in general, there is usually a poor correlation between Cu and Mo levels in mine soils, either total or extractable, and levels in reclamation

Table 3. Arithmetic means of total and extractable copper and molybdenum concentrations in mine soil, Big Sky Mine, Montana (*61*).

	Mean Concentration (ppm)			
	Cu		Mo	
Mine Soil	Total	Extractable*	Total	Extractable†
Spoil	21	1.4	2.3	0.46
Spoil overlain with 0.6m topsoil	15	1.0	1.5	.23
Spoil overlain with 0.6 m topsoil and 2.4 m of sandy overburden	10	.46	1.6	.31

*DTPA extractant.
†Ammonium oxalate extractant.

Table 4. Total copper and molybdenum concentrations in reclamation species from selected western coal mines (*25*).

	Concentration (ppm dry matter)			
	Cu		Mo	
Plant Species, Mine, State (number of samples)	Geometric Mean	Observed Range	Geometric Mean	Observed Range
Slender wheatgrass (*Agropyron trachycaulum*)				
Absaloka, Montana (10)	0.78	.58-1.0	0.20	.10-.61
Big Sky, Montana (10)	1.1	.70-2.0	.60	.19-1.5
Decker, Montana (10)	.87	.40-1.2	.33	.12-2.2
Crested wheatgrass (*Agropyron cristatum*)				
Dave Johnston, Wyoming (10)	1.3	.88-2.2	.22	.13-.88
Seminoe II, Wyoming (10)	3.4	1.5-3.8	.23	.15-.40
Intermediate wheatgrass (*Agropyron intermedium*)				
Energy Fuels, Colorado (10)	1.0	.76-1.3	.25	.12-1.7
Seneca II, Colorado (10)	1.1	.69-3.5	.36	.19-1.2
Husky, North Dakota (10)	1.1	.51-1.7	.31	.14-.58
Beulah South, North Dakota (10)	1.0	.69-1.5	.27	.18-.66
Velva, North Dakota (10)	1.5	1.2-2.2	.36	.19-2.4
Alkali sacaton (*Sporobolus airoides*)				
San Juan, New Mexico (6)	2.4	2.0-2.8	.70	.48-.90
Alfalfa (*Medicago sativa*)				
Energy Fuels, Colorado (10)	5.6	3.9-7.2	2.0	.36-3.1
Seneca II, Colorado (10)	5.3	1.1-8.0	1.2	.50-1.9
Big Sky, Montana (10)	7.6	4.7-10	4.3	2.3-12
Husky, North Dakota (10)	5.3	4.2-7.2	1.5	.88-5.5
Beulah South, North Dakota (10)	6.3	4.9-9.0	1.6	.43-5.8
Velva, North Dakota (10)	8.7	6.1-12	4.2	1.9-9.0
Fourwing saltbush (*Atriplex canescens*)				
Decker, Montana (10)	4.5	3.0-6.6	.46	.29-1.0
Jim Bridger, Wyoming (10)	4.9	4.4-7.0	.72	.23-5.1
San Juan, New Mexico (6)	9.7	8.4-12	.73	.53-1.3
Smooth brome (*Bromus inermis*)				
Dave Johnston, Wyoming (10)	1.1	.53-2.4	.24	.13-.72
Sandfoin (*Onobrychis viciaefolia*)				
Absaloka, Montana (10)	2.0	1.5-2.0	1.6	.36-5.5

plant species. Recent studies also have demonstrated that, except for soils that are unusually high in Mo, there is little difference between the Cu and Mo levels in identical species growing on native soils or on adjacent mine soils (*18, 23, 26, 54*). Furthermore, the variability in Cu and Mo levels in plants growing within several meters of each other in either native soils or mine soils is likely to be as great or greater than the variability in concentrations between plants growing on native versus mine sites.

Table 4 lists examples of Cu and Mo concentrations in a variety of reclamation species collected at 12 different surface coal mines in five western states. These data represent collections by one individual over a relatively short period of time. The same preparation and analytical procedures were used throughout the study. The data, therefore, are directly comparable and useful in evaluating differences in the chemical composition of reclamation species among states and mines.

Relations between soils and plants. The ability of plants to extract Mo from soils depends upon the plant species, the nutrient status of the plants, parent material, pH, organic matter content, and soil moisture.

Soil pH is one of the most important, single factors affecting Mo uptake. The solubility of most Mo compounds in soils is enhanced as pH increases (*31*). Molybdenum occurs as the molybdate ion, MoO_4^{2-} (*46*), and Mo uptake by plants is inhibited by SO_4^{2-}. Other factors affecting Mo mobilty are Fe and Al oxides, which tend to remove Mo from the aqueous solution, and soluble soil P, which enhances uptake of Mo by plants (*11, 43*).

Table 5 lists extraction solutions stipulated by regulating agencies in six western states as of 1983 for use in determining Cu and Mo in soils and overburden. Because of the different analytical methods, interstate comparison of Mo data in soils and overburden is difficult.

Molybdenum. In 1977 the Montana Department of State Lands set a suspect level of 0.3 ppm extractable Mo (*34*). The guideline apparently was based on studies conducted in Ireland (*80, 81*). Using the acid-ammonium oxalate procedure for available Mo, two researchers conducted pot experiments using different soils and *Trifolium subterraneum,* a clover that is a known accumulator of Mo (*80*). Levels of Mo in the soils ranged from 0.42 to 3.50 ppm. Responses to Mo fertilization were obtained in soils containing less than 0.12 ppm extractable Mo. The pH of these soils ranged from 4.9 to 5.5. Soils that produced herbage containing excessive quantities of Mo varied in extractable Mo from 0.30 to 0.57 ppm and had pH values from 6.5 to 7.8. This early work demonstrated the dependence of solubility and uptake of Mo on pH by one species. If an accumulator species had not been used, the conclusions regarding soil deficiency and toxic levels may have been different. The proposed Montana guideline (*35*) is 0.5 to 1.0 ppm (Table 5). The actual maximum acceptable Mo level depends upon the plant species proposed in the revegetation plan and their potential for Mo accumulation.

Wyoming's suspect level for Mo in soils and overburden is 1.0 ppm, using ammonium carbonate or AB-DTPA as extractants. This level apparently was established on the basis of studies of Mo uptake by plants growing on Colorado pasture soils (*78, 79*). Alfalfa plants grown in alkaline soils on Colorado's Front Range were evaluated in terms of total Mo in the plants and the soil Mo extractable by ammonium carbonate. In a related study alfalfa was grown in the greenhouse on the same soils amended with Mo. The rate of Mo accumulation by the alfalfa in the field was significantly lower than the rate of accumulation in the greenhouse. This difference was attributed to differences in water regimes. Greenhouse soils were watered daily, and uptake of Mo increased with greater water availability. The range of Mo concentration in alfalfa sampled in the field was limited to nontoxic levels for cattle. Extrapolation of these field data resulted in a critical level for ammonium-carbonate-extractable soil Mo of 1.7 ppm (*79*). This level correlates with a Mo concentration in the alfalfa of 10 ppm. The ability to extrapolate beyond actual data was based on the researchers' greenhouse study, in which extractable Mo in the amended soils with a pH greater than 7 was linearly correlated ($r = 0.977$) with uptake by alfalfa. The Wyoming suspect level of 1.0 ppm extractable Mo (equivalent of 6.7 ppm in alfalfa) is assumed to be a more conservative value based on dietary levels for cattle.

Considerable evidence indicates that Mo uptake is species specific, with legumes accumulating more Mo than grasses growing in the same field or in solution culture (*1, 4, 44, 79*). In a study of Cu and Mo uptake by legumes growing on topsoiled coal mine soils in Montana, six legume species were

Table 5. Analysis of copper and molybdenum in soil and overburden required by some western states.

State	Copper		Molybdenum	
	Extractant Solution	Suspect Level	Extractant Solution	Suspect Level
Colorado	DTPA* or AB-DTPA†	unspecified	DTPA* or AB-DTPA†	unspecified
Montana	analysis not required		Acid ammonium oxalate§	0.5-1.0 ppm‡
Wyoming	analysis not required		ammonium carbonate‖ or AB-DTPA†	1 ppm
North Dakota	analysis not required		analysis not required	
New Mexico	unspecified	unspecified	unspecified	unspecified
Utah	analysis not required		analysis not required	

*(*21*)
†(*72*)
‡Maximum value depends upon plant species and their potential for Mo accumulation.
§(*66*)
‖(*78*)

seeded in triplicate on three areas prepared for revegetation using different procedures (*61*). The species included yellow and white sweetclover (*Melilotus officinalis* and *M. alba*), white and purple prairieclover (*Petalostemon candidum* and *P. pupureum*), alfalfa (*Medicago sativa*), and cicer milkvetch (*Astragalus cicer*). Soils were characterized for several physical and chemical parameters, including pH and acid ammonium-oxalate-extractable Mo, at five depth increments: surface, 20 cm, 40 cm, 80 cm, and 120 cm. Site 1 included mixed interburden and overburden with no topsoil; it contained a mean extractable Mo level of 0.46 ppm. Site 2 included mixed interburden and overburden with topsoil; mean extractable Mo level was 0.23 ppm. Site 3 included segregated interburden with topsoil; mean extractable Mo level was 0.31 ppm. The native soils, site 4, had a mean extractable Mo level of 0.19 ppm. The pH of all materials varied from 7.0 to 7.9; there was little intersite variation. Vegetation was sampled and analyzed for total Cu, Mo, and S.

Molybdenum accumulation in this study was species specific. White sweetclover and white prairieclover accumulated the highest concentrations. Mo uptake was also site specific. Plants grown on soils with high Mo concentrations accumulated higher Mo concentrations. Figure 1 shows the mean concentrations in the vegetation and mean extractable soil levels. Extrapolation of data beyond 0.46 ppm Mo in soil is statistically invalid. However, using linear regression, the extractable level of Mo in these soils would need to be 1.9 ppm to correlate with 10 ppm Mo in alfalfa. This value is similar to the critical soil level of 1.7 ppm for alfalfa in Colorado soils. For the other species in this study to contain 10 ppm, the associated extractable levels would be 0.7 ppm for white sweetclover, 1.3 for yellow sweetclover, 0.9 for white prairieclover, 1.9 for cicer milkvetch, and 1.2 ppm for purple prairieclover.

Because Mo uptake by vegetation on mine soils appears to be species specific, the setting of a limiting numerical value, or suspect level, for extractable Mo necessarily depends upon which species are used in the revegetation seed mix. Regulatory policies that require Cu and Mo determinations in soils and overburden do not prevent imbalances of these metals from occurring in vegetation after seeding.

Copper. Plant species vary considerably in their Cu uptake from soils. Copper concentrations in most plants, however, seldom exceed 30 ppm, even at high rates of Cu fertilization (*50*). Plant-metabolic mechanisms appear to limit the amount of Cu in plant tissues to about 20 ppm (*56*), although higher levels do occur under certain conditions and in some Cu-tolerant species.

Some states require that Cu levels be determined in soils and overburden using DTPA extraction. In one test using DTPA the researchers estimated the levels of Cu that define deficient and nondeficient soils (*21*). A level of 0.2 ppm Cu indicated a need for treatment. The researchers reported a range of DTPA-extractable Cu in 37 Colorado soils of 0.2 to 3.2 ppm. Similar Cu levels in soils have been reported in the northern Great Plains (*55, 71*). A study of 313 soils in Montana showed DTPA-extractable levels ranging from 0.3 to 3.0 ppm, with one sample at 10 ppm (*32*). Levels of extractable Cu in mine spoils within the region (Tables 2 and 3) are similar to those from undisturbed land.

Toxicity of copper and molybdenum

Vegetation. Under natural conditions, Cu mineralization is toxic to native and cultivated plants. The degree of resistance to Cu toxicity by different species has been used as a geobotanical means of mineral exploration. Areas in Norway where vegetation had been poisoned (as much as 895 ppm Cu in dry plant material) by naturally occurring high Cu levels in soils (as much as 7,400 ppm) could be detected by examining satellite photographs (*6*). Deficiencies of Cu are much more common than Cu excesses, however. Fertilization with Cu, especially on highly organic or very sandy soils, is a common practice. Total and extractable Cu levels in mine soils from selected western coal mines (Tables 1, 2, and 3) are far below phytotoxic levels—probably by two orders of magnitude. Therefore, Cu toxicity is probably of little concern and should only be suspected in mine soils that are very acid.

No occurrence of Mo toxicity to plants in the field has been recorded. Such toxicity has been noted only under extreme experimental conditions. In solution culture, for example, Mo concentrations of 1,000 to 2,000 ppm were required to develop chlorotic leaves in tomatoes. Concentrations of several hundred ppm (dry material) Mo in plants have been reported without noticeable toxic symptoms (*31*). Values of less than 1 ppm are common in nonleguminous reclamation species, and legumes (alfalfa and sandfoin) can have Mo concentrations that are 10 times greater (Table 4).

Relations between plants and animals. Although elevated levels of Mo in vegetation are not injurious to the plants themselves, animals are sensitive to high concentrations in their diet. The effects of high Mo intake on grazing cattle depends upon the age of the animal; the amount and chemical form of the ingested Mo; the Cu status and intake; the inorganic SO_4^{2-} and S content of the diet, including levels of S-containing proteins; and other ingested metals (*27, 31*).

Molybdenosis is a Mo-induced, Cu-deficiency disease caused by high concentrations of Mo in forage. Molybdenum competitively reduces the amount of biologically available Cu (*38*). Monogastric animals are far less susceptible to Mo toxicity than are ruminants. Horses are the most tolerant, followed in order by pigs, rats, rabbits, guinea pigs, sheep, and cattle. Deer are far more tolerant than sheep or cattle (*82*), and younger animals are more severely affected than older animals (*4*). Molybdenum toxicity in cattle, therefore, is of greatest concern and has received the most study. Regulations sufficient to protect cattle foraging on western mined land will be adequate for all other grazers.

When Cu levels in feeds or forages are within normal limits—8 to 11 ppm—cattle generally are immune to Mo poisoning, even when Mo concentrations are as high as 5 to 6 ppm (*56*). At dietary levels of 1 to 2 ppm, Mo may be toxic to cattle if Cu levels are below 8 ppm. The critical Cu:Mo ratio in one test of animal feeds used in western Canada was 2:1 (*52*). In a study

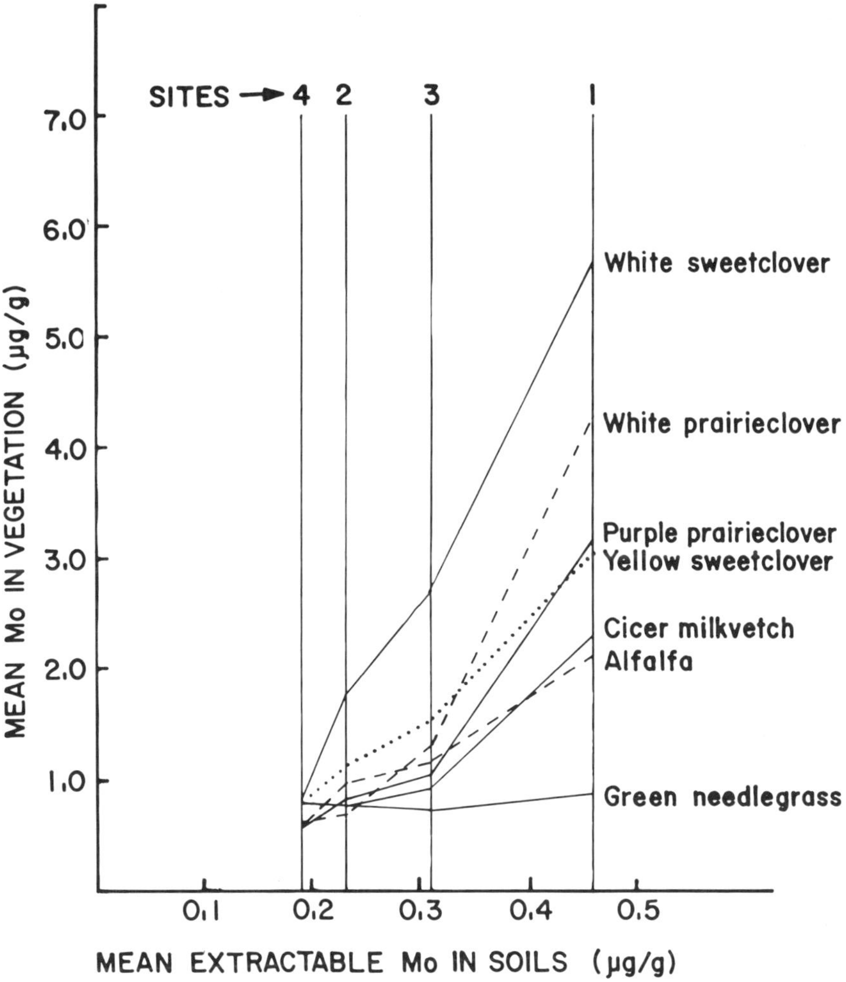

Figure 1. Mean molybdenum concentrations in legumes grown on mine soils in Montana (*61*).

of beef cattle, corn silage containing 7 ppm Cu with a Cu:Mo ratio of 6.6:1 did not warrant supplementation, and the ratio was considered favorable (*64*). When the ratio of Cu to Mo in feed is below 2:1, Cu deficiency can be expected in cattle unless the dietary Cu level exceeds 13 to 16 ppm (*56*).

Exactly how Mo and Cu interact is not known, but the formation of a Mo-Cu complex (lingren) may reduce Cu availability (*15*). Many unanswered questions remain as to the antagonism of these elements in rumen physiology (*38*). Fortunately, however, the ratio of Cu concentration in dry forage to Mo concentration is a fairly reliable and convenient method of assessing molybdenosis potential (*52*).

Researchers, then, generally agree that the critical Cu:Mo ratio is about 2:1 in cattle forage and about 5:1 in sheep forage, (*19*). Although far less reliable, concentrations of Cu and Mo in dry forage can be evaluated individually as indications of potential problems. Forage with levels of 20 to 30 ppm Cu or 5 to 10 ppm Mo should be considered physiologically suspect.

The toxicity of any particular level of dietary Mo is affected not only by the ratio of Cu to Mo but also by the S status of that diet. Mo and S can either increase or decrease the Cu status in an animal, depending upon their ingestion relative to that of Cu (*77*). Most investigators feel that SO_4^{2-} makes possible the metal interaction. But one researcher clearly demonstrated that total S rather than inorganic S is a more useful measurement of the Cu-Mo-S interrelationship (*75*). Normal limits of S concentration associated with disturbed metal metabolism in ruminants have not been well defined, but several investigators have indicated that levels above 0.35 percent are generally found in relation to molybdenosis (*36, 73, 83*). Suggested dietary intake of S is 0.1 percent for beef cattle (*57*) and 0.2 percent for dairy cattle (*58*). Excessive S ingestion may adversely affect the animal by interfering with the metabolism of other elements.

Recommendations for reclamationists

Several conclusions from Mo and Cu research lead to recommendations applicable to mined-land reclamation:

► Undisturbed areas in the West where Mo and Cu imbalances may result in forage levels harmful to ruminants can be expected in poorly drained, alkaline soils derived from granite, granitic conglomerates, and alkaline shale parent materials.

► Regulations comparing Mo and Cu levels in native soils to mine soils are inappropriate because rapid changes in chemical equilibria are expected when material that was confined to a reducing and biologically inactive zone is exposed by mining, which creates an oxidizing environment that is biologically active.

► Because many factors affect the availability of Mo in soils and the uptake of Mo by plants, soil tests often cannot predict Mo deficiencies and toxicities (*79*). Analysis of Mo concentrations in vegetation is one of the better indicators of Mo availability in soils (*31*); this relationship should be considered as a more reliable basis for reclamation regulations. Molybdenum uptake is species specific. Any regulations developed to limit extractable soil Mo levels must be based on the species used in the revegetation seed mixture.

► Management options prior to and after mining are available to control or mitigate the potential for molybdenosis developing in ruminants grazing revegetated mine soils. If areas to be mined include known or suspected sites of Mo toxicity, the dominant native vegetation can be assessed in terms of Cu, Mo, and S content. Special attention should be given to leguminous species growing in alkaline soils with high water tables. These analyses will determine if a potential post-mine problem may be expected with regard to the original topsoil.

Soil testing of the prepared seedbed for extractable Mo after mining is appropriate. The values should, however, be carefully assessed in terms of the vegetation to be seeded. Certain legumes should be avoided in the seed mixture because of their ability to absorb proportionally greater amounts of Mo. Evaluation of Cu and Mo concentrations and Cu:Mo ratios in the dominant vegetation should be performed sometime after revegetation and before grazing. The presence of Mo-accumulating plant species should be weighed in terms of their production on the revegetated area. A grazing animal's total dietary intake of Cu and Mo can be estimated from species-specific productivity data and species-specific Cu and Mo concentrations.

If unacceptably high Mo concentrations or unacceptably low Cu:Mo ratios are estimated for grazing cattle, a number of management options still remain available. These include feeding only dried and cured vegetation from the area, adding $CuSO_4$ to salt blocks or feeds, or subcutaneous injecton of Cu glycinate. These decisions should be guided by dietary levels of total Cu, Mo, and S and Cu:Mo ratios.

REFERENCES

1. Allaway, W. H. 1977. *Perspectives on molybdenum in soils and plants.* In W. R. Chappell and K. K. Peterson [editors] *Molybdenum in the Environment, Volume 2.* Marcel Dekker, Inc., New York, New York. pp. 317-339.
2. Aubert, H., and M. Pinta. 1977. *Trace elements in soils.* Elsevier Science Publications Co., Amsterdam, The Netherlands. 395 pp.
3. Baker, D. E. 1974. *Copper: Soil, water, plant relationships.* Federation Proceedings 33: 1,188-1,193.
4. Barshad, I. 1948. *Molybdenum content of pasture plants in relation to toxicity to cattle.* Soil Science 66: 187-195.
5. Bhella, H. S., and M. D. Dawson. 1972. *The use of anion exchange resin for determining available soil molybdenum.* Soil Science Society of America Proceedings 36: 177-179.

6. Bolviken, B., F. Honey, S. R. Levine, R.J.P. Lyon, and A. Prelat. 1977. *Detection of naturally heavy-metal-poisoned areas by Landsat-1 digital data.* Journal of Geochemical Experiment 8: 457-471.
7. Bortels, H. 1930. *Molydan als katalysator bei der biologischen stockstoffbundung.* Archiv für Mikrobiologie 1: 333-342.
8. Christianson, G. A., and G. A. Jacobson. 1971. *Report on molybdenosis in farm animals and its relation to uraniferous lignite ashing plant.* North Dakota State Department of Health, Bismarck. 9 pp.
9. Crock, J. G., and R. C. Severson. 1980. *Four reference soil and rock samples for measuring element availability in the western energy regions.* Circular 841. U.S. Geological Survey, Washington, D.C. 16 pp.
10. Davies, B. E. 1980. *Applied soil trace elements.* John Wiley & Sons, New York, New York. 842 pp.
11. Davies, E. B. 1956. *Factors affecting molybdenum availability in soils.* Soil Science 81: 209-221.
12. DeReuzo, E. C., E. Kaleita, P. G. Heytler, J. J. Oleson, B. L. Hutchings, and J. H. Williams. 1953. *Identification of xanthine oxidase factor as molybdenum.* Archives of Biochemistry Biophysics 45: 247-251.
13. Dick, A. T., and L. B. Bull. 1945. *Some preliminary observations on the effect of molybdenum on copper metabolism in herbivorous animals.* Australian Veterinarian Journal 21: 70.
14. Dollahite, J. W., L. D Rowe, L. M. Cook, D. Hightower, E. M. Bailey, and J. R. Kyzar 1972. *Copper deficiency and molybdenosis intoxication associated with grazing near a uranium mine.* Southwest Veterinarian 9: 47-50.
15. Dowdy, R. P., and Gennard Matrone. 1968. *A copper-molybdenum complex: Its effect and movement in the piglet and sheep.* Journal of Nutrition 95: 197-201.
16. Ebens, R. J., and J. A. Erdman. 1982. *Sweetclover and associated spoil materials from selected coal mines in the northern Great Plains.* In R. J. Ebens and H. T. Shacklette [editors] *Geochemistry of Some Rocks, Mine Spoils, Stream Sediments, Soils, Plants, and Waters in the Western Energy Region of the Conterminous United States.* Professional Paper 1237. U.S. Geological Survey, Washington, D.C. 173 pp.
17. Ebens, R. J., J. A. Erdman, G. L. Feder, A. A. Case, L. A. Selby. 1973. *Geochemical anomalities of a claypit area, Callaway County, Missouri, and related metabolic imbalance in beef cattle.* Professional Paper 807. U.S. Geological Survey, Washington, D.C. 24 pp.
18. Erdman, J. A., and R. J. Ebens. 1979. *Element content of crested wheatgrass grown on reclaimed coal spoils and on soils nearby.* Journal of Range Management 32: 159-161.
19. Erdman, J. A., R. J. Ebens, and A. A. Case. 1978. *Molybdenosis: A potential problem in ruminants grazing on coal mine spoils.* Journal of Range Management 31: 34-36.
20. Ferguson, W. S., A. H. Lewis, and S. J. Watson. 1938. *Action of molybdenum in nutrition of milking cattle.* Nature 141: 553.
21. Follett, R. H., and W. L. Lindsay. 1971. *Changes in DTPA-extractable zinc, iron, manganese, and copper in soils following fertilization.* Soil Science Society of America Proceedings 35: 600-602.
22. Gough, L. P., J. M. McNeil, and R. C. Severson. 1978. *Availability of elements in soils to native plants, Northern Great Plains—Assessment of new data.* In *Geochemical Survey of the Western Energy Regions.* Open File Report 78-1105. U.S. Geological Survey, Denver, Colorado. pp. 126-143.
23. Gough, L. P., and R. C. Severson. 1981. *Biogeochemical variability of plants at*

native and altered sites, San Juan Basin, New Mexico. Professional Paper 1134-D. U.S. Geological Survey, Washington, D.C. 36 pp.

24. Gough, L. P., and R. C. Severson. 1982. *Soils, mine spoils, and native plants at the San Juan mine, New Mexico.* In R. J. Ebens and H. T. Shacklette [editors] *Geochemistry of Some Rocks, Mine Spoils, Stream Sediments, Soils, Plants, and Waters in the Western Energy Region of the Conterminous United States.* Professional Paper 1237. U.S. Geological Survey, Washington, D.C. 173 pp.
25. Gough, L. P., and R. C. Severson. 1982. *Element concentrations in rehabilitation species from thirteen coal-strip mines in five western states and Alaska.* Open-file Report 81-182. U.S. Geological Survey, Washington, D.C. 110 pp.
26. Gough, L. P., and R. C. Severson. 1983. *Rehabilitation materials from surface-coal mines in western U.S.A. II. Biogeochemistry of wheatgrass, alfalfa and fourwing saltbush.* Reclamation and Revegetation Research 2: 103-122.
27. Gray, L. F., and G. H. Ellis. 1950. *Some interrelationships of copper, molybdenum, zinc and lead in the nutrition of the rat.* Journal of Nutrition 40: 441-452.
28. Grigg, J. L. 1953. *Determination of available soil molybdenum.* New Zealand Soil News 3: 37-40.
29. Gupta, U. C. 1979. *Copper in agricultural crops.* In J. O. Nriagu [editor] *Copper in the Environment. Part 1. Ecological Cycling.* John Wiley and Sons, Inc., New York, New York. pp. 255-288.
30. Gupta, U. C., and D. C. MacKay. 1965. *Extraction of water-soluble copper and molybdenum from Podzol soils.* Soil Science Society of America Proceedings 29: 323-327.
31. Gupta, U. C, and J. Lipsett. 1981. *Molybdenum in soils, plants, and animals.* Advances in Agronomy 34: 34-115.
32. Haby, V. A., and J. R. Sims. 1979. *Availability of micronutrient cations in Montana soils.* Bulletin 706. Montana Agricultural Experiment Station, Bozeman. 20 pp.
33. Haq, A. U., and M. H. Miller. 1972. *Prediction of available soil Zn, Cu and Mo using chemical extractants.* Agronomy Journal 64: 779-782.
34. Harrington, N. 1977. *Suspect levels of soil parameters.* Montana Department of State Lands, Helena.
35. Harrington, N. 1983. *Montana proposed soil and overburden guidelines.* Montana Department of State Lands, Helena.
36. Harve, G. N., and I. W. Dishington. 1962. *The mineral composition of pasture as influenced by various types of heavy nitrogen dressings.* Acta Agricultarea Scandinavica 12: 298-308.
37. Henning, S., and T. L. Jackson. 1973. *Determination of molybdenum in plant tissues using flameless atomic absorption.* Atomic Absorption Newsletter 12: 100-101.
38. Huisingh, J., and G. Matrone. 1976. *Interactions of molybdenum in animal nutrition.* In W. R. Chappell and K. K. Peterson [editors] *Molybdenum in the Environment, Volume I, the Biology of Molybdenum.* Marcel Dekker, Inc., New York, New York. 315 pp.
39. Jarrell, W. M., and M. D. Dawson. 1978. *Sorption and availability of molybdenum in soils of western Oregon.* Soil Science Society of America Journal 42: 412-415.
40. Kahn, S. U., R. O. Cloutier, and M. Hidiroglou. 1979. *Atomic absorption spectroscopic determination of molybdenum in plant tissue and blood plasma.* Journal of the Association of Official Analytical Chemists 62: 1,062-1,064.
41. Karimian, N., and F. R. Cox. 1979. *Molybdenum availability as predicted from selected soil chemical properties.* Agronomy Journal 71: 63-65.
42. Kubota, J. 1977. *Molybdenum status of United States soil and plants.* In W. R.

Chappell and K. K. Peterson [editors] *Molybdenum in the Environment. Volume 2.* Marcel Dekker, Inc., New York, New York. pp. 555-581.
43. Kubota, J., E. R. Lemon, and W. H. Allaway. 1963. *The effects of soil moisture content upon the uptake of molybdenum, copper and cobalt by alsike clover.* Soil Science Society of America Proceedings 27: 697-683.
44. Kubota, J., and W. H. Allaway. 1972. *Geographic distribution of trace element problems.* In J. J. Mortvedt, P. M. Giordamo, and W. L. Lindsay [editors] *Micronutrients in Agriculture.* Soil Science Society of America, Madison, Wisconsin.
45. Lavy, T. L., and S. A. Barber. 1964. *Movement of molybdenum in the soil and its effect on availability to plants.* Soil Science Society of America Proceedings 28: 93-97.
46. Lindsay, W. L. 1972. *Inorganic phase equilibria of micronutrients in soils.* In P. M. Mortvedt, P. M. Giordamo, and W. L. Lindsay [editors] *Micronutrients in Agriculture.* Soil Science Society of America, Madison, Wisconsin. pp. 41-57.
47. Lindsay, W. L. 1979. *Chemical equilibria in soils.* John Wiley and Sons, New York, New York. 449 pp.
48. Lindsay, W. L., and W. A. Norvell. 1969. *Development of DTPA micronutrient soil test.* Agronomy Abstracts X: 84.
49. Little, I. P., and P. C. Kerridge. 1978. *A laboratory assessment of the molybdenum status of nine Queensland soils.* Soil Science Society of America Proceedings 125: 102-106.
50. MacKay, P. C., W. E. Chipman, and U. C. Gupta. 1966. *Copper and molybdenum nutrition of crops grown on acid sphagnum peat.* Soil Science Society of America Proceedings 30: 755-759.
51. Martson, H. R. 1952. *Cobalt, copper and molybdenum in the nutrition of animals and plants.* Physiological Review 32: 66-121.
52. Miltmore, J. E., and J. L. Mason. 1971. *Copper and molybdenum ratio and molybdenum and copper concentration in ruminant feeds.* Canada Journal of Animal Science 51: 193-200.
53. Munshower, F. F., and D. R. Neuman. 1978. *Elemental concentrations in native range grasses from the Northern Great Plains of Montana.* Journal of Range Management 31: 145-148.
54. Munshower, F. F., and D. R. Neuman. 1979. *Trace element concentration in plants from revegetated strip mined lands and native range in southeastern Montana.* In M. K. Wali [editor] *Ecology and Coal Resource Development, Volume 2.* Pergamon Press, New York, New York. 1,091 pp.
55. Munshower, F. F., E. J. DuPuit, and D. R. Neuman. 1978. *The effects of stack emissions on the range resource in the vicinity of Colstrip, Montana.* Research Report 126. Montana Agricultural Experiment Station, Bozeman. 196 pp.
56. National Academy of Sciences. 1977. *Copper.* Washington, D.C. 115 pp.
57. National Research Council, Subcommittee of Beef Cattle Nutrition. 1976. *Nutrient requirements of beef cattle.* National Academy of Science, Washington, D.C. 56 pp.
58. National Research Council, Subcommittee on Dairy Cattle Nutrition. 1971. *Nutrient requirements of dairy cattle.* National Academy of Science, Washington, D.C. 42 pp.
59. Neuman, D. R. 1975. *Analysis of molybdenum in soil extract by nonflame atomic absorption spectroscopy.* In B. R. Culver [editor] *Analytical Methods of Carbon Rod Atomizers.* Varian Techtron Springvale, Victoria, Australia.
60. Neuman, D. R., and F. F. Munshower. 1981. *Rapid determination of molybdenum in botanical material by electrothermal atomic absorption spectrometry.* Analytica Chimica Acta 123: 325-328.

61. Neuman, D. R., and F. F. Munshower. 1984. *Copper and molybdenum uptake by forages grown on coal minesoils.* Journal of Range Management 31: 517-520.
62. Nicholas, D.J.D. 1975. *The function of trace elements in plants.* In D.J.D. Nicholas and Adrian R. Egan [editors] *Trace Elements in Soil-Plant-Animal System.* Academic Press, New York, New York. pp. 181-198.
63. Pathak, A. N., H. Shankar, and R. V. Misra. 1969. *Methods of molybdenum analyses in soils.* Journal of Indian Society of Soil Science 17: 151-153.
64. Pringle, W. L., W. K. Dawley, and J. E. Miltmore. 1973. *Sufficiency of Cu and Zn in barley, forage and corn silage rations as measured by responses to supplements by beef cattle.* Canada Journal of Animal Science 53: 497-502.
65. Purvis, E. R., and N. K. Peterson. 1956. *Methods of soil and plant analyses for molybdenum.* Soil Science 81: 223-228.
66. Reisenauer, H. M. 1965. *Molybdenum.* In C. A. Black [editor] *Methods of Soil Analysis, Part 2.* American Society of Agronomy, Madison, Wisconsin. pp. 1,050-1,058.
67. Reisenauer, H. M. 1967. *Soil testing and plant analysis, Part 1.* Special Publication No. 2. Soil Science Society of America, Madison, Wisconsin. pp. 71-102.
68. Sauchelli, V. 1969. *Trace elements in agriculture.* Van Nostrand Reinhold Co., New York, New York. 248 pp.
69. Severson, R. C., and L. P. Gough. 1981. *Chemical character of mine soils at one Alaskan and twelve western conterminous United States coal-stripmines.* Open-file Report 81-243. U.S. Geological Survey, Washington, D.C. 80 pp.
70. Severson, R. C., and L. P. Gough. 1983. *Rehabilitation materials from surface-coal mines in western U.S.A. I. Chemical characteristics of spoil and replaced cover-soil.* Reclamation on Revegetation Research 2: 83-102.
71. Severson, R. C., J. M. McNeil, and L. P. Gough. 1978. *Total and extractable element composition of some Northern Great Plains soils.* In *Geochemical Survey of the Western Energy Region.* Open File Report 78-1105. U.S. Geological Survey, Denver, Colorado. pp. 87-116.
72. Soltanpour, P. N., and A. P. Schwab. 1977. *A new soil test for simultaneous extraction of macro- and micro-nutrients in alkaline soils.* Communications in Soil Science and Plant Analysis 8: 195-207.
73. Spais, A. G. 1959. *Le cuivre et pathologie ovine et bovine.* Recueil de Medecine Veterinaire 135: 161-194.
74. Stone, L. R., J. A. Erdman, G. L. Feder, and H. D. Holland. 1983. *Molybdenosis in areas underlain by uranium-bearing lignites in the Northern Great Plains.* Journal of Range Management 36: 280-285.
75. Suttle, N. F. 1975. *The role of organic sulfur in the copper-molybdenum-S relationship in ruminant nutrition.* British Journal of Nutrition 34: 411-420.
76. Swanson, V. E., and C. Huffman. 1976. *Guidelines for sample collecting and analytical methods used in the U.S. Geological Survey for determining chemical composition of coal.* Circular 735. U.S. Geological Survey, Washington, D.C. 11 pp.
77. Underwood, E. J. 1977. *Molybdenum in animal nutrition.* In W. R. Chappell and K. K. Peterson [editors] *Molybdenum in the Environment, Volume 1.* Marcel Dekker, Inc., New York, New York. pp. 9-32.
78. Vlek, P.L.G. 1975. *The chemistry, availability and mobility of molybdenum in Colorado spoils.* Ph.D. Thesis. Colorado State University, Fort Collins.
79. Vlek, P.L.G., and W. L. Lindsay. 1977. *Molybdenum contamination in Colorado pasture soils.* In W. R. Chappell and K. K. Peterson [editors] *Molybdenum in the Environment.* Marcel Dekker, Inc., New York, New York. pp. 619-650.
80. Walsh, T., and M. Neenan. 1953. *High molybdenum levels in herbage on acid soils.* Nature 171: 1,120.

81. Walsh, T., M. Neenan, and L. B. O'Moore. 1952. *Molybdenum in relation to cropping and livestock problems under Irish conditions.* Nature 170: 149-150.
82. Ward, G. M., and J. G. Nagy. 1976. *Molybdenum and copper in Colorado forages, molybdenum toxicity in deer, and copper supplementation in cattle.* In W. R. Chappell and K. K. Peterson [editors] *Molybdenum in the Environment, Volume 1. The biology of molybdenum.* Marcel Dekker, Inc., New York, New York. 315 pp.
83. Whitehead, D. C., and E. C. Jones. 1969. *Nutrient elements in the herbage of white clover, red clover, lucerne and sainfoin.* Journal of Science and Food Agriculture 20: 585-591.
84. Williams, C., and J. Thornton. 1972. *The effect of soil additives on the uptake of molybdenum and selenium from soils from different environments.* Plant and Soil 36: 345-406.
85. Wilson, D. O. 1979. *Determination of molybdenum in wet-ashed digests of plant material using flameless atomic absorption.* Communications in Soil Science and Plant Analysis 10: 1,319-1,330.

Chapter 11

Acid-base potential and sulfur forms

Andrew A. Sobek, Lloyd R. Hossner, Darwin L. Sorensen, Patrick J. Sullivan, and Dennis F. Fransway

The acid-generating potential of sulfur (S) minerals in overburdens is commonly accepted as resulting from the oxidation of Fe sulfides, such as pyrite, pyrrhotite, marcasite, calcopyrite, and sphalerite. Numerous weathering and chemical tests for measuring acid-base potential have been applied to overburden materials at surface mines. Many of these tests are direct adaptations of U.S. Department of Agriculture methods, U.S. Geological Survey practices, or standard engineering procedures with ASTM designations; certain methods have also been proposed directly for overburden analysis. Often, the exact method(s) for conducting these analyses are not stipulated, leading to conflict between mining companies, state regulators, and environmental groups.

The geochemical data base that defines overburden acid-generation characteristics is important to management. These data are taken from soils, geology, and engineering literature developed in eastern and midwestern coal fields. There is not an adequate data base on acid generation for western coal fields. Researchers must use the data base developed on undisturbed agriculture and range soil chemical properties and extrapolate it to reclaimed mine soils to supplement the limited studies on acid generation of western overburden. Extrapolation of data must be tempered with the understanding that diagenesis has taken place in overburden but not in surface soils.

Sulfur forms

Acid-base potential laboratory procedures are based on the oxidation of sulfide compounds in exposed coal and coal overburden materials. The theory of this reaction has been elaborated upon by many investigators (*27, 35, 60, 63*):

$$2FeS_2 + 7\frac{1}{2}O_2 + 7H_2O \rightarrow 2Fe(OH)_3 + 2H_2SO_4$$

Using the stoichiometry of the reaction for pyrite oxidation, calculations for a material containing 0.1 percent S (all as pyrite) indicate that complete oxidation will yield a quantity of sulfuric acid requiring 3,125 tons of $CaCO_3$ to neutralize 1,000 tons of material. Where sulfur in overburden is present exclusively as pyrite, the total S content accurately quantifies the acid-producing potential. Removal from the acid-base potential calculation of sulfates and organic S, naturally present in some overburdens or resulting from weathering of pyritic materials, allows increased accuracy in predicting the acid-producing potential of materials containing mixed S species (*50*).

Much confusion surrounds the issue of organic S causing acid problems because many of the overburdens in western coal basins contain more organic S than pyritic S. Pyrite is an unstable, high-energy compound that will readily react in an oxidizing acidic environment to form acid sulfate salts; however, organic S is much more stable under such conditions. There is no theoretical basis in the literature to substantiate claims that organic S will cause acid problems under natural environmental conditions. Difficulties in accurately determining sulfides and organic S in overburden adds to the confusion on this matter. Insuring that all sulfate and reduced S has been removed from the sample prior to organic S determination is tenuous because no extraction procedure currently used removes individual S forms selectively.

Sulfates. Most Fe^{+2}, H^+, and SO_4^{-2} released by pyrite oxidation undergo further reactions. Hydrogen ions are largely inactivated by ion exchange reactions and weathering reactions; ferrous Fe is oxidized to ferric Fe; and most of the sulfate remains in solution and is leached out of the system. The remainder of the sulfate is either precipitated as jarosite and as basic Al sulfate or adsorbed. Gypsum and soluble sulfates may form when evaporation exceeds rainfall (*67*).

Jarosite [$KFe_3(SO_4)_2(OH)_6$] in most acid sulfate soils commonly occurs as earthy fillings of voids or as mottles in the soil matrix. Potassium jarosite is by far the most common (*67*). In agreement with theoretical stability relations (Figure 1), jarosite is formed only in acidic (pH 2-4), oxidizing (Eh $>$400 mV) environments. In acid sulfate soils it is only slightly stable and will eventually hydrolyze to geothite. However, yellow jarosite mottles sometimes persist for decades at pH values above 4 in limed soils (*68*). In some acid sulfate soils, particularly those high in organic matter (*32*), jarosite is not found.

Eventually, most pyritic Fe oxidizes to ferric oxides. Fine-grained geothite is formed one of two ways: quickly by direct oxidation of dissolved ferrous sulfate or slowly by hydrolysis of jarosite. Both reactions are acidic (Figure 2) and part of the sulfurization process:

$$Fe^{2+} + SO_4^{2-} + 1/4O_2 + 3/2H_2O \rightarrow FeOOH + 2H^+ + SO_4^{2-}$$
$$\text{jarosite} \rightarrow 3FeOOH + 2SO_4^{2-} + K^+ + 3H^+$$

Gypsum and jarosite are the most commonly mentioned sulfates in conjunction with coal overburden. Gypsum, a neutral salt, does not contribute to spoil acidity, while jarosite is quite insoluble ($Ksp = 10^{-98.6}$) (*70*), and may not undergo much hydrolysis, especially in materials with a pH near neutrality. Any acid generated from jarosite weathering would not be large when compared to that produced by pyrite oxidation. Carson and others (*10*) picked jarosite (87 percent pure) from a soil and determined its acidity. The titratable acidity was 18.7 meq/100 g compared to the theoretical acidity value of 522 meq/100 g. Melanterite [$FeSO_4(H_2O)_7$] and coquimbite [$Fe_2(SO_4)_3(H_2O)_9$] have also been found in association with acid mine drainage (*39*).

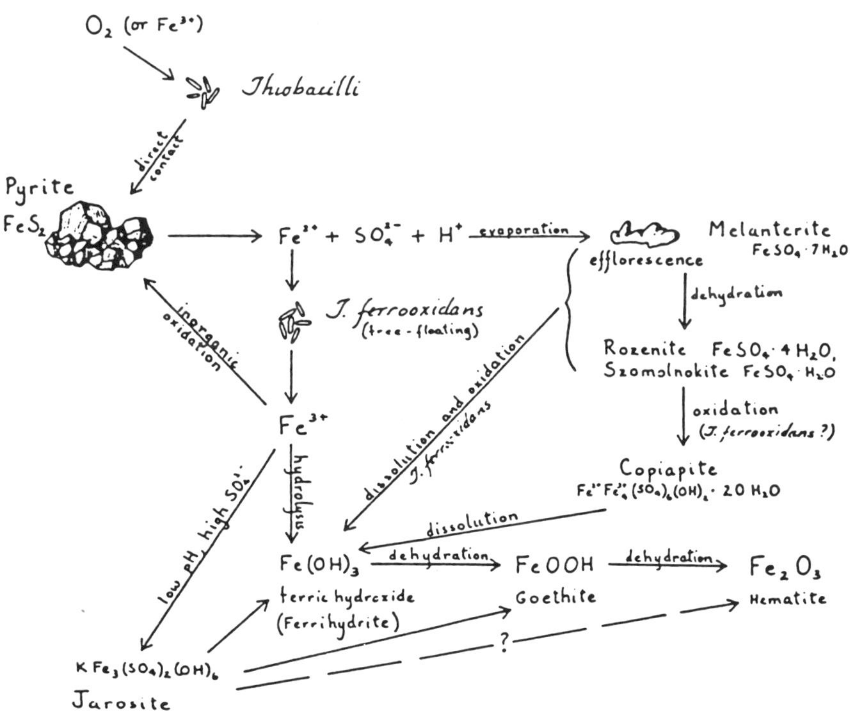

Figure 1. The overall sequence of mineral reactions for pyrite oxidation showing the relationships between oxidizing agents, catalysts, and mineral products (*38*).

Sulfides. A quantitative determination of iron sulfide weathering is a complex problem because weathering can involve both geochemical and biochemical components. For example, the geochemical oxidation of pyrite may be characterized by the following reactions (*57*):

$$FeS_2 + 7/2\,O_2 + H_2O = Fe^{2+} + 2SO_4^{2-} + 2H^+ \quad [1]$$

$$Fe^{2+} + 1/4\,O_2 + H^+ = Fe^{3+} + 1/2\,H_2O \quad [2]$$

$$Fe^{3+} + 3H_2O = Fe(OH)_3\,(s) + 3H^+ \quad [3]$$

$$FeS_2(s) + 14Fe^{3+} + 8H_2O = 15Fe^{2+} + 2SO_4^{2+} + 16H^+ \quad [4]$$

After pyrite is oxidized (equation 1), ferric Fe is produced extremely slowly (equation 2). This second reaction, however, can be accelerated by microbial catalysis to increase the overall rate of the oxidation of ferrous Fe (*4*). With the generation of ferric Fe, insoluble ferric hydroxide can then form under proper pH conditions (equation 3). Ferric Fe can also be reduced further by pyrite, resulting in the generation of more acidity and ferrous Fe (equation 4).

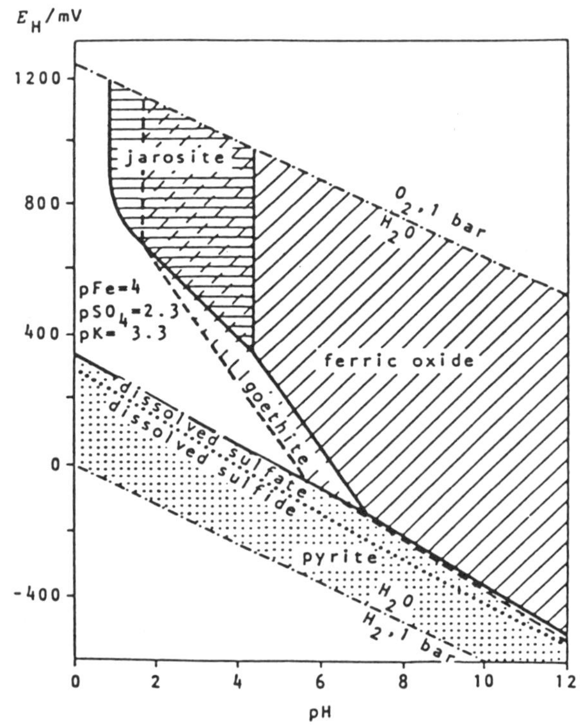

Figure 2. E_H − pH diagram of ferric oxide (pK_{so} = 39.9), jarosite, and pyrite at 25 °C. Solid phases are indicated by shading. The broken, hatched shading shows the extension of the Fe_2O_3 field at pK_{so} = 46.2 (geothite) (*66, 67*).

Chemical oxidation. The rate of chemical oxidation of pyrite is influenced by a number of factors, including pH, O_2 concentration, temperature, moisture content, and the ferrous/ferric ion equilibria of the system. These factors are highly interdependent; therefore, their individual contribution is not easily predicted.

Chemical oxidation of pyrite is quite sensitive to the pH of the reaction media. A major reason for this appears to be the influence of pH on the ferrous/ferric ion equilibria. The ferric ion may serve as an oxidant of Fe^{2+} in pyrite. Oxidation of Fe^{2+} is a rate-limiting step in the chemical oxidation scheme (*25*).

Pyrite agglomerates are unstable in the presence of O_2. Absence of oxygen will inhibit regeneration of Fe^{+3} from Fe^{+2}. The acidity of drainage water from coal mines is proportional to the partial pressure of O_2 in the gas phase in contact with the pyrite. In an inert atmosphere acidity production is dependent on dissolved O_2 concentration; therefore, exclusion of O_2 prevents oxidation of pyrite (*65, 71*). Grim and Hill (*22*) reported that the formation of acid could be decreased by a limited O_2 supply.

Water plays a major role in the oxidation of pyrite. Complete agreement, however, is lacking on whether water is a reaction medium or a reactant. Researchers at Harvard University considered water a reaction medium rather than a reactant in the oxidation of Fe pyrite (*25*). Morth and associates (*37*) found that oxidation rate increased with humidity and concluded that water was a reactant as well as a reaction medium. Furthermore, water removes the products of oxidation from the reaction site. Smith and Shumate (*46*) reported that the basic function of water was to remove the reaction products from the pyrite surface.

Biological oxidation. Microorganisms have frequently been associated with the oxidation of pyrite. Their action may be a direct attack on pyrite or an indirect procedure whereby the organisms generate Fe^{+3} from Fe^{+2}. Both reactions can generate sulfuric acid (*2, 22, 28, 43*).

Iron-oxidizing bacteria are indigenous to environments containing sulfides, O_2, and acid conditions. *Thiobacillus ferrooxidans* and *Thiobacillus thiooxidans* are the principal autotrophic microorganisms that obtain their energy from the oxidation of Fe and S compounds. These organisms are most active at pH 2.0-5.0 (*34*). *Thiobacillus ferrooxidans* can use reduced S as its sole energy source and oxidize elemental S to sulfate, with sulfite as an intermediate compound (*19, 22, 44, 45, 69*).

Thiobacillus thiooxidans obtain their energy by oxidizing elemental S. These bacteria regenerate ferric Fe. The ferric Fe can chemically oxidize pyrite. The rate of ferrous Fe oxidation increased from 10-fold to 50-fold when these bacteria were added to a pyrite system (*2, 37*). The population of bacteria limits the oxidation rate when the water-to-pyrite ratio is low; when the bacteria are in excess, the surface area of the pyrite determines the rate (*28, 46, 64*).

Organic sulfur. Coal and carbonaceous rock and/or sediments in coal overburden have two major stages of formation: (1) a diagenetic or peat-forming stage in which biochemical processes are important and (2) a metamorphic stage in which heat and pressure are brought to bear on the sediment. Casagrande and associates (*16*) found a relationship between the depositional environment and pyrite and organic S concentrations. Carbon-bonded S comprised 70 percent of the total S in freshwater peat and 50 percent of the total S in marine peat. More than 15 percent of the total S in marine environments was in pyrite combination. In freshwater peat pyrite levels were an order of magnitude lower.

Altschuler and colleagues (*1*) found the diagenesis of S and the organic matter in the Everglades were closely linked. The S cycle contained the following elements: (1) organic fixation of aqueous sulfate in upper peat zone; (2) reducible organic S to sulfide (pyrite) by sulfate-reducing bacteria, in the course of their use of organic matter; (3) reducible organic S to pyrite with conversion of some C-bonded sulfide in the basal peat zone; and (4) the progressive remineralization of organic S to pyrite with depth.

Soils. Organic S accounts for 90 percent of the total S in the surface soils of humid regions (*6*). Soil organic S can be divided into fractions: (1) C-bonded S (C-S), as in S-containing amino acids; and (2) non-C-bonded S, as in the ester sulfates ($R\text{-}O\text{-}SO_3$). Carbon-bonded S accounts for 12 to 35 percent of the total S in mineral soils compared with 47 to 50 percent in organic soils of Quebec (*33*). The non-C-bonded fraction of organic S (phenolic sulfates, choline sulfates, and sulfate esters of carbohydrates) comprises about 50 percent of the total S in soils (*62*).

Soil organic matter can serve either as a source or a sink for S. The incorporation of plant residues low in S content into the soil can cause S deficiencies in crops growing on these soils. The soil microorganisms grow and multiply during decomposition of this plant material. They may tie up much of the available S supply, creating a S deficiency in the growing crop.

The release of organically bound S in soils depends largely upon microbiological decomposition. The S associated with the organic fraction of soils must be converted to inorganic sulfate-S or to relatively low molecular weight uncombined organic S compounds, such as cystine and methionine, before it can be oxidized. These amino acids are decomposed very rapidly in soils (*8*).

The microbial decomposition of humic material is a relatively slow process. Thus, the sulfate-S release during incubation of a soil is very small in comparison with the total soil content of organic S. The limiting factor appears to be the breakdown or decomposition of the organic matter rather than the subsequent oxidation of the decomposition products (*26*).

Coal. The S content of coals may vary from below 1 percent to approximately 10 percent. Rasa coal (Istria) of Yugoslavia contains about 11 percent sulfur, while some Assam coals in India contain 6 to 8 percent sulfur, most of it in both cases occurring as organic S. Coals, like the Wyodak-Anderson seam in Wyoming, formed under freshwater conditions have a low S content, whereas coals formed under marine or estuarine conditions, like the Kittanning coal seam in Pennsylvania, contain a very high percentage of S. Organic S compounds originally present as proteins in vegetation give rise to the organic S in coal. During this process, concentration of the organic S takes place because nearly 3 m of plant remains become compacted to produce, ultimately, 0.3 m of coal seam. It has been reported that organic S could be present in molecular combination with cystin, a derivative of plant albumin in coal. The organic S compounds found in Assam coals in India reportedly consist of thiols, alkyl sulfides and disulfides, and partly as ring compounds (Table 1) (*17, 36*).

The origin of the organic S distributed in the organic matrix of coal is believed to be S contained in the original plant material. However, many high-S coals contain large concentrations of organic S or organic and inorganic S, leading to the conclusion that some S was trapped by the organic matrix during the coalification process. The "trapping" of S could have in-

Table 1. Physical forms and appearance of organic and inorganic S in coal (*3*).

	Organic S	Inorganic S
Major compounds	Thiophenes Aromatic and aliphatic Sulfides Φ-S-Φ R-S-R Thiols and thiophenois RSH, ΦSH	FeS_2 (pyrite), FeS_2 (marcasite), $CaSO_4$, $FeSO_4$. Exosure of coal to air converts FeS_2 to $FeSO_4$. Predominant specie: iron pyrite
Relation to the organic structure	Uniformly distributed in the organic matrix	Present in the form of inorganic patches, often crystalline, embedded in or on the surface of the organic matrix
Accessibility to chemical reagents	Reagents have to diffuse through the organic matrix to reach the organic sulfur	The surface inorganic sulfur is accessible. Reagents have to diffuse in the organic matrix to reach embedded sulfur. Crushing makes embedded sulfur accessible
Density	Density of organic matrix about 1.3 g/cm³	Pyrite, 5.0 g/cm³, marcasite, 4.87 g/cm³
Particle size in washed coal		<200 μm. Most in the range of 5-65 μ
Desulfurization potential	By chemical reaction only	By chemical reaction or by physical separation

volved an aerobic bioreduction of sulfates to H_2S and exchange of O functional groups by S groups. For example:

$$SO_4^{2-} + 5H_2 \xrightarrow{\text{(Bacteria)}} H_2S + 4H_2O \quad [5]$$

$$ROH + H_2S \rightarrow RSH + H_2O \quad [6]$$

Whenever minerals were present that could form stable inorganic sulfides, they competed with the organic groups for the H_2S. For example, siderite ($FeCO_3$) can be found only in low-S coals. Attar (*3*) stated that, because the rate of reaction of siderite with H_2S is so much larger than that of the organic functionalities, trapping of H_2S by the organic matrix could have occurred only after all the siderite has been converted to stable FeS_2 and could react with no more H_2S.

Acid-base account method

The toxicity associated with acid generation is due largely to the oxidation of Fe disulfides. This process takes place when earth-disturbing activities, such as mining (*27, 63*) and highway construction (*35*), expose Fe disulfides to the atmosphere. Because of public concern about acid drainage and barren mine spoils resulting from the surface exposure of acid-toxic or potentially acid-toxic materials, a basis to evaluate overburden materials and avoid this occurrence had to be established.

Acid-base accounting is a dependable method for evaluating coal overburden materials. An acid-base account consists of five components: (a) acid potential (b) neutralization potential, (c) pH, (d) rock type, and (e) depth below the land surface. The most important part of the method balances maximum potential acidity (from immediately titratable sources, plus sulfuric acid equivalent, calculated from total S) against total neutralizers (from alkaline carbonates, exchangeable bases, weatherable silicates, or other rock sources capable of neutralizing strong acids in an acid digestion) (*5, 30, 47, 51, 52*).

The potential acidity of materials is accurately quantified by the total S content when all S is present as a pyritic mineral. The presence of sulfate minerals, such as gypsum, jarosite, coquinite, or basalenite, in a sample of weathered overburden indicates that part of the total S is nonacid-forming. Samples high in organic C usually contain organic S. Therefore, the maximum potential acidity, as calculated, will be too high when these nonacid-forming forms of S are present. In such cases a pyritic S determination can be used to indicate acid potential more accurately.

The neutralization potential of overburden materials, the second component of the acid-base account, measures the amount of neutralizers present in the overburden. This value is found by treating a sample with a known amount of standardized HC1, heating it to assure complete reaction, and titrating with a standardized base. The result is then expressed as $CaCO_3$ equivalent. When balanced against acidity from the total (or pyritic) S measurement, a net acid-base account can be made (*23, 31*).

The third component of an acid-base account is pH. This property is indicative of the present chemical state of the sample. It does not predict the potential acid or alkaline condition of a sample; however, it does reflect the chemical changes made by the weathering process. Thus, a sample with a pH of less than 4.0 (in a pulverized rock slurry made with distilled water) is defined as being acid-toxic regardless of the acid-base balance (*23*).

Rock type, the fourth component, is incorporated into the acid-base account because knowledge of the rock type can provide an indication of the texture and base status of a future mine soil, as well as a measure of the stability of rock fragments (*5, 48, 49*). For example, sandstones containing moderate amounts of pyrite and lacking sufficient neutralizers will become active acid producers when exposed to the atmosphere. Likewise, carbonaceous shales contain pyrite and are a storehouse for quantities of heavy metals.

The final component of the acid-base account method is depth of the material below the land surface. This includes the sequence of samples from the land surface to the base of the lowest minable coal seam. Potentially acid-forming materials can be located and their thickness calculated in order to identify areas and volumes of materials that must be replaced well below the surface and away from groundwater (*51*).

In humid regions acid spoil and acid mine drainage are equally important to the mining industry. In semiarid to arid regions acid spoil formation is more important, and acid mine drainage is relatively rare in the western coal mining industry. However, acid mine drainage does pose a difficult problem to metal mines. Revegetation of acid-generating materials poses a significant problem to the mining industry. The acid-base accounting method (Figure 3) provides a useful tool for evaluating overburden in the coal fields of the United States because it is useless to look for more complex problems, like toxic concentrations of Al, until the possibility of acid mine soils is eliminated.

Acid potential methods. Formation of acid from the oxidation of reduced S compounds in exposed coal and coal overburden materials has been elaborated upon by many investigators (*21, 27, 37, 38, 46, 63*). Using the stoichiometry of the reaction for pyrite oxidation, calculations for a material containing 0.1 percent S (all as pyrite) indicate that complete oxidation will yield a quantity of sulfuric acid requiring 3.125 tons of $CaCO_3$ to neutralize 1,000 tons of material. Where S in overburden is present exclusively as pyrite, the total S content accurately quantifies the acid-producing potential. Removal of sulfates and organic S naturally present in some overburdens or resulting from weathering of pyritic materials allows increased accuracy in predicting the acid-producing potential of materials containing mixed S species.

There are several procedures used to determine acid potential. Many authors attribute the major source of strong mineral acidity in drastically disturbed lands to reduced forms of S, especially pyrite (*1, 10, 16, 54, 63*). Therefore, the methods used to predict acid potential fall into three categories: (a) determination of S content and use of a stoichiometric equation for pyrite oxidation to calculate acidity, (b) determination of Fe and calculation of acidity from a stoichiometric equation for pyrite oxidation, and (c) titration of acidity after oxidation of the sample by a strong chemical oxidant or a simulated weathering procedure.

The most popular and easiest methods are those that directly determine S content. Two such methods are the S furnace method and the Eschka method. The Eschka method (*6*) determines the sulfur fraction by wet chemistry ($BaSO_4$ precipitation) for each digestion extract, whereas the S furnace technique measures the total S left in the sample after removal of the digestion extract.

Scholz and Rathleff (*42*) found that the LECO S furnace agreed closely with the Eschka method for determining total S in coal and was rated as the preferable method because of its speed and accuracy. For overburden samples the S furnace is recommended because it eliminates the need to precipitate the sulfate in solution with barium (as in the Eschka method) and

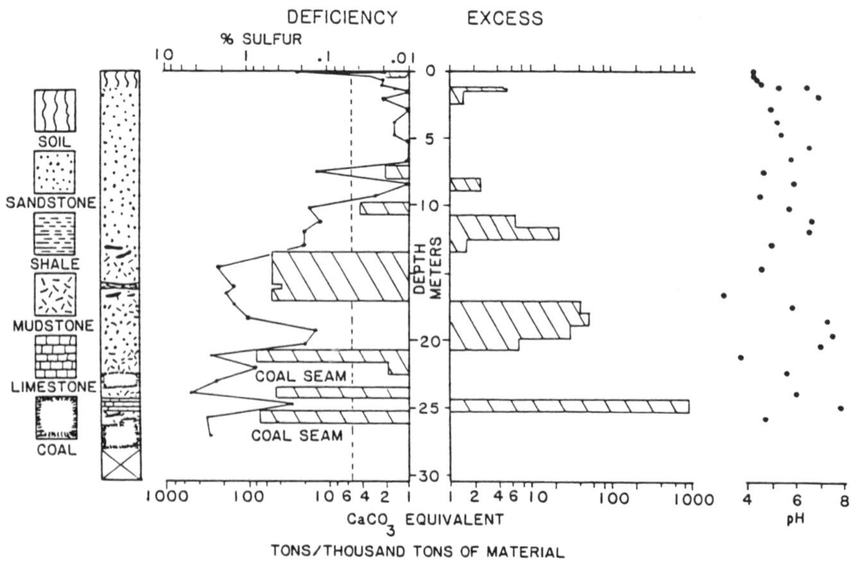

Figure 3. Typical acid-base account; upper and lower Bakerstown coal overburden, Mineral County, West Virginia.

gravimetrically determine the amount of precipitate. The Eschka method is adequate for coal, but the mixed mineralogy of overburden rocks complicates the technique and allows introduction of error, for example, coprecipitation of Fe^{3+} with $BaSO_4$ if the solution is not acidic enough to prevent formation of Fe^{3+}. The S furnace is quick, reproducible, and accurate enough for total S estimations. However, Tabatabai (*61*) cited studies where the LECO S analyzer gave inaccurate S analyses of soil and did not recommend the LECO instrument for research requiring accurate and precise determinations of total S.

Both of these methods use the ASTM D2492-77 procedure for determining different forms of S. Sulfate and sulfide S are determined by difference in sequential total S measurements after these S forms have been removed from the sample. The procedure is outlined as follows: (a) total S (TS_1) is measured; (b) sulfates are removed by washing using the ASTM procedure, followed by measurement of total S (TS_2) in the washed sample; and (c) sulfates and sulfides are removed by ASTM procedure, followed by measurement of total S (TS_3) in the residue. The fractions are sulfates = $TS_1 - TS_2$, sulfides = $TS_2 - TS_3$, organic sulfur = TS_3.

Sulfides may also be determined by generating H_2S from mineral sulfides, trapping the H_2S in an antioxidant buffer and measuring the sulfide with a sulfide ion electrode (*53*). This method should be considered when a S furnace is not available.

The second category of determining acid potential also uses the ASTM D2492-77 technique and assumes that the total Fe ($Fe^{2+} + Fe^{3+}$) left in the sample after digesting the sample with HC1 will come from pyrite, marcasite, and other iron sulfide minerals. The washed sample is digested in HNO_3 and the resulting solution is analyzed for total Fe; the acid potential is then calculated from the stoichiometric equation for pyrite oxidation. This technique works well for coal, but the mineralogy of coal overburden materials and the particle size of samples being digested complicates this procedure (*3, 56*). Reproducibility is a problem with this method.

Direct titration of a strongly oxidized sample (hydrogen peroxide method) or of leachings from a gently oxidized sample (simulated weathering test) is a third method of determining acid potential. Use of peroxide has been tested and found unreliable. Peroxide oxidation can be used for overburden samples with simple mineralogy; however, coal overburden contains few units with simple mineralogy. Even uncomplicated mineralogies, such as sandstone, have been analyzed by H_2O_2, with erroneous results occurring because unstabilized H_2O_2 was not used or because the sample was not heated enough to destroy the excess H_2O_2 before titration. Sobek and associates (*51*) did not recommend this procedure for routine use with all overburden samples. Potential acidity determinations of east Texas overburden sediments of Eocene age using a modification of the Sobek method (*51*) were found to be accurate and reproducible (*40*). Standard additions of pyrite were quantitatively recovered when the volume of H_2O_2 for pyrite oxidation was increased, the excess H_2O_2 was completely destroyed using Cu^{+2} catalyst, and the acidity generated from pyrite was separated from the soil prior to titration.

The more gentle oxidation process (simulated weathering) is a good method; however, the humidity cell weathering technique (*14*) requires four to eight weeks to obtain a trend line, and data interpretation remains a problem with this method. Originally, Hanna and Brant (*24*) developed a sample cell supplied with constant saturated water vapor. Samples placed into the cell could be leached with water at a constant rate from an overhead funnel. A similar method was used to evaluate acid mine drainage production (*11*). Caruccio (*11*) placed samples in humidified plastic containers (Figure 4) and flushed them at regular intervals with deionized water. This method provides for simple control over air, temperature, moisture, and microbes. After this initial development, numerous studies used the humidity cell to characterize leachate from coal and overburden materials (*12, 13, 15, 20, 48*). Laboratory procedures for using the humidity cell are not explicitly defined in most of this literature.

New weathering methods, such as the modified soxhlet technique, show promise for predicting pyrite weathering reactions, although more testing is still required (*59*). Use of the modified soxhlet by Sullivan and Sobek (*58*) has shown that soluble acid sulfates contribute a significant concentration of hydrogen ion from weathered refuse or spoil. Thus, weathering studies should leach acid sulfates, such as iron sulfates or jarosites, prior to assessing the acid-generating potential of overburden materials by the weathering process.

Sullivan and associates (*59*) have also shown that long-term leaching of weathered pyritic waste will influence soluble acid sulfate release as well as changes in leachate geochemistry. In this study a pyritic coal waste was leached continuously under saturated conditions for 6 months at 20 °C. Waste samples were removed after 3 months and 6 months of leaching. These samples, in addition to a sample with no leaching, were used to make

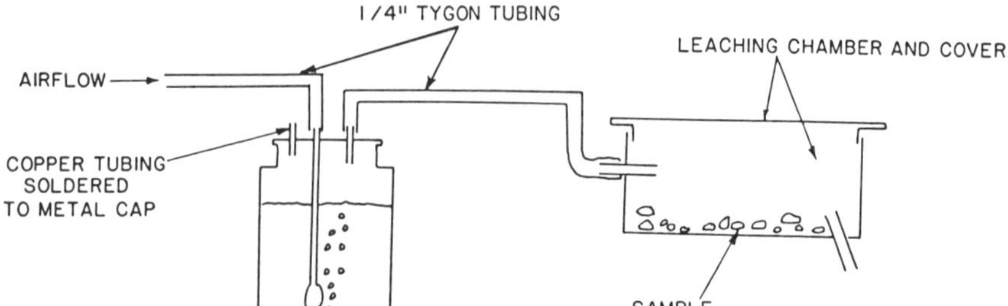

Figure 4. Humidity cell apparatus.

up equilibrium solutions. The leaching data showed a continuous release of Fe and sulfate species and low acidity under anaerobic conditions. As leaching progressed the relatively soluble sulfate, carbonates, and amorphous phases were removed, leaving the control of ionic activities to less soluble primary and secondary minerals. Data from the equilibrium solutions were used to calculate the saturation index for common secondary minerals that could form in the waste/water system. The saturation index data show that (1) the common Fe mineral phases under anaerobic and acid conditions will continually dissolve; (2) the mineral halotrichite may be controlling the activities of Fe^{2+}, Al, and sulfate; and (3) the solubility of illite may limit the activities of Mg, K, and Si. All of these data indicate changes in the solid phase of the waste material with leaching and the potential change in the long-term geochemistry of the waste/water system.

A relatively short-term, accurate bioassay or weathering technique is needed to help resolve questions about acid-producing potential in situations where acid-base accounting using chemical techniques fails to provide an adequate answer. Duncan and Walden (*18*) developed a slurry flask bioassay for overburden acid-producing potential. In this procedure the spoil or overburden is considered to have acid-producing potential if the amount of acid produced by *Thiobacillus ferrooxidans* in the shaken slurry exceeds the neutralizing capacity of an equal weight of spoil or overburden. Bruynesteyn and Hackl (*9*) recognized that this procedure does not simulate the open percolation system common in nature and recommended use of a once-through leaching column to determine acid-producing potential. Efforts are continuing to improve bioassay procedures by using leaching procedures, inoculation with mixed cultures of Thiobacilli, and the addition of surfactants to the leaching solution (*55*).

Neutralization methods. The inherent base content of overburden material, including basic cations present as carbonates or exchangeable cations on clays, is the second important component of the acid-base account. The amount of bases in unweathered overburden may be sufficient to equal or overwhelm the potential acidity of the material. The methods commonly used to characterize this rock property are digestion of sample with hot acid (*51*) and digestion of sample with cold acid (*13*).

The neutralization potential method is a modification of the procedure used to measure the neutralization equivalence of agricultural limestone (*23, 29*). This involves treating the pulverized sample with a known quantity of standardized HCl, heating to ensure complete dissolution of bases, and back-titration with standard NaOH. Calculation of the proportion of added acid that has been neutralized by the sample components allows ready conversion to equivalent tons of $CaCO_3$ per 1,000 tons of material. This procedure is simple, rapid, and reproducible. Application of heat allows the reaction to go to completion more rapidly than allowing the sample and acid to remain at room temperature for periods of 24 to 48 hours with frequent stirring.[1] Also, this procedure allows a maximum neutralization potential to be measured, including that contributed by silicates and dolomite (*5*).

[1]Freeman, J. R. 1974. Unpublished research on the neutralization potential method. West Virginia University, Morgantown.

Table 2. Acid-base account for overburden section, northern Powder River Basin, Wyoming.

					Tons $CaCO_3$ Equivalent/1,000 tons Material	
Sample No.	Thickness (ft)	Paste pH	Rock Type	Total S (%)	Amount Present (neutralization potential)	Excess (+) or Deficiency (−)
1	1.8	7.5	Siltstone	0.000	127.0	+127.0
2	0.56	7.9	Ironstone	0.019	433.0	+432.0
3	1.0	7.3	Siltstone	0.005	106.0	+106.0
4	2.1	8.6	Siltstone	0.005	111.0	+111.0
5	3.0	8.4	Sandstone	0.010	373.0	+373.0
6	0.33	8.7	Mudstone	0.010	164.0	+164.0
7	1.6	8.6	Siltstone	0.025	147.0	+146.0
8	4.1	8.5	Siltstone	0.029	173.0	+172.0
9	1.1	8.1	Mudstone	0.029	6.2	+5.3
10	0.33	8.4	Siltstone	0.005	97.9	+97.7
Covered Zone	5.0					
11	5.0	7.6	Mudstone	0.029	141.0	+140.0
12	5.0	7.6	Mudstone	0.025	141.0	+266.0
13	0.82	8.1	Siltstone	0.025	498.0	+497.0
14	2.8	7.2	Siltstone	0.010	135.0	+135.0
15	0.82	7.0	Shale	0.005	160.0	+160.0
16	5.0	7.7	Sandstone	0.140	207.0	+203.0
17	5.0	7.2	Siltstone	0.090	152.0	+149.0
18	5.0	4.6	Siltstone	0.229	121.0	+114.0
19	5.0	7.2	Siltstone	0.045	293.0	+149.0
20	5.0	7.3	Siltstone	0.055	151.0	+149.0
Covered Zone	5.0					
21	5.0	8.7	Sandstone	0.080	151.0	+149.0
22	5.0	7.6	Sandstone	0.048	196.0	+195.0
23	5.0	7.5	Sandstone	0.045	160.0	+159.0
24	5.0	7.8	Sandstone	0.038	164.0	+163.0
25	5.0	8.2	Sandstone	0.015	162.0	+162.0
26	5.0	7.2	Sandstone	0.030	180.0	+179.0
27	5.0	7.2	Siltstone	0.055	185.0	+184.0
28	5.0 9.5	7.0	Mudstone coal (Armstrong)	1.105	2.5	−32.1
29	4.0	2.5	Siltstone	0.303	1.0	−8.5
30	4.7	3.2	Coal & shale	2.325	0.2	−72.4
31	5.0	4.0	Siltstone	0.283	1.7	−7.1
32	5.0	7.1	Siltstone	0.019	17.3	+16.7
33	5.0	8.0	Siltstone	0.045	13.0	+11.6
34	5.0	7.1	Siltstone	0.098	13.3	+10.6
35	5.0	8.3	Mudstone	0.035	25.0	+23.9
36	5.0	8.1	Mudstone	0.030	21.8	+20.9
37	5.0	8.3	Siltstone	0.039	24.5	+23.3
38	5.0	8.0	Siltstone	0.205	40.0	+33.6
39	5.0	7.2	Mudstone	0.163	27.3	+22.2
40	5.0	7.5	Shale & coal	0.025	15.3	+14.5

Net for section $\frac{[(+17{,}282.13)+(-569.9)]}{163.56}$ = + 102.18 tons/1,000 tons excess material.

The cold alkalinity method is also simple and rapid, but it is not always reproducible.[2] There are several cold methods currently being used in the United States and Canada. They differ in solution-to-acid ratio, concentration of acid used, and contact time between sample and solution (2.5-4 hours) (*14*). These methods, when modified, are highly correlated to the hot acid technique of Sobek and associates (*51*).

pH methods. The most common measure of soil or rock characteristics is pH. Six factors affect the measurement of pH: (a) drying the overburden sample, (b) sample:water ratio used, (c) soluble salts content, (d) seasonally influenced CO_2 content, (e) amount of sample grinding, and (f) electrode junction potential (*29*). The pH of a sample characterizes the present condition of the sample at one point in time. It reflects past conditions by measuring active acidity, but it is not predictive of future acid or neutralization potential. Determination of pH can be helpful because many phytotoxic materials (heavy metals) increase in solubility when pH is less than 4.0 (where acidity is dominated by mineral acids in solution). No other measurement is more useful.

The ease of obtaining a 1:1 (volume:volume) pH measurement has made it the most popular technique used; however, the more accurate measurement of pH is the paste method (*41*). Easy preparation of paste pH is achieved with a 2:1 soil:water ratio, which gives a satisfactory paste with most samples. More water can be added to fine-textured material or more sample added to coarse-textured materials if necessary to develop the desired paste for pH measurements. For a more complete discussion of the subject, see chapter 1.

Rock type and depth. Rock type (description of overburden materials), as defined by Sobek and associates (*51*), indicates the mineralogy of the sample and is useful in predicting the chemical properties and reactions of the sample when exposed to oxidizing conditions (*5, 49*). Combined with depth, which indicates the sequence of samples from the land surface to the base of the pit and the thickness of the relative units, rock type provides a morphological picture of depositional history that can be used to make certain broad general assumptions about the chemical nature of the sample before laboratory testing begins.

Examples of acid-base accounting

Table 2 presents an acid-base account for an overburden section in Wyoming's northern Powder River Basin. Note that when a net deficiency or excess of bases is calculated for the section as a whole, there is a large excess

[2]Ibid.

of bases. The influence of a small section of pyritic, carbonaceous, fine-grained clastics adjacent to the mineable Armstrong coal would probably be negligible if this material were replaced well below the surface. Total S determinations for this section were made by LECO furnace. This type of result is fairly representative for many western mines where a thick section of alkaline-producing material exists and where local groundwater tends to be alkaline and/or saline due to dissolution of native alkaline/saline material and long regional flow paths.

Table 3 presents a similar acid-base account for a mine in Campbell County, Wyoming, with one important distinction: the acid-base account was calculated twice—once with S values obtained by Fisher S furnace and once with S values obtained by the peroxide method. Note the wide discrepancy in values obtained and the fact that the method used influences the net deficiency or excess of bases for the section as a whole. That is, the Fisher furnace values for S indicated a net excess of bases for the section, while the peroxide method values indicated a net deficiency of bases for the section.

Table 3. Comparative acid-base accounts by two methods for Campbell County, Wyoming, Mine (Wyodak Coal).

				Tons $CaCO_3$ Equivalent/1,000 Tons Material				
				Maximum Acid Potential		Maximum Neutralization	Excess (+) or Deficiency (−)	
Sample No.	Thickness (ft)	Paste pH	Rock Type	(Fisher)*	(H_2O_2)	Potential	(Fisher)	(H_2O_2)
1	10	7.3	Clay & sand	1.3	3.7	17.0	+16.0	+13.3
2	10	5.5	Silt	0.9	13.0	2.1	+1.2	−11.0
3	10	7.2	Clay, silty	6.9	53.0	6.2	−0.7	−47.0
4	10	7.6	Clay	3.4	20.0	11.0	+7.6	−9.0
5	10	7.6	Clay, silty	1.6	16.0	8.2	+6.6	−7.8
6	10	7.6	Silt	1.6	25.0	7.9	+6.3	−17.0
7	10	6.8	Clay & sand	8.8	51.0	18.0	+9.2	−33.0
8	10	6.8	Clay	5.0	64.0	6.6	+1.6	−57.0
9	10	6.1	Clay, carbonaceous; coal	5.9	44.0	7.7	+1.8	−36.0
10	60	No Sample	-	-	-	-	-	-
11	10	5.5	Clay, carbonaceous	5.6	18.0	4.4	−1.2	−13.6
12	10	7.4	Clay	5.3	74.0	20.0	+15.0	−54.0
13	10	6.0	Clay, silty	2.5	25.0	4.6	+2.1	−20.0
14	10	7.4	Clay	1.6	29.0	3.4	+1.8	−26.0
15	5	6.9	Clay & silt	3.8	46.0	11.0	+7.2	−35.0

*Based on (pyritic + organic) S.

$$\text{Net for section (Fisher)} = \frac{[(+764) + (-19)]}{135} = 5.52 \text{ tons } CaCO_3 \text{ equivalent/1,000 tons material.}$$

$$\text{Net for section } (H_2O_2) = \frac{[(+133.0) + (-3{,}664)]}{135} = -27.14 \text{ tons } CaCO_3 \text{ equivalent/1,000 tons material.}$$

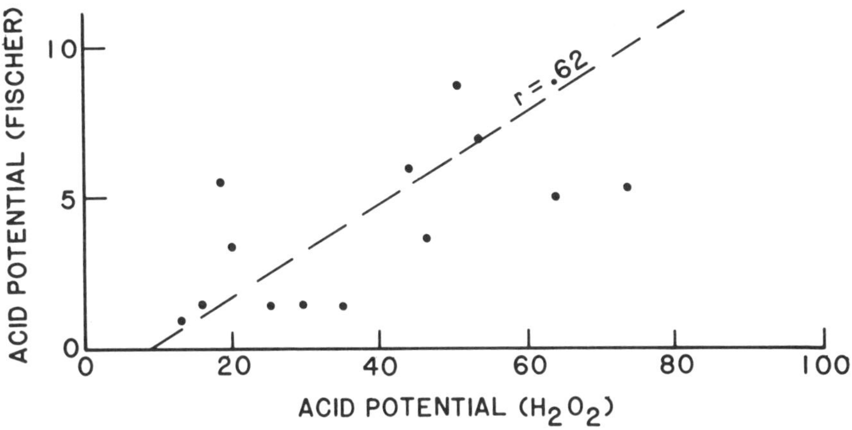

Figure 5. Plot of acid potential from Fisher sulfur furnace and H_2O_2 method for samples from drillhole in Campbell County, Wyoming.

Figure 5 shows a plot of the acid potential for the two methods. The two data sets yield a correlation coefficient (Pearson r) of 0.64. When viewed in relation to actual water quality data for this site (Table 4), the acid-base account done by S furnace, which shows a net excess of bases, becomes the more applicable of the two methods. In general, the water quality data suggest the presence of spoils that are neutral to alkaline with high dissolved solids, high sulfate, low Fe, and moderately high Ca. The low values for Fe and Mn in comparison with the relatively high values for alkalinity and Ca suggest an alkaline rather than an acid drainage. This example illustrates an extreme case of method-dependent results.

Table 5 is an acid-base account for the Lower Kittanning coal seam. In spite of a pyritic shale above the coal and an acid underclay, the section as a whole shows a net excess of bases. It is important to stress that the acid-base account is a measure of maximum potential excess or deficiency of bases. Thus looking at table 6, which gives ranges for Fe, Mn, and SO_4 in monitoring wells in reclaimed spoils, there is a wide range of values. These variations are due to placement of individual lithologic units in the spoil sequence and to the complex relationship of the freshly placed spoils to local and regional flow systems, where pre-existing water quality may be influencing spoil water quality.

For comparison with the acid-base account, some alternative leaching tests were also done with overburden samples from the same section shown in table 6. A "Caruccio Plot" (Figure 6) gives the overburden data illustrated according to the patented method of Caruccio and Caruccio (*14*). In this method individual samples are assigned a designation of acid or alkaline based on net acidity or alkalinity of leachate after 6 weeks of leaching in

Table 4. Water quality data for a spoil monitoring well about 0.7 mile from drill hole, Campbell County, Wyoming, Mine (Wyodak Coal). Data from seven samples taken between April 6, 1982, and September 29, 1983, at 2- to 4-month intervals.

Parameter	Minimum Value	Maximum Value
pH	7.0	8.1
Conductivity (μmho/cm)	5,300.0	6,400.0
Total dissolved solids (mg/liter)	5,150.0	6,140.0
Alkalinity (mg/liter)	429.0	530.0
Sulfate (mg/liter)	3,350.0	3,650.0
Ca (mg/liter)	467.0	533.0
Fe (mg/liter)	0.07	0.38
Mn (mg/liter)	0.32	1.25

Table 5. Acid-base account for Kittanning coal mine.

					Tons $CaCO_3$ Equivalent/1,000 Tons Material		
Sample No.	Thickness (ft)	Paste pH	Rock type	Total Sulfur (%)	Maximum from Percent Total S* (acid potential)	Amount Present (neutralization potential)	Excess (+) or Deficiency (−)
1	18.0	6.1	Siltstone	<0.01	-	2.6	+2.6
2	23.4	7.7	Sandstone	<0.01	-	4.8	+4.8
3	8.6	6.2	Shale	0.09	2.8	5.7	+2.9
4	16.5	7.3	Shale	0.26	8.1	9.3	+1.2
5	6.3	7.3	Shale, (Middle Kittanning Coal Horizon)	0.36	11.2	23.2	+12.0
6	6.9	7.8	Sandstone	0.15	4.7	24.7	+20.0
7	10.0	8.1	Shale	0.03	1.0	22.0	+21.0
8	25.2	7.6	Shale	0.53	16.6	20.0	+3.4
9	5.8	7.4	Shale	1.90	59.4	58.0	−1.4
10	No Sample	-	Coal, (Lower Kittanning)	-	-	-	-
11	5.3	7.5	Claystone	.95	29.8	7.8	−22.0

$$\text{Net for section} = \frac{[(+713.14) + (-124.72)]}{126.0} = +4.67 \text{ tons } CaCO_3 \text{ equivalent/1,000 tons material.}$$

Table 6. Water quality data for Kittanning coal mine; samples taken from 10 locations in reclaimed spoils on April 14, 1983.

Parameter	n	Minimum Value	Maximum Value
Total Fe (mg/liter)	10	<0.05	28.9
Total Mn (mg/liter)	10	<0.01	46.4
SO_4 (mg/liter)	10	21.0	1,453.0

plastic humidity cells. The values for alkaline production potential, which is twice the neutralization potential, and total S are then plotted against each other (Figure 6); each data point is designated as acid or alkaline, based on the 6-week leaching. Finally, a conservative line is drawn to connect the lowest alkaline samples. This method is designed to split the data field into alkaline and acid sectors. In this case only one sample (No. 1) fell outside the alkaline sector, that is, below the "A Line" (Figure 6).

Results of the ASTM leaching test for the same samples (Figure 7) are similar to the results of the Caruccio Plot, that is, only sample 1 generally fell into the acid-producing category. It might be argued that both of these leaching tests are either relatively gentle (Caruccio) or of such short duration (ASTM) that the results indicate only the removal of already weathered or degraded phases without weathering of the primary pyrite. Thus, these tests tend to indicate a minimum measure of potential acidity or alkalinity, as opposed to the acid-base account, which gives a maximum measure.

This argument is supported by two lines of evidence. First are the values for saturated-paste pH for this section (Table 5). Only samples 1 and 3 had a pH below 7.0, suggesting that a simple pH determination as a measure of immediately available acid-producers or neutralizers, frequently yields essentially the same information as the more time-consuming leaching tests.

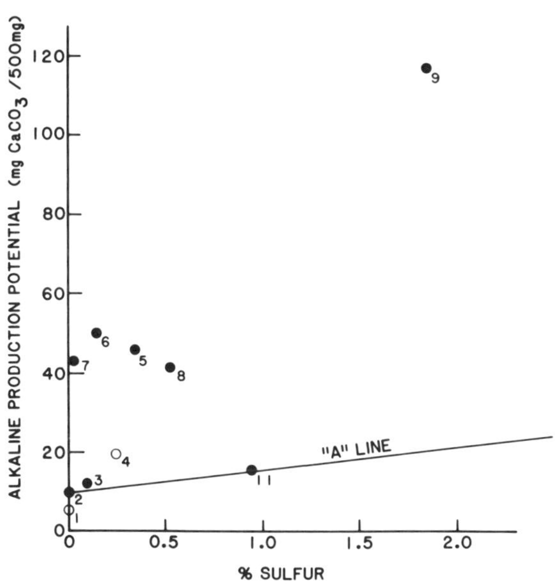

Figure 6. Caruccio plot for Kittanning overburden coal. Closed circles indicate alkaline units and open circles indicate acid units based on net acidity of alkalinity after 6 weeks leaching in humidity cells. Numbers refer to sample numbers in table 5.

The second line of evidence is a plot of modified soxhlet leaching for a highly weathered pyritic gob from Staunton, Illinois (Figure 8). The soxhlet leaching, in a side-by-side comparison with humidity cell leaching (*58*), consistently showed higher cumulative values for all variables achieved in a shorter period of time. Note in figure 7 that values for SO_4, acidity, and conductivity rapidly declined while weathered sulfates were being removed from the samples. This was followed by an increase in these variables (and an associated decline in pH) when chemical weathering of pyrite was initiated in the 144-hour leachate. Thus, one must carefully evaluate what a leaching test does in relation to what information is desired.

Management recommendations

Pyritic S and elemental S are probably the only S forms needed when determining the acid-generation characteristic of overburden sediments. Oxi-

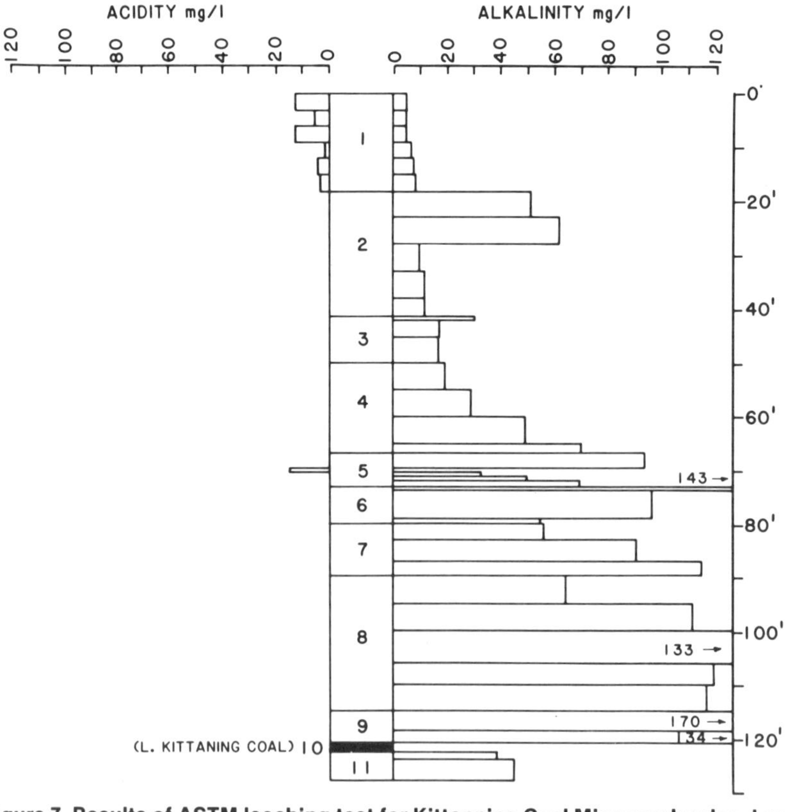

Figure 7. Results of ASTM leaching test for Kittanning Coal Mine overburden. Leaching test done on subsamples of same core reported in table 6. See table 5 for descriptions of numbered samples.

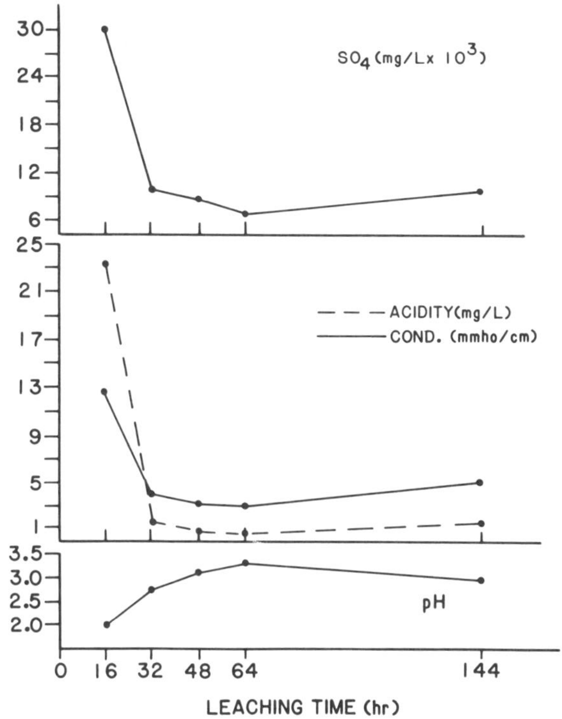

Figure 8. Modified Soxhlet leaching data for Staunton gob.

dation of pyrite is the main source of acidity in most mine spoils and acid mine drainage. Pyrites are oxidized in natural systems by both chemical and biological reactions, and acidity produced from these reactions overwhelms all other sources of acid in an oxidizing environment. In general, biological oxidation is considered most important because its rate of acid formation is much higher. Biological oxidation requires the presence of specific strains of S- and Fe-oxidizing bacteria.

The contribution of organic S to spoil acidity is probably negligible. Boon and Smith (*7*) cited other work indicating micron size pyrite is embedded in carbonaceous matter and is protected from HNO_3 extraction by the organic matrix. Thus, acid generated by these sulfides can be mistaken for acid generated by organic S. Recent experiments to define the amount of acid produced from organic S or carbonaceous overburden material yielded mixed results.[3] Some samples have yielded amounts of titratable acidity in excess of the theoretical maximum calculated from the stoichiometric equation for pyrite oxidation. Other samples that were washed free of all inorganic S

[3]Dollhopf, D. Personal communication, 1984.

compounds, following the ASTM procedure D2492-77 and artificially weathered, produced no acidity, although the sample contained 3 percent organic S. Recent work indicates the problems and potential errors in determining organic and pyritic S accordng to the ASTM method are a result of analyzing a sample that is not ground fine enough (*3, 56*). Therefore, weathering data based on the quantities of pyritic versus organic S must be viewed carefully or a wrong conclusion can be inferred.

Although organically bound S includes both soil organic S and coal organic S, conclusions drawn as to the ability of organic S to generate acidity quickly under normal atmospheric conditions must be taken more from coal research than from soils research. The reason is that coal overburden and coal are approximately the same age, while recent organic matter has not undergone the diagenetic processes that coal overburden and coal have experienced.

Finally, there are certain conclusions that can be drawn from the preceding discussion:

► The outcome of overburden analysis is method-dependent; therefore, the person evaluating data must know the capabilities and limitations of each methodology.

► The acid-base account method is the best available technique for evaluating acid potential of overburden materials.

► Weathering methods should be used to help resolve marginal acid-producing potential determined by acid-base accounting.

► The organic S content of sediments will not cause an acid generation problem under natural environmental conditions. There is no theoretical basis in the literature to substantiate claims that organic S will cause acid problems. In fact, most recent evidence indicates mircon-size sulfides embedded in the organic matric of carbonaceous materials are responsible for the perceived acid generation from organic S.

ACKNOWLEDGEMENT

The authors thank Jean E. Bogner of the Argonne National Laboratory and Larry Ligocki of Kiewit Mining & Engineering Company for their constructive review of this manuscript. Ms. Bogner was especially helpful in providing timely suggestions and ideas.

REFERENCES

1. Altschuler, Z. S., M. M. Schenepfe, C. C. Silver, and F. O. Simons. 1983. *Sulfur diagenesis in Everglades peat and origin of pyrite in coal.* Science 221 (4607): 221-227.
2. Arkesteyn, G.J.M.W. 1979. *Pyrite oxidation by* Thiobacillus ferrooxidans *with special reference to the sulfur moiety of the mineral.* Antonie van Leluwenholk Journal of Microbiology Serology 45: 423-425.
3. Attar, A. 1979. *Sulfur groups in coal and their determinations.* In C. Karr, Jr.

[editor] *Analytical Methods for Coal and Coal Products.* Academic Press, New York, New York. pp. 585-624.

4. Baker, M. 1975. *Inactive and abandoned underground mines-water pollution prevention and control.* EPA 440/9-75-007. U.S. Environmental Protection Agency, Washington, D.C.
5. Banaszak, K. J. 1981. *Predicted changes in the mineralogy of a spoil as a function of net neutralization potential and rate of flushing.* In *Symposium on Surface Mining Hydrology Sedimentology and Reclamation.* University of Kentucky, Lexington. pp. 459-462.
6. Bardsley, C. E., and J. D. Lancaster. 1960. *Determination of reserve sulfur and soluble sulfates in soils.* Soil Science Society of America Proceedings 14: 265-268.
7. Boon, D. Y., and P. J. Smith. 1985. *Carbonaceous materials problems associated with reclamation.* In Proceedings, Second Annual Meeting of the American Society of Surface Mine Reclamation. American Society of Surface Mine Reclamation, Princeton, West Virginia. pp. 358-256.
8. Bremmer, J. M. 1965. *Soil organic nitrogen in soil nitrogen.* Monograph No. 10. American Society of Agronomy, Madison, Wisconsin.
9. Brunynesteyn, A., and R. P. Hackl. 1982. *Evaluation of acid production potential of mining waste materials.* Minerals and the Environment 4: 5-8.
10. Carson, C. D., D. S. Fanning, and J. B. Dixon. 1982. *Alfisols and ultisols with acid sulfate weathering features in Texas.* In *Acid Sulfate Weathering.* Soil Science Society of America, Madison, Wisconsin. pp. 127-146.
11. Caruccio, F. T. 1968. *An evaluation of factors affecting acid mine drainage production and the ground water interactions in selected areas of western Pennsylvania.* In Proceedings, Second Symposium on Coal Mine Drainage Research, Bituminous Coal Research, Monroeville, Pennsylvania. pp. 107-152.
12. Caruccio, F. T. 1975. *Estimating the acid production potential of coal mine reuse.* In Chadwick and Goodman [editors] *The Ecology of Resource Degradation and Renewal.* Blackwell Scientific Publication, London, England. pp. 197-205.
13. Caruccio, F. T., and G. G. Caruccio. 1980. *An evaluation of the proposed Holly Grove Mine (Canaan, W. Va.) to impact the Little Kanawha River with acid waters.* Special Report prepared for the U.S. Environmental Protection Agency. GARGEID, Columbia, South Carolina. p. 89.
14. Caruccio, F. T., and G. G. Caruccio. 1982. *Method of determining weathering characteristics of rock formations in earth moving operation.* United States Patent 4,328,001. U.S. Government Printing Office, Washington, D.C.
15. Caruccio, F. T., J. C. Ferm, J. Horne, G. Geidel, and B. Baganz. 1977. *Paleoenvironmental of coal and its relation to drainage quality.* EPA-600/7-77-067. U.S. Environmental Protection Agency, Washington, D.C.
16. Casagrande, D. J., K. Siefert, C. Berschinski, and N. Sutton. 1977. *Sulfur in peat-forming systems of the Okefenokee Swamp and Florida Everglades: Origins of sulfur in coal.* Geochimica et Cosmochimica Acta 41: 161-167.
17. Chakrabarti, J. N. 1979. *Analytical procedures for sulfur in coal desulfurization products.* In C. Karr, Jr. [editor] *Analytical Methods for Coal and Coal Products.* Academic Press, New York, New York. pp. 179-332.
18. Duncan, D. W., and C. C. Walden. 1975. *Prediction of acid generation potential.* Report Proposed by B. C. Research Ltd. Water Pollution Control Directorate, Environmental Protection Service, Environment Canada, Ottawa, Ontario.
19. Eccleston, M., and D. P. Kelly. 1978. *Oxidation kinetic and chemostat growth kinetics of* Thiobacillus ferrooxidans *on tetrathionate and thiosulfate.* Journal of Bacteriology 134: 718-727.

20. Geidel, G. 1978. *Alkaline and acid production potentials of overburden material: The rate of release.* Reclamation Review 2: 101-107.
21. Grim, E. C., and R. D. Hill. 1974. *Environmental protection in surface mining of coal.* EPA 67012-74-093. National Environmental Research Center, Office of Research and Development, Environmental Protection Agency, Washington, D.C.
22. Groudev, S. 1979. *Mechanism of bacterial oxidation of pyrite.* Mikrobiologija 16: 75-87.
23. Grube, W. E., Jr., R. M. Smith, R. N. Singh, and A. A. Sobek. 1973. *Characterization of coal overburden materials and minesoils in advance of surface mining.* In *Research and Applied Technology Symposium on Mined-Land Reclamation.* Bituminous Coal Research, Inc., Monroeville, Pennsylvania. pp. 134-152.
24. Hanna, G. P., and R. A. Brant. 1962. *Stratigraphic relations to acid mine water production.* In Proceedings, 17th Industrial Waste Conference, Series 112. Engineering Extension, Purdue University, Lafayette, Indiana.
25. Harvard University. 1970. *Oxygenation of ferrous iron.* Project 14010. Federal Water Quality Administration, U.S. Department of the Interior, Washington, D.C.
26. Harward, M. E., T. T. Chao, and S. C. Fang. 1962. *The sulfur status and sulfur supplying power of Oregon soils.* Agronomy Journal 54: 101-106.
27. Hill, R. D. 1970. *Elkins mine drainage pollution control demonstration project.* In *Third Symposium on Coal Mine Drainage Research.* Bituminous Coal Research Inc., Monroeville, Pennsylvania. pp. 284-303.
28. Hutchins, S. R., M. S. Davidson, J. A. Brierley, and C. L. Brierley. 1986. *Microorganisms in reclamation of metals.* Annual Rev. Microbiology 40: 311-336.
29. Jackson, M. L. 1958. *Soil chemical analysis.* Prentice Hall, Englewood Cliffs, New Jersey.
30. Knabe, W. 1973. *Development and application of the "Domsdorf Ameleorative Treatment" on toxic spoil banks of lignite opencast mines in Germany.* In R. J. Hunik and G. Davis [editors] *Ecology and Reclamation of Devastated Land.* Gordon and Breach Science Publication Inc., New York, New York. pp. II: 273-295.
31. Koller, P., and W. Knabe. 1962. *The reclamation of lands stripped for brown coal.* Forestry Department Series. Ohio Agricultural Experiment Station, Wooster.
32. Kosaka, J. 1971. *Report on acid sulfate soils in the Muda Irrigation area.* (Japanese, English summary). Bulletin 17:29. Tropical Agricultural Research Center (Japan).
33. Lowe, L. E., and W. A. DeLong. 1960. *Carbon bonded sulfur in selected Quebec soils.* Soil Science Society of America Proceedings 24: 337-340.
34. Manning, H. L., and T. M. Cook. 1972. *Physicology of acidophilic bacteria of acid mine water.* Technical Report No. 8. Water Resources Research Center, University of Maryland, College Park.
35. Miller, W. L., C. L. Godfrey, W. C. McCully, and G. W. Thomas. 1976. *Formation of soil acidity in carbonaceous soil materials exposed by highway excavations in east Texas.* Soil Science 121: 162-169.
36. Meyers, R. A. 1977. *Coal desulfurization.* Marcel Dekker, Inc., New York, New York.
37. Morth, A., E. E. Smith, and K. S. Shumate. 1972. *Pyritic systems: A mathematical model.* EPA-R2-72-002, Project No. 14010 EAH. U.S. Environmental Protection Agency, Washington, D.C.
38. Nordstrom, D. K. 1982. *Aqueous pyrite oxidation and the consequent formation of secondary iron minerals.* In *Acid Sulfate Weathering.* Soil Science Society of

America, Madison, Wisconsin. pp. 37-56.
39. Nordstrom, D. K., E. A. Jenne, and J. W. Ball. 1981. *Redox equilibria of iron in acid mine waters.* In E. A. Jenne [editor] *Chemical Modeling in Aqueous Systems, Speciation, Sorption, Solubility, and Kinetics.* American Chemical Society, Washington, D.C. pp. 51-79.
40. O'Shay, T. A. 1982. *The determination of potential acidity in overburden sediments.* Masters Thesis. Texas A&M University, College Station.
41. Peech, M. 1965. *Hydrogen-ion activity.* In C. A. Black [editor] *Methods of Soil Analysis.* American Society of Agronomy, Madison, Wisconsin. pp. 914-920.
42. Scholz, A. G., and D. Rathleff. 1982. *A rapid method to determine total sulphur.* Journal of Coal Quality 2(1): 16-19.
43. Silver, M., and A. E. Torma. 1974. *Oxidaton of metal sulfides by* Thiobacillus ferrooxidans *grown on different substrates.* Canadian Journal of Microbilogy 20: 141-147.
44. Silver, M., and D. G. Lundgren. 1968. *Sulfur-oxidizing enzyme of* Ferrobacillus ferrooxidans (Thiobacillus ferrooxidans). Canadian Journal of Biochemistry 46: 457-461.
45. Silver, M., and D. G. Lundgren. 1968. *The thiosulfate-oxidizing enzyme of* Ferrobacillus ferrooxidans (Thiobacillus ferrooxidans). Canadian Journal of Biochemistry 46: 1,215-1,220.
46. Smith, E. E., and K. S. Shumate. 1970. *Sulfide to sulfate reaction mechanism.* Grant No. 14010 FPS. Federal Water Quality Administration, U.S. Department of the Interior, Washington, D.C.
47. Smith, R. M., and A. A. Sobek. 1978. *Physical and chemical properties of overburdens, spoils, wastes, and new soils.* In P. Sutton [editor] *Reclamation of Drastically Disturbed Lands.* American Society of Agronomy, Madison, Wisconsin. pp. 149-172.
48. Smith, R. M., W. E. Grube, Jr., R. N. Singh, and A. A. Sobek. 1974. *Rock types and laboratory analysis as a basis for managing minesoils.* Tenth Forum on Geology of Industrial Minerals Proceedings. Ohio Department of Natural Resources, Columbus. pp. 47-52.
49. Smith, R. M., W. E. Grube, Jr., T. Arkle, Jr., and A. Sobek. 1974. *Mine spoil potentials for soil and water quality.* EPA 670/2-74-070. U.S. Environmental Protection Agency, Washington, D.C.
50. Sobek, A. A., and J. E. Bogner. 1984. *Using the acid-based account to predict acid potential of coal overburden.* In Proceedings, Third Biennial Symposium on Surface Coal Mining and Reclamation in the Great Plains. Montana State University, Bozeman.
51. Sobek, A. A., W. A. Schuller, J. R. Freeman, and R. M. Smith. 1978. *Field and laboratory methods applicable to overburdens and mine soils.* EPA 600/2-78-054. U.S. Environmental Protection Agency, Cincinnati, Ohio.
52. Sobek, A. A., R. M. Smith, and T. Arkle, Jr. 1976. *Planning and classification of highly disturbed soils.* In Proceedings, International Hill Land Symposium, West Virginia University, Morgantown. pp. 535-539.
53. Sorensen, D. L., W. A. Kneib, and D. B. Porcella. 1979. *Determination of sulfide in pyritic soils and minerals with a sulfide ion electrode.* Analytical Chemistry 51: 1,870-1,872.
54. Sorensen, D. L., D. B. Porcella, and B. Z. Richardson. 1980. *Determining the lime requirement for the Blackbird mine spoil.* Journal of Environmental Quality 10: 162-166.
55. Sorensen, D. L., J. E. McLean, and E. E. Farmer. 1985. *Development of a bioassay for acid producing potential in mine spoils and overburden.* In Proceedings, Annual Meeting, American Society for Surface Mining and Reclamation.

American Society for Surface Mining and Reclamation, Princeton, West Virginia.

56. Stanton, R. W., and J. J. Renton. 1981. *Organic and pyritic sulfur in coal: Potential errors in determination.* Geological and Economic Survey Circular No. C-22. West Virginia University, Morgantown.
57. Stumm, W., and J. J. Morgan. 1970. *Aquatic chemistry.* Wiley, New York, New York.
58. Sullivan, P. J., and A. A. Sobek. 1982. *Laboratory weathering studies of coal refuse.* Minerals and the Environment 4: 9-16.
59. Sullivan, P. J., A. A. Sobek, and J. P. Rybarczyk. 1986. *Evaluating mineral dissolution in laboratory weathering experiments using leachate equilibria.* Soil Science Society of American Journal 50: 251-254.
60. Sullivan, P. J., S. V. Mattigod, and A. A. Sobek. 1986. *Dissolution of iron sulfates from pyritic coal waste.* Environmental Science and Technology 20: 1,018-1,016.
61. Tabatabai, M. A. 1982. *Sulfur.* In A. L. Page [editor] *Methods of Soil Analysis* (part 2): *Chemical and Microbiological Properties.* American Society of Agronomy, Madison, Wisconsin pp. 501-538.
62. Tabatabai, M. A., and J. M. Bremmer. 1972. *Forms of sulfur and carbon, nitrogen and sulfur relationships in Iowa soils.* Soil Science 114: 380-386.
63. Temple, K. L., and W. A. Koehler. 1954. *Drainage from bituminous coal mines.* Research Bulletin 25. West Virginia University Engineering Station, Morgantown.
64. Tuovinen, O. H., and D. P. Kelly. 1972. *Biology of* Thiobacillus ferrooxidans *in relation to the microbiological leaching of sulphide ores.* Zeitschrift fuer Allg. Mikrobiologie 12: 311-346.
65. Van Breemen, N. 1973. *Soil forming processes in acid sulfate soils.* Publication 18, Volume I. In H. Dost [editor] *Acid Sulfate Soils.* Internatonal Land Reclamation Institute, Wageningen, The Netherlands. pp. 66-130.
66. Van Breemen, N. 1976. *Genesis and solution chemistry of acid sulfate soils in Thailand.* Agricultural Research Report 848. PUDOC, Wageningen, The Netherlands. p. 263.
67. Van Breemen, N. 1982. *Genesis, morphology and classification of acid sulfate soils in coastal plains.* In *Acid Sulfate Weathering.* Soil Science Society of America, Madison, Wisconsin. pp. 95-108.
68. Verhoeven, B. 1973. *Acid sulfate soils in the Wieringermeerpolder, a good 40 years after reclamation.* Publication 18, Volume II. In H. Dost [editor] *Acid Sulfate Soils.* International Land Reclamation Institute, Wageningen, The Netherlands. pp. 114-128.
69. Vestal, J. R., and D. G. Lundgren. 1971. *The sulfite oxidation of* Thiobacillus ferrooxidans. Canadian Journal of Biochemical 4: 1,125-1,130.
70. Vlek, P.L.G., Th. J. M. Blom, J. Beek, and W. L. Lindsay. 1974. *Determination of solubility product of various iron hydroxides and jarosite by the chelation method.* Soil Science Society of America Proceedings 38: 429-432.
71. Zuberer, D. A. 1971. *Microbial oxidations of pyritic mine spoil materials.* M.S. thesis. West Virginia University, Morgantown.

Chapter 12

Rock fragments

Larry Munn, Neil Harrington, and Dennis R. McGirr

The quantity, size, lithology, and distribution of rock fragments affect the productivity and distribution of native plant communities on rangeland and forest land and the productivity of cropland. These properties also affect the success of reclamation efforts on mined land.

Rock fragments result from the weathering of residual parent material. Fragments remain near their point of origin or are transported and deposited through colluvial, alluvial, and glacial processes. Rock fragments in the soil reflect both the initial rock fragment content of the soil parent material and the weathering and redistribution that has occurred within the soil over time. Some soils contain essentially no rock fragments, while other soils consist mostly of loose rock with little fine material.

A soil's rock fragment content affects many soil physical and chemical properties, such as water-holding capacity, infiltration rate, permeability, aeration, pore space, erodibility, soluble salt content, fertility, and bulk density. Each of these properties has certain influences on the effectiveness of the soil as a plant growth medium.

Definitions and descriptions

A rock fragment must meet a slaking test to qualify as a rock fragment. The slaking test identifies those fragments that will break down when shaken overnight in a sodium hexa-metaphosphate solution (*90, 91*). Scientists use this test to distinguish between a "field" rock fragment—what appears to be a rock fragment in the field—and a "laboratory-qualified" rock fragment.

Because of the interest in rock fragment weathering observed in reclaimed mine soils, researchers have sought to improve rock fragment identification. Sobek and associates developed three slaking procedures of different intensity (*88*). A mild water slaking test indicates the percentage of material that will weather quickly to provide fine earth in mine soils. A more intensive slaking procedure involves shaking the material for 16 hours in a dispersing agent followed by sieving. This method presumably represents the proportion of material that will weather under intense environmental influences at the spoil surface. The third method, intermediate in intensity, uses a lower strength dispersing agent, overnight submergence, and limited hand-shaking. This method closely approximates that anticipated in soil surveys for determining "true" rock fragment content (*91*).

Soil scientists generally evaluate rock fragments in soils in terms of volume relative to the whole soil volume, in terms of size and shape, and in terms of the spacing of surface fragments. The classification developed by the Soil Conservation Service (*91*) is the method commonly used to evaluate soil rock fragments during land inventories in the United States. Materials with a diameter smaller than 2 mm are referred to as fine earth materials (*90*). Materials greater than 2 mm in diameter are described as rock fragments. Materials between 2 mm and 25 cm in diameter are called coarse fragments, although this term appears to often be used in a more generic sense without adherence to this definition in the scientific community. Table 1 shows size limits and terminology for rock fragments of various shapes.

Soil scientists describe the lithology of rock fragments during soil profile descriptions. But the description of rock type often is not complete, particularly in alluvial soils and others with rocks of mixed lithology.

Table 1. Classification of rock fragments by shape and size (*91*).

Fragment	Diameter of Rounded, Subrounded, Angular, or Irregular Fragments*	Length of Flat Fragment*
	cm	
Gravel†	0.2-7.6	
Fine gravel	0.2-0.5	
Medium gravel	0.5-2	
Coarse gravel	2-7.6	
Cobble	7.6-25	
Stone	25-60	
Boulder	>60	
Channer		0.2-15
Flagstone		15-38
Stone		38-60
Boulder		>60

*If shape is significant to classification or interpretations, it is indicated as, for example, "angular gravel" or "irregular boulders."
†A single gravel fragment is called a pebble.

In the Soil Conservation Service (SCS) system the quantity of rock fragments in a soil is designated by adding a modifying term to the textural class name, as follows (*91*):

► *Less than 15 percent rock fragments by volume:* No special term is used or "nongravelly" and "nonstony" are used in writing for contrast with soils having more than 15 percent pebbles, cobbles, stones, or boulders.

► *Fifteen to 35 percent rock fragments by volume:* The adjective term of the dominant kind of rock fragment is used as a modifier of the textural term, for example, "gravelly loam," "stony loam," or "bouldery loam."

► *Thirty-five to 60 percent by volume:* The adjective term of the dominant kind of rock fragment is used with the word "very" as a modifier of the textural term, for example, "very gravelly loam" or "very bouldery loam."

► *More than 60 percent rock fragments by volume:* If enough fine earth is present to determine the texture class (about 5 percent or more by volume), the adjective term of the dominant kind of rock fragment is used with the word "extremely" as a modifier of the textural term, for example, "extremely gravelly loam" or "extremely bouldery loam." If there is too little fine earth to determine the texture class (less than about 5 percent by volume), the terms "gravel," cobbles," "stones," and "boulders" are used in place of fine earth texture.

► The class limits apply to the volume of the layer occupied by all pieces of rock larger than 2 mm. Total volume of rock fragments in each layer is estimated and the size class that makes up the greatest volume is used as the modifier.

The SCS classification system groups soils on the basis of exposed bedrock (percent coverage) as well as on the basis of stoniness or boulderiness (surface coverage and distance between fragments) (*91*). There are seven classes of rock outcrop coverage and six classes of stoniness or boulderiness (Tables 2 and 3). The percentage limits used as boundaries between classes are different for bedrock outcrop than for loose rock fragments.

Another well-known rock fragment classification is that of the American Association of State Highway Officials (*3*). This system designates two general categories for particle sizing: granular materials—35 percent or less passing No. 200 (75-μm) sieve—and silt-clay materials—more than 35 percent passing No. 200 sieve. The granular grouping is subdivided into four classes: boulders, retained on a 75-mm sieve; gravel, will pass a 75-mm sieve, retained on a 2-mm sieve; coarse sand, will pass a 2-mm sieve, retained on a 425-μm sieve; and fine sand, will pass a 425-μm sieve, retained on a 75-μm sieve.

Another engineering classification, the Unified Classification of the American Society for Testing and Materials (*4*), evaluates material smaller than 75 mm in diameter. In this system material is "gravel" if 50 percent or more of the material that is retained on the No. 200 (75-μm) sieve is large enough to be retained on the No. 4 (4.75-mm) sieve. The system subdivides gravels into a number of groups of clean gravels or of gravels with fines, depending upon the particle size of the fine materials (less than 0.75-mm diameter) and the uniformity of the gravels.

Rock fragments in soil classification

Rock fragment content is a determinant of particle size classes (family taxonomic level) for mineral soils and for some (terric) subgroups of organic soils (*90*). Soils consisting of stones, cobbles, gravel, and very coarse sand particles in the textural control section with too little fine earth material to fill the interstices larger than 1 mm in diameter are called fragmental. If 60 percent or more of the whole soil (weight basis) consists of volcanic ash, cinders, and pumice and 35 percent or more (volume basis) is composed of cinders that have a diameter of 2 mm or greater, the soil is called cindery. If there is enough fine earth material to fill the interstices greater than 1 mm in diameter and rock fragments make up 35 percent or more by volume, the term skeletal is used in conjunction with a term describing the particle size distribution of the fine earth material, for example, "clayey-skeletal" or "loamy-skeletal."

SCS manuals recognize that some soils have adjacent horizons with markedly contrasting textures and rock fragment contents (*90*). Where the transition between horizons is less than 12.5-cm thick, some 40 combinations of strongly contrasting particle-size classes are identified, including many based in part on rock fragment content. Examples include "loamy-skeletal over sandy" and "fine-silty over sandy or sandy-skeletal." The intent of these designations is to recognize changes in pore-size distribution that seriously affect movement of water within the profile.

The term lithic contact is defined as the boundary between soil and coherent underlying material, that is, hard rock. A lithic contact is diagnostic at the subgroup taxonomic level if it is within 50 cm of the mineral soil surface (*90*). Similarly, a paralithic contact is recognized between the soil

Table 2. Classes of area of rock outcrop, including soil supporting little or no vegetation (*91*).

Class	Percent of Ground Surface of Exposed Bedrock
1	<0.1
2	0.1 to 1
3	2 to 10
4	10 to 25
5	25 to 50
6	50 to 90
7	>90

Table 3. Classes of stoniness and boulderiness in relation to surface coverage and spacing between rock fragments (*91*).

Class	Percentage of Surface Covered	Distance Between Stones or Boulders (m) by Diameter of Fragments		
		25 cm	60 cm	120 cm
1	<0.01	>25	>60	>120
2	0.01-0.1	8-25	20-60	37-120
3	0.1-3	1-8	3-20	6-37
4	3-15	0.5-1	1-3	2-6
5	15-50	0.1-0.5	0.2-1	0.5-2
6	>50	<0.1	<0.2	<0.5

and underlying coherent but softer material. One criterion given for distinguishing materials below lithic and paralithic contacts is that material below a paralithic contact "when moist, can be dug with difficulty with a spade" (*90*). Other criteria are associated with cracks and root penetration. It often is difficult to identify paralithic contacts in the field (*70*). Lithic contacts usually are identified easily, except in small holes where rock fragments are large.

Measuring rock fragment content

Evaluating the rock fragment content of soil or overburden is useful only if the methodologies, definitions, and terminology used are clearly specified or evident. The key to soil description or sample collection in the field, using any methodology, is to follow procedures in a consistent, methodical manner.

Scientists generally report the rock fragment content of soils as volume percentages and percent surface coverage (*89, 92*). In practice, soil scientists commonly estimate volume percentages visually (*89, 92*) or derive those values from weight percentages obtained by sieving and weighing the 2-mm to 76-mm fraction. If a soil contains appreciable amounts of fragments greater than 76 mm in diameter, weighing becomes difficult. If percentages obtained by weighing (2 mm to 76 mm) are combined with percentages obtained by visual estimates, care should be taken not to imply unwarranted precision.

Daniels and associates described a point-count method for evaluating soil rock fragment content (*31*). Transecting and visual estimates can be used to evaluate surface rock coverage and stoniness (*17, 92, 93*). Evaluating soils containing large rock fragments is complicated by the need to have a sampling site (soil pit) at least twice the size of the largest rock fragment (*53*). Sampling rocky soils without a backhoe or similar equipment often results in a limited number of observations and/or inadequate sampling pit size. A significant complication is the problem of distinguishing between field and laboratory-qualified rock fragments discussed earlier. In addition, the presence of rock fragments complicates the evaluation of bulk density (*26, 30, 71, 86*) and soil hydraulic properties (*5, 72, 78*).

Rock fragment effects on soil properties

Assessment of the effects of rock fragment content on the quality of soil and spoil materials in terms of reclamation potential (*63, 107*) apparently is based upon a number of assumptions about the behavior of rock fragments in soils. Negative attributes of rock fragments include a reduction in rooting volume, a decline in total water-holding capacity, a reduction in total soil nutrients, and elevated surface temperatures as a result of surface rock coverage. Surface rock coverage also can cause problems with agricultural equipment during seeding and seedbed preparation on rangeland and pasture sites and during other cultural practices on sites planted to agronomic crops. Positive attributes of rock fragments in soils include reduced susceptibility to erosion and compaction and increased macroporosity.

Rock fragments in the soil reduce the volume of fine soil material available to plant roots. This generally results in lower cation exchange capacity, porosity, total nutrients, and water storage capacity than for similar soils without rock fragments (*25, 27, 71*). However, the assumption that rock fragments are inert material often is not true. Various researchers have reported important contributions by rock fragments to plant-available water (*25, 28, 83*), nutrient content (*104*) and cation exchange capacity (*25, 79*). Rock fragments effectively decrease root density, which increases the effectiveness of individual roots in obtaining water and nutrients (*97, 98, 99*).

Small rock fragments (2 to 5 mm in diameter) generally make a greater contribution to available water storage than larger fragments. Studies summarized by Childs (*25*) showed that available water-holding capacities varied from 3 to 12 percent by weight in rocks of varying lithologies. Pumice was reported to have an available water-holding capacity of 91 to 124 percent (*25*). The water-holding capacity of rock fragments is a function of rock type and weathering state, which influences porosity. Smaller rock fragments usually are more porous than larger fragments in natural soils (*43*). Coile (*28*) found that rock fragments hold from 5 to 25 percent water by weight at saturation.

Rock fragments also enhance the water-holding capacity of the soil on steep slopes by reducing erosion of the organic matter rich surface layer (*15, 25, 33*). Organic matter is distributed over a greater depth in stony soils (*25*), reflecting both altered root distribution and reduced erosion.

Rock fragments also influence soil water content and distribution of water within the profile. Rock fragments generally enhance water infiltration into soils (*33, 41*). A given quantity of water will penetrate to a greater depth in rocky soils compared with nonrocky soils of similar fine-earth texture (*13, 54*). Where rock-fragment distribution contributes to distinct soil layering, water movement downward may be slowed (*60, 102, 103*). Researchers have found that water concentrates under surface stones (*58*) and at points of contact between rock fragments within the soil (*6*). Hillel and Tadmore (*47*) showed that rocky soils have a lower water storage capacity (volume of water per unit depth). Birkeland (*13*) and Nettleton and associates (*65*) found that carbonates and clay normally are leached to greater depths in coarse-textured, gravelly soils than in fine-textured soils in arid and semiarid regions. In arid regions the deeper penetration of a limited amount of precipitation would reduce upward movement of water in response to evaporation (*49, 61, 80*). In general, upland native range sites in the drier portions of the Great Plains show evidence of limited leaching of water below the root zone (*37*).

While scientists generally attribute the cation exchange capacity of soils to their clay and organic matter fractions, there is evidence that the sand and silt fractions (*101*) and rock fragments (*79*) may contribute significantly to total cation exchange capacity. Rivard and DeKimpe (*79*) reported that particles greater than 2 mm in diameter contributed up to 69 percent of the total cation exchange capacity of soils derived from sandstones and shales. Shale and sandstone fragments often break down rapidly in the surface layers of reclaimed mine soils (*83*), resulting in an increase in fines. Such fragments may not meet the definition of true rock fragments as described earlier.

A number of researchers have noted the effect of stones in reducing soil erosion (*15, 24, 25, 33, 35, 51, 81, 87*). Researchers have proposed using rock fragment layers to control erosion on construction sites (*59*) and to conserve water in row crops (*1, 36*). Sencindiver and associates (*84*) found that a rock layer at the surface of mine spoil reduced erosion. For otherwise similar soil surface layers, the K factor (inherent soil erodibility) of the universal soil loss equation (USLE) is lower when rock fragments are present (*16, 93*). Based on the USLE, erosion from a soil surface layer containing 50 percent rock fragments by volume would be only 30 percent of that from a similar soil without rock fragments (*55*).

Research has shown beneficial effects of a stone mulch on agronomic crops, attributed to lower evaporation losses (*14, 36*). But negative effects ascribed to higher soil temperatures also have been reported (*1*). Researchers in central Montana reported increased spring wheat yields in fields with a stone mulch (*27*).

The specific heat of rock fragments generally is higher than that of fine earth particles (*25*). Soils with significant rock contents, therefore, have higher heat capacities than dry, rock-free soils. A higher heat capacity would result in slower warming and cooling of the soil. However, rock fragments also increase the rate of heat transfer in soils (*25, 57*). This may result in earlier initiation of biological activity in rocky soils in the spring, but also may increase evaporative losses of soil water.

Rock fragment effects on productivity and plant distribution

Quantifying the effect of rock fragments on vegetative productivity is difficult to document because, on natural landscapes, the rock fragment content of soils does not vary independently of other soil properties, such as texture, organic matter, mineralogy, exchangeable cations, and topographic position. This is further complicated by vegetative community differences associated with varying rock contents and topographic positions. Nevertheless, numerous studies have linked rock fragment distribution and content with plant productivity. The relationship varies according to climate and vegetation.

In forestry, site quality commonly correlates with available soil moisture and effective rooting volume (*7, 21, 22, 23, 28, 39, 40, 108*). Carmean (*21, 22, 23*) and Hannah (*42*) reported relationships between site index for oaks (*Quercus* spp.) and soil horizon texture and stone content in the eastern United States. Site index—tree height at a specific age—was lower on sites with appreciable subsoil rock fragment content (*22, 23*). In northern Idaho, Brown and Loewenstein (*18*) found a positive but low correlation ($r = 0.51$) between site index for mixed conifers (primarily Douglas-fir, *Pseudotsuga menziesii* var. *glauca,* and grand fir, *Abies grandis*) and the ratio of fine earth to rock content. They attributed this to the influence of rock on water-holding capacity. These findings are from areas of higher precipitation (76-127 cm), where the effect of rock fragments usually is negative. Site preparation for planting often is more difficult on skeletal soils (*48, 52*).

Magier and Ravina (*54*) found that soil rock content had positive effects on apple (*Malus sylvestris*) orchards in Israel under drip irrigation, within the limits of the study. They concluded, from their own work and from the data of other workers, that rock fragments greater than 19 mm in diameter and within the range of 5 to 30 percent of the soil volume, were correlated positively with numbers of roots, trunk circumference, and apple yield. They also found a positive relationship between aerated porosity at field capacity and soil rock fragment content; moreover, rock fragment contents of 30 to 60 percent reduced the effects of compaction on orchard soils. In contrast, soils with layers of densely packed small rock fragments and soils with less than 10 percent rock fragments produced poor tree growth. They considered rock fragment contents of 25 to 30 percent as optimum for apple production on these specific soils.

Using soil interpretation records from SCS computer files in the northeastern United States, Hidlebaugh (*46*) examined 34 records of yield ratings for crops and pastures and site indexes for trees in the same soil phase, with and without rock fragment modifiers. Only one record showed a higher crop yield for the rock-fragment-containing phase, while 16 records showed lower crop yields and 8 showed lower pasture yields. Other records showed yields were the same—9 for crops, 12 for pasture. Four records showed higher site indexes for tree growth in the phases containing rock fragments. The other 30 records showed yields were the same. Many tree species commonly are associated with rocky soils (*44*). Some sagebrush (*Artemesia spp.*) species (*10, 32*) and mountain mahogany (*Cercocarpus montanus*) (*56, 62*) are restricted to rocky or shallow sites.

The relationship between soil rock fragment content and range site has been used for years. Examples are the gravelly range site, gravelly loamy range site, and various shallow range sites in the detailed soil survey for the High Plains Grasslands Research Station (*96*). That soil survey reports plant community composition and productivity potential for each range site. Productivity on the gravelly loamy range site is 1,233 kg/ha compared to 1,569 kg/ha on a loamy range site in a 38- to 43-cm precipitation zone.

Table 4 shows data from a number of published soil surveys that illustrate the trend in productivity relationships observed for soils with and without significant rock fragment content, along a moisture gradient from humid to arid climates. In humid climates crop productivity generally is higher on fine-textured, non-rocky soils. As precipitation declines, the relationship generally continues to hold, but the difference in productivity potential diminishes and eventually is reversed in areas of very low precipitation (less than 30 cm). In arid and semiarid regions the deeper penetration of limited precipitation and greater relative availability of water at low moisture contents in coarse-textured soils and soils containing rock fragments in surface horizons often result in these soils being more productive than finer textured soils in comparable upland topographic positions. This relationship often is true in the case of loamy and sandy range sites compared with clayey range sites in semiarid and arid regions. Potential yields on stony range sites often are limited by topographic position, for example, ridge crests, and reduced infiltration of water due to snow redistribution and runoff.

Studies of range productivity often show correlations between subsoil moisture content and productivity (*45, 76, 106*). Rock fragments will affect subsoil moisture. Noy-Meir (*67*) concluded that the dominant plant species on each site in arid and semiarid regions will be the species with the greatest capability to use available soil moisture. Klemmedsen and Smith (*50*), describing edaphic relationships of plant communities in Arizona and New Mexico, attributed many vegetation distribution patterns to water-holding capacity, soil depth, and rock content. Weaver (*105*) found similar relationships in Montana. Shantz (*85*) suggested that improved plant-soil water relations resulting from deep water penetration in coarse-textured and

Table 4. Productivity ratings for similar soils with and without rock fragments in different climates.

Soil Survey Area	Soil Series (Phase)	Soil Classification	Yield Rating*
Goshen County, Wyoming, Northern part 34-38 cm annual precipitation (*64*)	Alice	Aridic Haplustolls Coarse-loamy, mixed, mesic	Sandy range site (670-2,010 kg/ha)
	Dix	Torriorthentic Haplustolls Sandy-skeletal, mixed, mesic	Gravelly range site (340-1,010 kg/ha)
	Willowman	Aridic Argiustolls Loamy-skeletal, mixed, mesic	Coarse-upland range site (560-1,570 kg/ha)
	Wolf	Aridic Argiustolls Fine-loamy, mixed, mesic	Loamy range site (670-2,010 kg/ha)
Johnson County, Wyoming, Southern part 38-48 cm annual precipitation (*95*)	Natrop	Argic Cryoborolls Loamy-skeletal, mixed	Coarse-upland range site (38-48 cm) (670-1,230 kg/ha)
	Passcreek	Argic Cryoborolls Fine-loamy, mixed	Loamy range site (38-48 cm) (1,230-2,800 kg/ha)
Broadwater County, Montana 30-43 cm annual precipitation (*69*)	Ameshia	Borollis Calciorthids Coarse-loamy, mixed	Dryland wheat (19.5 quintals/ha) Potatoes, irrigated (392 quintals/ha) Sugar beets, irrigated (35.9 t/ha)
	Ameshia (cobbly)		Dryland wheat (18.2 quintals/ha) Potatoes, irrigated (364 quintals/ha) Sugar beets, irrigated (35.9 mt/ha)
	Passcreek (Channery silt loam)	Argic Cryoborolls Fine-loamy, mixed	Dryland wheat (23.6 quintals/ha)
	Bridger (Silt loam)	Argic Cryoborolls Fine-mixed	Dryland wheat (26.9 quintals/ha) Alfalfa, irrigated (10.1 t/ha)
	Bridger (Cobbly loam)	Argic Cryoborolls Clayey-skeletal varient	Dryland wheat (26.9 quintals/ha) Alfalfa, irrigated (10.1 t/ha)
	Chinook (Sandy loam)	Aridic Haploborolls Coarse-loamy, mixed	Spring wheat, irrigated (37.0 quintals/ha) Sugar beets, irrigated (38.1 mt/ha)
	Chinook (Sandy loam, gravelly subsoil)	Aridic Haploborolls Coarse-loamy over sandy or sandy skeletal, mixed	Spring wheat, irrigated (30.3 quintals/ha) Sugar beets, irrigated (54.9 mt/ha)
Douglas County, Washington 20-28 cm annual precipitation (*11*)	Dougville (Loam)	Aridic Haploxerolls Coarse-loamy, mixed, mesic	Loamy range site (780-1,570 kg/ha) (1,180 normal)
	Heyton (Very stony loam)	Aridic Haloxerolls Loamy-skeletal, mixed, mesic	Loamy range site (730-1,460 kg/ha) (1,180 normal)

	Zen (Silt loam)	Aridic Haploxerolls Fine-silty, mixed, mesic	Loamy range site (670-1,340 kg/ha) (900 normal)
Dona Ana County area, New Mexico 20 cm annual precipitation (*19*)	Dona Ana (Fine sandy loam)	Typic Haplargids Fine-loamy, mixed, thermic	Loamy range site (200-760 kg/ha) (530 normal)
	Pinaleno (Very gravelly sandy loam)	Typic Haplargids Loamy-skeletal, mixed, thermic	Gravelly loam range site (140-950 kg/ha) (670 normal)
Ashland County, Ohio 94 cm annual precipitation (*77*)	Chili (Loam)	Typic Hapludalfs Fine-loamy, mixed, mesic	White oak SI_{50} - 25.9 Northern red oak SI_{50} - 27.5
	Conotton (Gravelly-loam)	Typic Hapludalfs Loamy-skeletal, mixed, mesic	White oak SI_{50} - 22.6 Northern red oak SI_{50} - 22.6
	Berks	Typic Dystrochrepts Loamy-skeletal, mixed, mesic	Corn (44.0 quintals/ha) Oats (18.0 quintals/ha) Northern red oak SI_{50} - 22.6
	Lordstown	Typic Dystrochrepts Coarse-loamy, mixed, mesic	Corn (50.2 quintals/ha) Oats (23.3 quintals/ha) Northern red oak SI_{50} - 19.4
Crook County, Wyoming 46 cm annual precipitation (*34*)	Lakoa	Typic Eutraboralfs Fine-loamy, mixed	Ponderosa pine SI_{50} - 23.3
	Grizzly	Typic Eutraboralfs Loamy-skeletal, mixed	Ponderosa pine SI_{50} - 19.4
	Nevee	Ustic Torriorthents Coarse-silty, mixed (calcareous), mesic	Loamy range site (1,680-3,360 kg/ha) (2,460 normal)
	Nihill	Ustic Torriorthents Loamy-skeletal, mixed (calcareous), mesic	Loamy range site (1,680-3,360 kg/ha) (2,460 normal)
Carroll County, Illinois 88 cm annual precipitation (*73*)	Port Byron (Silt loam)	Typic Hapludoll Fine-silty, mixed, mesic	Oats (23.3 quintals/ha) Alfalfa (9.4 t/ha)
	Rodman (Gravelly loam)	Typic Hapludoll Sandy-skeletal, mixed, mesic	Oats (9.3 quintals/ha) Alfalfa (4.3 t/ha)
	Sparta (Loamy sand)	Entic Hapludoll Sandy, mixed, mesic	Oats (14.4 quintals/ha) Alfalfa (5.6 t/ha)

*SI_{50} is site index, average height of dominant and codominant trees, of 50 years in meters. Ratings for range production is given as kg/ha forage (range for dry and wet years) and, in some cases, a single value for normal years.

rocky soils allowed humid region plants to extend their range into semiarid regions and semiarid-region plants to extend into arid regions.

Rock fragment criteria in mine soils

In some states or regions soil rock fragment data have been collected for premining resource inventory and reclamation planning. In those cases assessing the suitability of various soils for reclamation, in terms of rock fragment content, is possible. Table 5 shows recent rock fragment guidelines that have been recommended or proposed by various state and federal agencies and researchers.

The guidelines range from fairly simplified or general criteria to a detailed categorization of suitability. Other differences are characterized by the implicit inclusion of the 2-mm to 7.6-cm size fraction (gravels) by Wyoming (*107*), Montana (*63*), Colorado (personal communication from James Herron, Mine Land Reclamation Division, Colorado Department of

Table 5. Rock fragment suitability criteria in soil and soil substitutes.

Source	Measurement Specifications	Rock Fragment Suitability Criteria Good or Suitable	Fair	Poor	Unsuitable
Wyoming Department of Environmental Quality (*107*)	% by volume	<25	Marginal 25-35		>35
Montana Department of State Lands (*63*)	Lift 1 Soil Lift 2 Soil (% by volume)	- -	- -		Suspect Level >20* >35*
Colorado Department of Natural Resources†	% by volume	-	>20		>35
Bureau of Land Management (*20*)	>7.6 cm (% by volume)	<15	15-35	>35	
Office of Surface Mining (*68*)	7.6-25.4 cm	15	Marginal 15-35		35
	>25.4 cm (% by volume)	3	3-10		10
Soil Conservation Service (*93*)	7.6-25.4 cm >25.4 cm (% by weight)	≤15 < 3	15-35 3-10	>35 >10	-
Berg et al. (*12*)	% by volume	0-10	10-20	20-35	>35
Shafer (*82*)	% by volume	<15	15-35	>35	-
Colorado School of Mines Research Institute (*29*)‡	% by weight	-	>30-35		

*These values may vary depending upon the plant species proposed for revegetation in specific locations. For example, a soil with a very high rock fragment content throughout its profile may be completely salvaged if used for certain shrub or tree plantings.

†Personal communication from James Herron, Mine Land Reclamation Division, Colorado Department of Natural Resources.

‡These guidelines are intended to apply to overburden or general plant growth media.

Natural Resources), Berg and associates (*12*), Schafer (*82*), and the Colorado School of Mines (*29*), but the exclusion of this fraction by others. Another difference involves the units of measurement; SCS (*93*) and the Colorado School of Mines (*29*) base recommendations on a percent by weight of coarse fragments, whereas all others are on a percent by volume basis.

Some recommendations are based on the size fractions of the fragments. In other cases no such distinction has been made. Where such a distinction has been made, the limits of rock fragment content in each suitability class decreases as the size class of the fragments increases.

Montana (*63*) has made different recommendations for first-lift and second-lift soil materials. Those regulations require that, in general, soils must be segregated in the salvage and replacement operations. The first or surface lift includes the A and upper B horizons and the second lift includes lower B and C horizons. A lower tolerance limit of rock fragment content has been suggested for the surface material.

Despite the variability and differences among the recommendations, a thread of similarity runs through the various categories. For example, the average upper limit of rock fragment content of soils in the good or suitable category ranges from 15 to 25 percent. The lower limit for materials in the poor to unsuitable categories is about 35 percent. In addition, the Office of Surface Mining (OSM), SCS, Bureau of Land Management (BLM), and Schafer class limits are quite similar, although the latter two do not categorize rock fragment sizes in the same way as OSM and SCS.

These similarities probably result from the fact these guidelines likely originate in whole or in part from guidelines that have been included in draft manuscripts of the *National Soils Handbook* (*93*) for several years. However, the one major difference already mentioned that ironically exists between the standards in table 5 and most others listed is whether the fragment content is based on a weight or a volume basis. This appears to be a serious discrepancy that should be investigated and resolved. For example, in a soil with a fine-earth bulk density of 1.3 g/cm^3, a rock fragment content of 35 percent by volume would be about 52 percent by weight, using a rock density of 2.65 g/cm^3.

The guidelines in table 5 are based on the limiting properties of rock fragments discussed earlier. In general, substantial quantities of fragments reduce the nutrient-supplying and water storage capabilities of soils, both of which are important to plant growth (*38*). The extent of reduced water-holding capacity depends upon the abundance and porosity of the fragments (*93*). Rock fragment content should be taken into account to assess correctly the nutrient status of soils (*12*). Other limitations of importance include the physical restrictions to planting of seed and emergence of seedlings and reduced seed-soil contact because of high quantities of rock fragments (*29*).

Schafer (*82*) outlined a capability classification of mine soils based on the

SCS system. One of his criteria for developing the classes was rock fragment content. According to Schafer's outline, Class IV soils must have a lithic (hard) rock fragment content of less than 35 percent by volume and a soft rock fragment content of less than 50 percent by volume. True rock fragments were defined earlier as maintaining their coherence "after overnight shaking in a weak solution of sodium hexa-metaphosphate" (*91*).

Researchers also have reported positive influences of rock fragments. Some percentage of rock fragments in a soil aids aeration and drainage (*29*) and helps protect the fine earth fraction from erosive forces (*90*). The effect of rock fragments on K values (inherent erodibility) in the USLE has been accounted for by procedures in the *National Soils Handbook* (*93*). Rock fragments generally reduce the K factor in proportion to their volumetric content in soil. These considerations may account in part for the apparent lack of concern for the gravel size fraction in SCS and BLM guidelines. On the other hand, this simply may reflect a belief that this fraction has no detrimental effect on the suitability or use of soils for reclamation purposes.

There appears to be a direct relationship between the average lower limit of rock fragment content of poor to unsuitable materials (35%) in table 5 with the differentiating characteristic of skeletal (>35% by volume fragments of 2 mm or more in diameter) versus nonskeletal particle size classes in the naming of soil families (*90*). There does not appear to be a stated rationale for the levels in the other suitability categories, except that they seem to relate to the SCS (*93*) criteria that have existed for several years.

Little consideration is given to evaluating rock fragments in overburden or regraded spoils, except insofar as these materials might be used as substitutes for soil (*29, 107*). Schafer (*82*) recommended that if overburden is used as cover soil the rock type and slaking properties should be evaluated. Fair to good material would include weakly consolidated sandstones and siltstones and unconsolidated surface deposits. Material in this category should also meet the test that at least 20 percent of the material can pass through a 5-mm sieve after 30 minutes in water. Barrett and associates (*9*) concluded that overburden material can be used as cover soil if at least 65 percent of material in a core sample passes through a 5-mm sieve after it has been wet under vacuum and then shaken for 15 minutes in a horizontal shaker.

As with soil, all of these overburden recommendations are based on the principle that a certain minimum percentage of fines is necessary to support adequate vegetative cover. However, there are a variety of methods for determining weathering potential and also different interpretations of this characteristic for reclamation suitability.

Rock fragment effects on vegetative productivity in mine soils

Redente and associates (*75*) found that the proportion of shrubs, forbs, and grasses on reclaimed sites in Colorado varied with soil rock fragment content. The highest productivity of fourwing saltbush (*Atriplex canescens*) occurred on a site with 85 percent rocks in the surface horizon. Highest grass production occurred on the site with lowest (21 percent) surface rock content. The researchers attributed the productivity differences to competition for water. On irrigated sites they reported no differences in productivity attributable to different soil rock contents. Extreme rockiness in the surface layer resulted in poor seed-to-soil contact, they concluded.

In another study, Redente and associates (*74*) reported a significant relationship ($r = .78$) between the content of rock fragments greater than 4.76 mm in reclaimed soils in Colorado and the Shannon-Weiner Diversity Index of the vegetation. They also observed that as irrigation increased soil moisture, the diversity of vegetation on the study sites decreased, with grass dominating the sites.

Ashby and associates (*6*) reviewed studies about the effects of rock fragment content on vegetative performance in surface mine soil materials. They presented examples of crops, forage, and trees in the eastern and midwestern United States that performed quite well on rocky mine soils exhibiting a wide range of rock fragment content, derived principally from overburden. In some cases the fragments were soft and probably behaved partially as soil in supplying water and nutrients (*8*). In other cases the nature of the fragments was not clear. In certain situations described the fragments may have benefited soil physical properties, water relations, and drainage. Nevertheless, the researchers concluded that further research on mine soil properties with rock fragment content as a single variable was needed.

Schafer and associates (*83*) noted the presence of soft sandstone and siltstone fragments in mine soils at the Rosebud Mine near Colstrip, Montana. They indicated that such fragments could supply water to plant roots, although roots had not penetrated the rock fragments in the early stages of reclamation.

In a Pennsylvania study that compared corn yields on natural soils and adjacent mine soils, Nielsen and associates (*66*) found that yields were 9 to 12 percent lower on the mine soils in three of four comparisons. In two of the study soil pairs on which yields were lower, the rock fragment content of the subsoil was much higher on the mine soils (90 percent) as compared to the natural soils (0 to 30 percent).

We discussed the armoring and self-mulching properties of rocky soil materials earlier. Increased site stability due to surface layer rock fragments may contribute to vegetative establishment on sites subject to erosion (*55*).

There is little research on western region mineland reclamation documenting the effects of rock fragments on revegetative performance. Casual observation of native rangeland in the region suggests that rock fragments are common on many range sites. Future research no doubt will reveal edaphic-plant community relationships as interesting and as complex on reclaimed sites as on native range.

Soil salvage and seeding problems

The salvage of rocky soils could necessitate additional equipment, such as ripper dozers, or cause excessive wear on scrapers used for soil removal and replacement. However, such problems depend upon the rock fragment content and the size and the hardness of the fragments. To date, such problems have not been documented.

Rock fragments of sufficient quantity, hardness, and size may influence the effectiveness of seeding and planting techniques and may adversely affect seeding equipment. Rock fragments may inhibit proper seeding depths for various species, proper soil-seed contact (*75*), and seedling emergence (*29*), but there has been little documentation of these effects on reclaimed areas in the West. Data simply are not available about soil rock fragment content and size in relation to vegetative establishment on mined land.

Reclamation goals and postmining land use

The guidelines in table 5 generally apply to grassland, hayland, cropland, and shrubland vegetation. Other vegetation is equally important from vegetative diversity and wildlife habitat points of view. These include some species that appear to have specific topographic-edaphic environmental requirements to compete successfully and maintain themselves.

In portions of the northern Great Plains, for example, near Colstrip, Montana, ponderosa pines often occur on soils with significant rock contents, although this species is found on other soils as well (*100*). The reclamation plan for the Absaloka Mine near Hardin, Montana, attempted to take into account the characteristics of native rocky soils on which ponderosa pine frequently occurs in this area (*2*). Specifically, cover soil material was obtained from a gravelly, clinker-derived soil; placed on a regraded, steep, northeast-facing slope; and planted with pine seedlings. A rocky terrace was constructed above the area receiving the gravelly cover soil. The rocky surface materials on this slope were intended to reduce runoff and encourage infiltration from snowmelt and rainfall (*2*).

Stark (*94*) proposed somewhat different procedures for reconstructing ponderosa pine soils. She recommended that rock, of unspecified size, be placed at a 50- to 60-cm depth below a sandy rooting medium or that spoil surfaces below the replaced soils be compacted. These procedures would restrict downward water movement, maximizing water availability to roots. Stark also recommended that additional rock fragments 20 cm in diameter be placed on the soil surface and extending partially into the soil to provide additional moisture to roots through distillation and condensation.

Fisher and Deutsch (*38*) concluded that ponderosa pine reestablishment on mined land is difficult because of the preference of this species for fractured sandstone bedrock situations. Ponderosa pine can obtain water from fissures in the bedrock, which gives it a marked competitive advantage in such situations. The researchers suggested that use of boulder-size rock fragments (>152 cm in diameter) would be necessary to reconstruct this kind of rooting environment.

Although the procedures for reconstructing a ponderosa pine rooting medium are not well defined, the possible beneficial use of rock fragments has been recognized and is being investigated.

Summary

The effect of rock fragments on mine soil properties and revegetation performance has received limited attention in the West. Guidelines that have been published appear to have their origin in the disturbed area soil reconstruction standards of the *National Soils Handbook* (*93*). However, the discrepancy between units of measurement, volume versus weight, should be resolved so that discussion of rock fragments in the reclamation community can be conducted on a uniformly accepted basis. In addition, in any oral and written communication or discussion, rock fragments should be described in terms of their size range, hardness, porosity, and other properties that affect their contribution to soil properties.

Rock fragments in soils reduce water-holding capacity, but also increase the rate and depth of infiltration. The rock component of soils is not inert, but may contribute in varying degrees to water-holding capacity, nutrient availability, and heat capacity. The presence of rock may alter root distribution and organic matter accumulation. Rocks on the surface and in the surface layer protect the soil from wind and water erosion. Rocks in the soil alter the biological potential of the site, both in terms of biomass and plant species composition. These effects are a function of the size, type, and quantity of rock fragments, and their influence on plant communities depends on climatic conditions.

Research has shown that maximum grass production on unirrigated, reclaimed sites in Colorado occurred on soils with the lowest rock fragment content (*75*). This is consistent with most published guidelines. On the other hand, the opposite relationship was reported in Colorado for four-wing saltbush (*75*) and vegetative diversity (*74*). The researchers in these latter cases also reported that all of the relationships broke down under irrigation or increased soil moisture.

Rocky soils have been used or proposed for use in reestablishing ponderosa pine (*2, 94*). However, the long-term performance of ponderosa pine relative to rock fragment characterisitcs of mine soils is not known.

To determine the effectiveness, or lack thereof, of existing guidelines on rock fragment content for mine soil reconstruction, there must be concerted efforts to monitor revegetative performance in terms of establishment, growth, diversity, and landscape stability. Postmining land use and reclamation objectives must be considered when selecting soils that may be particularly suited to the establishment of the desired postmining plant communities. Using gravelly and rocky soils may be beneficial in reestablishing certain plant species or communities, thereby enhancing diversity on reclaimed mine sites.

REFERENCES

1. Adams, J. E. 1967. *Effects of mulches on soil temperature and grain sorghum development.* Agronomy Journal 57: 471-474.
2. Amendola, F. A., M. D. Mitchell, and D. W. Simpson. 1984. *Reclamation of tree and shrub species at the Absaloka Mine.* In F. F. Munshower and S. E. Fisher, Jr. [editors] *Proceedings, Third Biennial Symposium on Surface Coal Mining and Reclamation in the Great Plains.* Montana Agricultural Experiment Station. Montana State University, Bozeman, pp. 237-257.
3. American Association of State Highway Officials. 1966. *Standard recommended practice for the classification of soils and soil-aggregate mixtures for highway construction purposes.* In *Standard Specifications for Highway Materials and Methods of Sampling and Testing. Part I, Specifications.* Washington, D.C. pp. 41-47.
4. American Society for Testing and Materials. 1979. *Standard test method for classification of soils for engineering purposes.* In *Annual Book of ASTM Standards, Part 19: Natural Building Stones; Soil and Rock; Peats, Mosses, and Humus.* ANSI/ASTM D2487-69. Philadelphia, Pennsylvania. pp. 374-378.
5. Arya, L. M., D. A. Farrell, and G. R. Blake. 1975. *A field study of soil water depletion patterns in presence of growing soybean roots: I. Determination of hydraulic properties of the soil.* Soil Science Society of America Proceedings 39: 424-430.
6. Ashby, W. C., W. G. Vogel, C. A. Kolar, and G. R. Philo. 1984. *Productivity of stony soils on strip mines.* In J. D. Nichols, P. L. Brown, and W. J. Grant [editors] *Erosion and Productivity of Soils Containing Rock Fragments.* Special Publication Number 13. Soil Science Society of America, Madison, Wisconsin. pp. 31-44.
7. Auten, J. T. 1945. *Prediction of site index for yellow-poplar from soil and topography.* Journal of Forestry 43: 662-668.
8. Barnhisel, R. I. 1977. *Reclamation of surface mined coal spoil.* Technical Report 600/7-77-093. U.S. Environmental Protection Agency, Cincinnati, Ohio.
9. Barrett, J., P. C. Deutsch, F. G. Ethridge, W. T. Franklin, R. D. Heil, D. B. McWhorter, and A. D. Youngberg. 1980. *Procedures recommended for overburden and hydrologic studies of surface mines.* General Technical Report INT-71. Intermountain Forest and Range Experiment Station, Forest Service, U.S. Department of Agriculture, Ogden, Utah.
10. Beetle, A. A., and K. J. Johnson. 1982. *Sagebrush in Wyoming.* Bulletin 779. Wyoming Agricultural Experiment Station, University of Wyoming, Laramie.
11. Beieler, V. E. 1981. *Soil survey of Douglas County, Washington.* Washington State University Agricultural Research Center and Soil Conservation Service, U.S. Department of Agriculture, Pullman, Washington.
12. Berg, W. A., S. D. Merrill, and D. J. Dollhopf. 1979. *Soils and overburden anal-*

ysis and sampling techniques. In *User Guide to Soils—Mining and Reclamation in the West.* General Technical Report INT-68. Intermountain Forest and Range Experiment Station, Forest Service, U.S. Department of Agriculture, Ogden, Utah. pp. 13-27.

13. Birkeland, P. W. 1984. *Soils and geomorphology.* Oxford University Press, New York, New York. pp. 179-181.
14. Bond, J. J., and W. O. Willis. 1971. *Soil water evaporation: Long term drying as influenced by surface residue and evaporation potential.* Soil Science Society of America Proceedings 35: 984-987.
15. Box, J. E., Jr. 1981. *The effects of surface slaty fragments on soil erosion by water.* Soil Science Society of America Journal 45: 111-116.
16. Box, J. E., Jr., and L. D. Meyer. 1984. *Adjustment of the universal soil loss equation for cropland soils containing coarse fragments.* In J. D. Nichols, P. L. Brown, and W. J. Grant [editors] *Erosion and Productivity of Soils Containing Rock Fragments.* Special Publication Number 13. Soil Science Society of America, Madison, Wisconsin. pp. 83-90.
17. Breeding, C. H. 1971. *Classes and phases of stoniness and rockiness.* In *Proceedings of the National Technical Work-Planning Conference of the Cooperative Soil Survey.* Soil Conservation Service, U.S. Department of Agriculture, Washington, D. C. pp. 73-78.
18. Brown, H. G., and H. Loewenstein. 1978. *Predicting site productivity of mixed conifer stands in northern Idaho from soil and topographic variables.* Soil Science Society of America Journal 42: 967-971.
19. Bulloch, H. E., Jr., and R. E. Neher. 1980. *Soil survey of Dona Ana County area, New Mexico.* Soil Conservation Service, U.S. Department of Agriculture; Bureau of Land Management, U.S. Department of the Interior; and New Mexico Agricultural Experiment Station, Las Cruces, New Mexico.
20. Bureau of Land Management, U.S. Department of the Interior. 1977. *Guidelines for reclamation areas.* EMRIA Handbook. Denver, Colorado.
21. Carmean, W. H. 1961. *Soil survey refinements needed for accurate classification of black oak site quality in southeastern Ohio.* Soil Science Society of America Proceedings 25: 394-397.
22. Carmean, W. H. 1965. *Black oak site quality in relation to soil and topography in southeastern Ohio.* Soil Science Society of America Proceedings 29: 308-312.
23. Carmean, W. H. 1970. *Site quality for eastern hardwoods.* In *The Silviculture of Oaks and Associated Species.* Research Paper NI-144:36-56. Northeastern Forest Experiment Station, Forest Service, U.S. Department of Agriculture, Broomall, Pennsylvania.
24. Chepil, W. S., N. P. Woodruff, F. H. Siddoway, D. W. Fryrear, and D. V. Armbrust. 1963. *Vegetative and nonvegetative materials to control wind and water erosion.* Soil Science Society of America Proceedings 27: 86-90.
25. Childs, S. 1981. *The nature of skeletal soils in steep terrain.* In S. D. Hobbs and O. T. Helgerson [editors] *Reforestation of Skeletal Soils.* Forest Research Laboratory, Oregon State University, Corvallis. pp. 11-20.
26. Chilkosz, E. J., R. L. Cunningham, G. W. Petersen, and R. C. Conce. 1983. *Characteristics, interpretations, and uses of Pennsylvania minesoils.* Progress Report 381. Pennsylvania Agricultural Experiment Station, Pennsylvania State University, University Park.
27. Choriki, R. T., J. C. Hide, S. L. Krall, and B. L. Brown. 1964. *Rock and gravel mulch aid in moisture storage.* Crops and Soils 16(9): 24.
28. Coile, T. S. 1952. *Moisture content of small stone in soil.* Soil Science 75: 203-207.

29. Colorado School of Mines Research Institute. 1981. *State of the art and guidelines for surface coal mine overburden sampling and analysis.* Phase 1 and 2 reports. Prepared for Office of Surface Mining, U.S. Department of the Interior, Washington, D.C.
30. Cunningham, R. L., and R. P. Matelski. 1968. *Bulk density measurements on certain soils high in coarse fragments.* Soil Science Society of America Proceedings 32: 109-111.
31. Daniels, R. B., E. E. Gamble, L. J. Bartelli, and L. A. Nelson. 1968. *Application of the point-count method to problems of soil morphology.* Soil Science 106: 149-152.
32. Daubenmire, R. 1970. *Steppe vegetation of Washington.* Technical Bulletin Number 62. Washington Agricultural Experiment Station, Washington State University, Pullman.
33. Edwards, W. M., P. F. Germann, L. B. Owens, and C. R. Amerman. 1984. *Watershed studies of factors influencing infiltration, runoff, and erosion on stony and nonstony soils.* In J. D. Nichols, P. L. Brown, and W. J. Grant [editors] *Erosion and Productivity of Soils Containing Rock Fragments.* Special Publication Number 13. Soil Science Society of America, Madison, Wisconsin. pp. 45-54.
34. Elwonger, W. G. 1983. *Soil survey of Crook County, Wyoming.* Soil Conservation Service and Forest Service, U.S. Department of Agriculture, and Wyoming Agricultural Experiment Station, University of Wyoming, Laramie.
35. Epstein, E., W. J. Grant, and R. A. Struchtemeyer 1966. *Effects of stones on runoff, erosion, and soil moisture.* Soil Science Society of America Proceedings 30: 638-640.
36. Fairbourn, M. L. 1973. *Effect of gravel mulch on crop yields.* Agronomy Journal 65: 925-928.
37. Ferguson, H., and T. Bateridge. 1982. *Salt status of glacial till soils as affected by the crop-fallow system of dryland farming.* Soil Science Society of America Journal 46: 807-812.
38. Fisher, S. E., and P. Deutsch. 1983. *The soil resource: Its importance in the west and its role in coal development and reclamation.* In D. Books, L. Branch, and C. Fisher [editors] *Coal Development: Collector Papers, Volume II. Proceedings of Coal Development Workshops.* Bureau of Land Management, U.S. Department of the Interior, Washington, D.C. pp. 845-984.
39. Gaiser, R. N. 1950. *Relation between soil characteristics and site index of loblolly pine in the Coastal Plain region of Virginia and the Carolinas.* Journal of Forestry 48: 271-275.
40. Gessel, S. P. 1949. *Correlation between certain soil characteristics and site for Douglas-fir in northwestern Washington.* Soil Science Society of America Proceedings 13: 333-337.
41. Grant, W. J., and R. A. Struchtemeyer. 1959. *Influence of the coarse fraction in two Maine potato soils on infiltration, runoff, and erosion.* Soil Science Society of America Proceedings 23: 391-394.
42. Hannah, P. R. 1968. *Topography and soil relations for white and black oaks in southern Indiana.* Research Paper NC-25. North Central Forest Experiment Station, Forest Service, U.S. Department of Agriculture, St. Paul, Minnesota.
43. Hanson, C. T., and R. L. Blevins. 1979. *Soil water in coarse fragments.* Soil Science Society of America Journal 43: 819-820.
44. Harlow, W. M., and E. S. Harrar. 1969. *Textbook of dendrology.* McGraw-Hill Book Company, New York, New York. 512 pp.
45. Herbel, C. H., F. N. Ares, and R. A. Wright. 1972. *Drought effects on a semidesert grassland range.* Ecology 53: 1,084-1,093.

46. Hidlebaugh, A. R. 1984. *Use of soil survey information to determine extent and effect of rock fragments on productivity.* In J. D. Nichols, P. L. Brown, and W. J. Grant [editors]. *Erosion and Productivity of Soils Containing Rock Fragments.* Special Publication Number 13. Soil Science Society of America, Madison, Wisconsin. pp. 7-12.
47. Hillel, D., and N. Tadmore. 1962. *Water regime and vegetation in the Central Negev Highlands of Israel.* Ecology 43: 33-41.
48. Hobbs, S. D. 1981. *Stocktype selection and planting techniques for Douglas-fir on skeletal soils in southwest Oregon.* In S. D. Hobbs and O. T. Helgerson [editors] *Reforestation of Skeletal Soils.* Forest Research Laboratory, Oregon State University, Corvallis. pp. 92-96.
49. Jackson, R. D. 1973. *Diurnal changes in soil water content during drying.* In R. R. Bruce, K. W. Flack, and H. M. Taylor [editors] *Field Soil Water Regime.* Special Publication Number 5. Soil Science Society of America, Madison, Wisconsin. pp. 37-55.
50. Klemmedsen, J. O., and E. L. Smith. 1978. *Vegetation: soil relationships of forests, woodlands, and grasslands of Arizona and New Mexico.* In C. T. Youngberg [editor] *Forest Soils and Land Use.* Colorado State University, Fort Collins. pp. 42-67.
51. Lamb, J., Jr., and J. E. Chapman. 1943. *Effect of surface stones on erosion, evaporation, soil temperature, and soil moisture.* Journal of the American Society of Agronomy 35: 567-578.
52. Lewis, R., and A. Abee. 1981. *Site preparation strategies for skeletal soils.* In S. D. Hobbs and O. T. Helgerson [editors] *Reforestation of Skeletal Soils.* Forest Research Laboratory, Oregon State University, Corvallis. pp. 40-49.
53. Lyford, W. H. 1964. *Coarse fragments in the Gloucester soils of the Harvard Forest.* Harvard Forest Paper Number 9, Harvard University, Petenham, Massachusetts. 16 pp.
54. Magier,, J., and I. Ravina. 1984. *Rock fragments and soil depth as factors in land evaluation of Terra Rosa.* In J. D Nichols, P. L. Brown, and W. J. Grant [editors] *Erosion and Productivity of Soils Containing Rock Fragments.* Special Publication Number 13. Soil Science Society of America, Madison, Wisconsin.
55. McCormack, D. E., K. K. Young, and G. M. Darby. 1984. *Rock fragments and the K factor of the universal soil loss equation.* In J. D. Nichols, P. L. Brown, and W. J. Grant [editors] *Erosion and Productivity of Soils Containing Rock Fragments.* Special Publication Number 13. Soil Science Society of America, Madison, Wisconsin. pp. 73-81.
56. Medin, D. E. 1960. *Physical site factors influencing annual production of true mountain mahogany,* Cercocarpus montanus. Ecology 41: 454-460.
57. Mehuys, G. R., L. H. Stolzy, and J. Letey. 1975. *Temperature distributions under stones submitted to a diurnal heat wave.* Soil Science 120: 437-441.
58. Mehuys, G. R., L. H. Stolzy, J. Letey, and L. V. Wees. 1975. *Effect of stones on the hydraulic conductivity of relatively dry desert soils.* Soil Science Society of America Proceedings 39: 37-42.
59. Meyer, L. D., C. B. Johnson, and G. R. Foster. 1972. *Stone and woodchip mulches for erosion control on construction sites.* Journal of Soil and Water Conservation 27: 264-269.
60. Miller, D. E. 1973. *Water retention and flow in layered soil profiles.* In R. R. Bruce, K. W. Flack, and H. M. Taylor [editors] *Field Soil Water Regime.* Special Publication Number 5. Soil Science Society of America, Madison, Wisconsin. pp. 107-117.
61. Miller, D. E., and J. S. Aarstad. 1971. *Available water as related to evapotran-*

spiration rates and deep drainage. Soil Science Society of America Proceedings 35: 131-134.

62. Miller, W. B. 1964. *An ecological study of the mountain mahogany community and related biotic associations of the Big Horn mountains.* M.S. thesis. Plant Science Division, Range Management Section, University of Wyoming, Laramie.
63. Montana Department of State Lands. 1983. *Draft soil and overburden guidelines.* Helena.
64. Nelson, F. L., F. Stephens, and H. B. Ravenhold. 1981. *Soil survey of Goshen County, Wyoming, Northern Part.* Soil Conservation Service, U.S. Department of Agriculture, and Wyoming Agricultural Experiment Station, University of Wyoming, Laramie.
65. Nettleton, W. D., J. E. Witty, R. E. Nelson, and J. W. Hawley. 1975. *Genesis of argillic horizons in soils of desert areas of the southwestern United States.* Soil Science Society of America Proceedings 39: 919-926.
66. Nielsen, G. A., R. L. Cunningham, P. P. Parlock, and D. R. Seibert. 1980. *Corn on strip-mined land can equal other soil yields.* Science of Agriculture 27(2): 12-13.
67. Noy-Meir, I. 1973. *Desert ecosystems: Environments and producers.* Annual Review of Ecology and Systematics 4: 25-51.
68. Office of Surface Mining, U.S. Department of the Interior. 1983. *Criteria for evaluation of topsoil and topsoil substitutes in the high desert ecosystem of southwestern United States.* Denver, Colorado.
69. Olsen, J. A., M. H. Haub, and L. C. Bingham. 1977. *Soil survey of Broadwater County area, Montana.* Soil Conservation Service, U.S. Department of Agriculture, and Montana Agricultural Experiment Station, Montana State University, Bozeman.
70. Paetzold, R. F., and M. J. Maurbach. 1984. *Hydraulic properties of some soils with paralithic contacts.* Soil Science Society of America Journal 48: 1,355-1,359.
71. Pedersen, T. A., A. S. Rogowski, and R. Pennock. 1980. *Physical characteristics of some minesoils.* Soil Science Society of America Journal 44: 321-328.
72. Ravina, I., and J. Magier. 1984. *Hydraulic conductivity and water retention of clay soils containing coarse fragment.* Soil Science Society of America Journal 48: 736-740.
73. Ray, B. W., R. Rehner, and J. B. Fehrenbacher. 1975. *Soil survey: Carroll County, Illinois.* Soil Conservation Service, U.S. Department of Agriculture, and Illinois Agricultural Experiment Station, University of Illinois, Champaign/Urbana.
74. Redente, E. F., J. M. Stark, M. E. Biodini, and T. A. Oliver. 1984. *Vegetation structure and succession as they relate to soil disturbance and retorted oil shale.* In E. F. Redente and C. W. Cook [editors] *Ecological Studies of Natural and Established Ecosystems on Energy Related Disturbances in Colorado.* Report to the U.S. Department of Energy, Contract DE-ASO2-76EV04018. Department of Range Science, Colorado State University, Fort Collins.
75. Redente, E. F., T. B. Doerr, C. B. Mount, C. E. Grygiel, T. E. Sievers, and M. Biodini. 1981. *Effects of plant species, soil material, and cultural practices upon plant establishment and succession.* In E. F. Redente and C. W. Cook [editors] *Revegetation Research on Oil Shale Lands in the Piceance Basin.* Report to the U.S. Department of Energy, Contract DE-ASO2-76EV04018. Range Science Department, Colorado State University, Fort Collins.
76. Redmann, R. E. 1975. *Production ecology of grassland plant communities in western North Dakota.* Ecological Monographs 45: 83-106.

77. Redmond, C. E., and D. L. Brown. 1980. *Soil survey of Ashland County, Ohio.* Soil Conservation Service, U.S. Department of Agriculture, and Ohio Department of Natural Resources, Columbus.
78. Reinhart, K. G. 1961. *The problem of stones in soil-moisture measurement.* Soil Science Society of America Journal 25: 268-270.
79. Rivard, R., and C. R. DeKimpe. 1980. *Proprietes de quelques sols riches en graviers dans la region de Quebec.* Canadian Journal of Soil Science 60: 263-273.
80. Rivers, E. D., and R. F. Shipp. 1972. *Available water capacity of sand and gravelly North Dakota soils.* Soil Science 113: 74-80.
81. Saini, G. R., and W. J. Grant. 1980. *Long-term effects of intensive cultivation on soil quality in the potato-growing areas of New Brunswick (Canada) and Maine (USA).* Canadian Journal of Soil Science 60: 421-428.
82. Schafer, W. M. 1979. *Guides for estimating cover-soil quality and mine soil capability for use in coal strip mine reclamation in the Western United States.* Reclamation Review 2: 67-74.
83. Schafer, W. M., G. A. Nielsen, D. J. Kollhopf, and K. Temple. 1979. *Soil genesis, hydrological properties, root characteristics, and microbial activity of 1-to 50-year-old strip mine spoils.* EPA-600/7-79-100. U.S. Environmental Protection Agency, Cincinnati, Ohio.
84. Sencindiver, J. C., J. T. Ammons, and R. M. Smith. 1979. *Overburden and minesoil potentials for land use.* Proceedings, West Virginia Academy of Science 51: 138-146.
85. Shantz, H. L. *Plants as soil indicators.* In *Soils and Men, the 1938 Yearbook of Agriculture.* U.S. Department of Agriculture, Washington, D.C. pp. 835-860.
86. Shipp, R. F., and R. P. Matelski. 1965. *Bulk density and coarse fragment determinations on some Pennsylvania soils.* Soil Science 99: 392-397.
87. Simonton, J. R., E. Rawitz, and E. D. Shirley. 1984. *Effects of rock fragments on erosion of semiarid rangeland soils.* In J. D. Nichols, P. L. Brown, and W. J. Grant [editors] *Erosion and Productivity of Soils Containing Rock Fragments.* Special Publication Number 13. Soil Science Society of America, Madison, Wisconsin. pp. 65-72.
88. Sobek, A. A., W. A. Schuller, J. R. Freeman, and R. M. Smith. 1978. *Field and laboratory methods applicable to overburdens and minesoils.* EPA-600/2-78-054. U.S. Environmental Protection Agency, Cincinnati, Ohio.
89. Soil Survey Staff. 1972. *Soil survey investigations report No. 1: Soil survey laboratory methods and procedures for collecting samples.* Soil Conservation Service, U.S. Department of Agriculture, Washington, D.C. 63 pp.
90. Soil Survey Staff. 1975. *Soil taxonomy—a basic system of soil classification for making and interpreting soil surveys.* Agriculture Handbook Number 436. Soil Conservation Service, U.S. Department of Agriculture, Washington, D.C.
91. Soil Survey Staff. 1981. *Soil survey manual.* Chapter 4. Working Draft (430-V-SSM). Soil Conservation Service, U.S. Department of Agriculture, Washington, D.C.
92. Soil Survey Staff. 1982. *Soil survey manual.* Agriculture Handbook Number 18. Soil Conservation Service, U.S. Department of Agriculture, Washington, D.C.
93. Soil Survey Staff. 1983. *National soils handbook. Part 603: Application of soil information.* Working Draft 430-VI-NSH. Soil Conservation Service, U.S. Department of Agriculture, Washington, D.C. pp. 603-633.
94. Stark, N. 1982. *The soils and ecology of establishing ponderosa pines on reclaimed land.* In *Proceedings, Symposium on Surface Coal Mining Reclamation in the Northern Great Plains.* Report Number 194. Montana Agricultural Experiment Station Research, Montana State University, Bozeman. pp. B-5-1 - B-5-10.

95. Stephens, J. R., Jr. 1975. *Soil survey of Johnson County, Wyoming, Southern part.* Soil Conservation Service, U.S. Department of Agriculture, and Wyoming Agricultural Experiment Station, University of Wyoming, Laramie.
96. Stevenson, A., R. E. Baumgartner, and G. E. Schuman. 1984. *High Plains Grasslands Research Station—detailed soil survey.* Wyoming Agricultural Experiment Station, University of Wyoming, Laramie.
97. Stevenson, D. S. 1967. *Effective soil volume and its importance to root and top growth of plants.* Canadian Journal of Soil Science 47: 163-174.
98. Stevenson, D. S. 1970. *Soil volume and fertilizer effects on growth and nutrient content of sunflower.* Canadian Journal of Soil Science 50: 353-360.
99. Stevenson, D. S., and L. Boersma. 1964. *Effect of soil water content on the growth of adventitious roots of sunflower.* Agronomy Journal 56: 509-512.
100. Stout, B. B. 1980. *Studies of tree root systems at Colstrip, Montana.* In *Proceedings, Symposium on Adequate Reclamation of Mined Lands?* Western Agriculture Experiment Stations Coordinating Committee on Mine Waste Reclamation and Soil Conservation Society of America, Ankeny, Iowa. pp. 18-1 - 18-8.
101. Tedrow, J.C.F. 1966. *Properties of sand and silt fractions in New Jersey soils.* Soil Science 101: 24-30.
102. Unger, P. W. 1971. *Soil profile gravel layers: I. Effect on water storage, distribution, and evaporation.* Soil Science Society of America Proceedings 35: 631-634.
103. Unger, P. W. 1971. *Soil profile gravel layers: II. Effect on growth and water use by a hybrid forage sorghum.* Soil Science Society of America Proceedings 35: 980-983.
104. VanLear, D. H. 1971. *Effects of spoil texture on growth of K-31 tall fescue.* Research Note NE-141. Forest Service, U.S. Department of Agriculture, Broomall, Pennsylvania.
105. Weaver, T. 1978. *Changes in soils along a vegetation-altitudinal gradient of the northern Rocky Mountains.* In C. T. Youngberg [editor] *Forest Soils and Land Use.* Colorado State University, Fort Collins. pp. 14-29.
106. Whitman, W. C. 1950. *Productivity of natural prairie vegetation.* Bulletin 356. North Dakota Agricultural Experiment Station, Fargo. pp. 29-30.
107. Wyoming Department of Environmental Quality. 1985. *Guideline No. 1—Topsoil and overburden.* Cheyenne.
108. Zahner, R. 1958. *Site-quality relationships of pine forests in southern Arkansas and northern Louisiana.* Forest Science 4: 162-175.

Chapter 13

Cadmium and zinc

Nirander M. Safaya, Joan E. McLean, and Gary A. Halvorson

Cadmium and zinc are naturally occurring trace metals that are often considered together because of their close geochemical association and similarities in chemical reactivity. The loss of two electrons from an atom of Cd or Zn imparts to each an electron configuration with completely filled d orbitals; this results in a highly stable 2^+ oxidation state. But Cd and Zn differ greatly in their significance to biological systems. Whereas Zn is an essential nutrient for plants, animals, and humans, Cd is best known for its toxicity to plants and as a causative agent of several disease syndromes in animals and humans.

Zinc is an essential constituent of many enzyme systems, and its role in metabolic processes critical to the development, division, and differentiation of cells is well established (*163, 164*). On the other hand, Cd has not been demonstrated to be essential for plants or animals, though in one study (*134*) Cd at a low concentration was found to improve growth in rats. Unlike Zn, Cd is toxic at extremely low concentrations. Cadmium toxicity is believed to be due in part to its ability to replace Zn in certain enzymatic reactions. Yamegata and Shigematsu (*179*) reported Cd poisoning in segments of the rural population in Japan who consumed rice grown on Cd-polluted soils. In some other studies, incidence of death from arteriosclerotic heart disease in 28 U.S. cities was found to be correlated with atmospheric Cd and Zn pollution (*26, 62*). Although Zn deficiency is more common and widespread than Zn toxicity (*104*), the toxic effects of Zn were recognized before it was shown to be an essential plant nutrient (*23*). Lee and Craddock (*86*) reported cases where protracted use of Zn resulted in Zn toxicity to agronomic species.

Inherent concentrations of Cd and Zn in soils vary greatly depending upon the geologic origin and degree of weathering of parent materials. However, concentrations of these trace elements in soils and water can increase as a result of atmospheric deposition, additions of Zn and phosphate fertilizers to soils, and the agricultural use or disposal of plant residues and industrial and municipal wastes (*37*). Cadmium and Zn are added to the environment every day from such sources of atmospheric pollution as mining and smelting of Zn, Pb, and Cu; steel-producing activities; manufacturing processes for plastic, batteries, pigments, alloys, and fertilizers; combustion processes involving coal, oil, wood, paper, and urban organic trash; and from tires and motor oils used in automobiles (*85, 149*).

Biological importance of cadmium and zinc

Cadmium. Growth reductions in plants due to Cd have been recorded in many studies (*1, 74, 112*). Allaway (*1*) found that plant growth decreased when Cd concentrations in plant tissues were 3 μg/g or more. Page and associates (*112*) observed that Cd concentrations in nutrient solution that were sufficient to reduce growth varied with plant species: 0.2 μg/ml for beets, beans, and turnips; 1 μg/ml for corn and lettuce; 5 μg/ml for tomatoes and barley; and 9 μg/ml for cabbage. However, under soil culture conditions, Traynor and Knezek (*156*) reported maximum reductions in corn growth when 281 μg/g Cd was applied to the soil; Cd concentration in the plant ash was 131 μg/g. Cadmium retards both shoot growth and root growth (*14, 63, 99, 158*). Several metabolic processes in plants are known to be affected by Cd (*63, 81, 112*). It increases respiration rates (*87*), affects mitochondrial oxidation-reduction reactions (*100*), and inhibits photosynthesis and nitrogen fixation (*74*). Wickliff and associates (*174*) found that Cd reduced growth, nodulation, and nitrogenase and nitrate reductase activities in red alder. In another study (*168*) nodule number and weight and acetylene reduction in Red-Mexican beans declined with increasing amounts of Cd in the roots and shoots, resulting from increased levels of Cd supply.

Mammals have low tolerance for Cd, which may be due to their inability to excrete it. With time, Cd accumulates in the liver and kidney. However, various studies have shown that absorption of Cd from the gastrointestinal tract is poor. Therefore, Cd poisoning in animals fed low Cd diet is relatively unlikely (*53*). When fed forages grown on Cd-contaminated soils, Cd toxicity in animals can occur, resulting in reduced growth and in some cases death (*21, 88, 161*). In laboratory studies, dietary Cd concentrations greater than 100 μg/g proved harmful to a variety of species (*12*). In humans, renal tubular damage and pulmonary emphysema are the two principal pathological lesions associated with excessive Cd exposure (*121*). When Cd accumulates in the renal cortex at concentrations ranging from 200 to 300 μg/g wet weight, renal dysfunction results. The first signs of the B_2 microproteinuria occur at the 200 μg/g level.

Food and tobacco smoke are the primary routes of Cd intake in nonoccupationally exposed populations. Most people tend to accrete Cd until about 50 years of age; thereafter a negative balance ensues. The average U.S. daily diet contains 23 to 92 μg Cd (*85*). A seven-year study indicated that the mean Cd intake of 15- to 20-year-old males in the United States is 38 ± 12 μg/day, which includes Cd in drinking water (*84*). Less than 10 percent of ingested Cd doses and 30 percent of inhaled doses are absorbed by the body (*85*). Cadmium toxicity has also been linked to erythrocyte destruction, testicular damage, and cardiovascular disorders. The "Itai Itai" disease in Japan resulted from high Cd levels in water, rice, and fish in the endemic area; the chronic source of the Cd pollution was a Zn-Pb-Cd smelter (*53*). Consumption of unpolished rice containing up to 3.4 μg/g of Cd caused Cd poisoning in segments of the rural Japanese population (*179*). The potentially toxic effects of high Cd levels in diet can be attenuated by Zn (*36, 85*), Se, compounds containing thiol groups, or EDTA-type sequestering agents (*85*). The Ca salt of EDTA markedly increases the excretion of a number of metals, including Cd.

Zinc. Severe metabolic disorders in plants, animals, and humans can be caused by the deficiency as well as the toxicity of Zn. Because Zn is an essential component of a number of enzymes—more than 80 Zn-dependent enzymes have been reported—Zn deficiency causes serious disorders in the growth, developmental, and reproductive processes. Effects of Zn deficiency on plant growth have been documented in numerous publications (*31, 104, 122, 125, 128, 166*). Zinc plays an important role in protein biosynthesis. Its deficiency is known to cause a sharp reduction in the RNA and ribosomal content of cells. Reductions in photosynthetic rate, chloroplast development and grana formation, and meristematic activity have also been associated with Zn deficiency in various studies. Safaya (*122*) reported that Zn plays a role in the regulation of P uptake by plants. Zinc-deficient plants exhibit chlorosis in older leaves, stunted growth, and drastic yield reductions. Some plant disease syndromes, such as white bud of maize, little leaf or rosette in fruit trees, mottle leaf or frenching in citrus, and bronzing of tung, are all manifestations of Zn deficiency.

In animals and humans Zn is necessary for proper skeletal and sexual development and maternal-fetal metabolism (*71, 115, 135, 161*). Taste acuity in humans depends in part on Zn nutrition (*71*). Zinc deficiency causes retarded growth, skin diseases, disorders of feathers or hair coat, reduced feed use efficiency, delayed sexual maturity or loss of fertility and, in acute cases, mortality. By controlling the activity of carbonic anhydrase, Zn plays an important role in maintaining the acid-base equilibrium in the body and in the release of CO_2 in the lungs. Carbonic anhydrase is also important in bone calcification and egg shell formation. Enzymes, such as aldolase, uricase, and kidney phosphatase, are Zn-dependent. Hence, its deficiency affects the activity of these enzymes.

The average concentration of Zn in humans is about 30 $\mu g/g$. Similar concentrations of Zn occur in most other animals. Zinc deficiencies occur in humans as well as other animals in many parts of the world. However, the occurrence of Zn deficiency in animals shows no clearcut relationship with the geographic pattern of its deficiency in plants (*82*). Zinc deficiency in animals often results from ineffective digestion and use of dietary Zn, rather than an absolute deficiency of this element. Monogastric animals with diets high in phytate and Ca may develop Zn deficiency even when the diet contains more than 20 $\mu g/g$ of Zn. However, animals on diets low in phytate and Ca may show optimum growth with less than 5 $\mu g/g$ of Zn in their diets (*82*).

Toxicity of Zn to plants or animals is less common than its deficiency. Under natural conditions Zn toxicity in plants occurs on some acid peats and on soils developed on mineralized zones and outcroppings. Zinc concentrations in such soils range from 0.43 to 12.5 percent (*120, 150*). Zinc toxicity also occurs in vegetation growing on metal mine tailings and on agricultural land that has received heavy applications of sewage sludge, protracted applications of Zn fertilizers, or where excessive amounts of Zn oxysulphate have been used as a pesticide. The presence of luxuriant ragweed growths in areas where other vegetation is stunted is a good indication of Zn mineral outcrops (*120*). Generally, plants show symptoms of Zn toxicity when they contain more than 400 $\mu g/g$ of Zn (*31*). But in some soils high in Zn a distinctly different flora may develop that is capable of growing and completing its life cycle while accumulating more than 400 $\mu g/g$ of Zn in tissues (*7*).

Zinc toxicity symptoms in plants include stunting, chlorosis, and development of light blue-green coloration at the tips, which spreads to the base, followed by leaf drying (*154*). The symptoms often resemble those of P toxicity.

High levels of Zn affect several physiological and biochemical processes; for example, Zn has been found to inhibit CO_2 fixation in isolated spinach chloroplasts (*65*), photosynthesis and respiration in *chlorella* (*42*), oxidation of succinate by isolated corn mitochondria (*16*), and phloem transport (*129*). One of the primary mechanisms of Zn toxicity may be due to its ability to alter membrane permeability. High Zn levels cause cell membranes to become 'leaky' to K (*41, 177*). Excessive Zn levels also interfere with Fe metabolism, presumably by reducing the amount of reductant necessary for Fe translocation (*3*).

Zinc toxicity in animals and humans is rare because of a wide margin of safety that exists between normal intake levels and those likely to produce deleterious effects (*161*). Also, there is a wide variation among animals in their tolerance to Zn. Most laboratory animals tolerate Zn intake levels up to 0.25 percent in their diets. Diets containing Zn concentrations greater than 0.25 percent, however, can cause reduced appetite, growth depression, and mortality. Pigs showed no ill effects when fed diets containing 1,000 μg/g of Zn. In sheep or cattle, dietary Zn levels of 500 μg/g do not cause toxicity symptoms. But diets with 900 μg/g of Zn reduced gains and lowered feed efficiency (*36*). In humans Zn poisoning generally results from intake of acidic foods stored in galvanized iron containers. Zn salts, when consumed at the rate of 50 to 500 mg/kg body weight, can prove lethal to adults (*155*).

Geochemical abundance of cadmium and zinc

Concentrations of Cd and Zn in the lithosphere average 0.18 μg/g (*53*) and 80 μg/g (*59*), respectively. According to Wallace and Berry (*169*), Cd and Zn usually appear in the ratio of 1:200 to 1:50.

Cadmium. Inherent Cd levels in soils vary with the parent rock; they are highest in soils developed on basaltic rock and lowest in soils on granitic rock (*8*). Under unpolluted conditions, the Cd content of soils ranges from 0.01 to 22 μg/g (*1*); a range of 0.06 to 0.5 μg/g is considered normal (*121*). According to Lindsay (*89*), total Cd in agricultural soils ranges from 0.01 to 0.7 μg/g, with an average of 0.06 μg/g. Concentrations of inherent Cd exceeding 1 μg/g generally occur in areas of geochemical enrichment. Boon (*19*) reported Cd concentrations of up to 230 μg/g in the mineralized area of Aspen, Colorado. Analysis of soil samples collected from 24 soil series in California's Santa Monica Mountains indicated that Cd (extracted by 4N HNO_3) was highest (mean, 7.5 μg/g) in residual soils developed from shale and lowest (mean, 0.84 μg/g) in soils developed from sandstone and basalt (*91*). Alluvial soils with parent material from mixed sources contained intermediate levels of Cd (mean, 1.5 μg/g). Some shale samples contained as much as 33 μg/g of Cd (*91*). This confirmed the earlier observations showing that Topanga and Monterey shales of Miocene age found in the Santa Monica Mountains are very high in Cd (*96, 180*). Some carbonaceous shales from a Wyoming coal mine contained 13.1 μg/g of Cd (*20*). However, all shales do not contain high Cd levels. For example, in Green River oil shale Cd concentrations ranged from 0.5 to 4.0 μg/g (*39*).

Analysis of soil samples collected from 21 locations in unglaciated areas of the northern Great Plains by Severson and associates (*137, 139*) showed A horizons contained relatively higher amounts of Cd compared to samples from C horizons. DTPA-extractable Cd ranged from less than 0.01 μg/g to 0.3 μg/g, but higher Cd levels, 0.2 to 3.6 μg/g, were recovered following extraction of soil samples with hydroquinone. In Southeastern Montana topsoils, DTPA-extractable Cd rarely exceeded 0.1 μg/g (*109*). But Cd concentrations greater than 0.7 μg/g have been observed in some northern Great

Plains soils (*107*). Severson and Gough (*138*) analyzed topsoil and spoil materials collected from 13 mine sites in North Dakota, Montana, Wyoming, Colorado, New Mexico, and Alaska. Lower values for Cd were found in soil and spoil samples from New Mexico and higher values in samples from Colorado. Differences in the Cd content of topsoils and spoils generally were marginal. Safaya and associates (*127*) conducted a seasonal monitoring study of trace metals on some surface-mined lands in North Dakota over a 4-year period. EDTA-extractable Cd in spoil and soil materials ranged from 0.27 to 0.63 μg/g, although some samples of glacial material at one site averaged as low as 0.04 μg/g. Moreover, Cd levels declined with time at some sites. These and other studies show that naturally occurring Cd concentrations in soils and spoil materials throughout the western United States are roughly equal to the average geochemical abundance of Cd in the earth's crust.

Zinc. Concentrations of total native Zn in soils range from 10 to 300 μg/g (*89*). These concentrations generally correlate with the total concentrations of Zn in the parent rocks. Soils derived from basalt and shale are often high in Zn, compared to soils derived from granite, limestone, or sandstone. Zinc concentrations in soils are generally somewhat lower than the average Zn concentration in the earth's crust.

In unglaciated regions of the northern Great Plains, Zn concentrations in soil samples ranged from 14 to 174 μg/g; the average was 62 μg/g (*139*). Differences in total Zn concentrations of samples from A and C soil horizons generally were small. But Zn fractions extracted by DTPA, EDTA, and $(NH_4)_2C_2O_4$ were generally lower in C-horizon samples than in samples from the A horizon. In surface and subsoil samples from the Power River Basin, Anderson and associates (*5*) found Zn concentrations in the range of 35 to 100 μg/g and 34 to 110 μg/g, respectively, with averages of 59 and 61 μg/g.

Analysis of 140 surface soils in Colorado's Piceance Creek Basin showed Zn concentrations ranging from 45 to 140 μg/g, with an average of 80 μg/g (*119*). Topsoil and spoil samples from 13 mine sites in North Dakota, Montana, Wyoming, Colorado, New Mexico, and Alaska, in most cases, contained higher amounts of Zn in spoil material than in topsoils (*138*). At the San Juan Mine in New Mexico, Severson and associates (*137*) measured less variability in Zn concentration in the replaced topsoils than in spoils in the reclaimed areas. The geometric mean of Zn concentrations in topsoil was 41 μg/g; 95 percent of the values ranged from 35 to 48 μg/g. But the mean Zn concentration in spoils was 56 μg/g, and 95 percent of the values ranged from 43 to 73 μg/g. Severson (*136*) found significant regional variations in Zn concentrations in the Bighorn Basin and concluded that in this case one summary statistic could not be used as a baseline estimate of Zn concentrations. In North Dakota, EDTA-extractable Zn levels in spoil and soil materials ranged from 1.0 to 11.7 μg/g (*127*). No significant changes in the Zn status of these materials were detected over a 4-year period.

DTPA-extractable Zn concentrations of less than 1.0 μg/g are considered deficient for many economically important crops grown in the north central United States (*173*). Many of the soils and geological materials sampled and analyzed in the above mentioned studies contained less than 1.0 μg/g of Zn and could therefore be considered Zn deficient. Many researchers have found marginally to potentially Zn-deficient mine soils and spoils in the western United States (*70, 123, 127, 132, 178*). In some field studies on reclaimed and undisturbed lands in western North Dakota, however, corn showed no response to Zn fertilization (*64*).

Chemistry and bioavailability of cadmium and zinc

Soil factors. As in the case of other heavy metals, the fate of Cd and Zn in soils is controlled by a complex and dynamic system of physical, chemical, and biological factors. At any given time, these metals are distributed among several pools in the soil: (a) in soil solution, (b) in exchangeable or adsorbed forms, (c) as precipitates, (d) incorporated into biological systems and residues, (e) held in secondary minerals and as insoluble hydrous oxides, and (f) held in primary minerals. The first four pools are in rapid and reversible equilibrium with one another. It is these sources of Cd and Zn that must be considered when describing the plant-available and leachable fractions of these metals in soils.

The solid phase of soil influences the concentration of metals, including Cd and Zn, in the soil solution. A number of processes, including inorganic and organic complexation, acid-base reactions, redox potential, precipitation/dissolution, and interfacial interactions, govern the concentration of heavy metals in the soil aqueous phase. Clay, organic matter, hydrous metal oxides, carbonates, and inorganic compounds in soil can sorb metals and reduce their concentrations in the soil solution.

Precipitates of Cd and Zn include oxides, hydrous oxides, carbonates, phosphates, and, in reducing environments, sulfide. These precipitates may exert a dominant control on the ionic concentrations of Cd and Zn in the soil solution. The concentration at which precipitation occurs depends upon the individual compounds in question and pH. As long as the solubility product is exceeded, the solid phase will continue to form, reducing the concentration of metal in the soil solution until apparent equilibrium is reached. If the concentration of metal ions drops below this critical level as a result of dilution, plant uptake, adsorption, or other causes, the solid phase will dissolve to maintain an equilibrium condition between the solid and solution phases.

Use of a solubility diagram is a convenient technique to show how the solubility of heavy metal compounds varies with pH. At the same time, it allows prediction of which solid phase regulates heavy metal activity in the soil solution. pCd-pH diagrams, independently constructed by Street and associates (*130*) and Santillan-Medrano and Jurinak (*153*), have shown that in a calcareous Nibley soil the solubility of Cd decreases with increasing pH. This is the usual trend for most heavy metals. These researchers also theorized that a Cd phosphate compound could be regulating the activity of Cd^{2+} ions in solution.

Soil pH, more than any other single factor, influences solution concentrations of Cd and Zn. The solubility of Cd and Zn compounds increases with decreasing soil pH. Soil pH also directly effects Cd and Zn sorption (*68, 83, 98*). Many sorption sites in soils are pH-dependent, in that as pH decreases the number of negatively charged sites also decrease. In acid media Cd and Zn compete with H^+ and Al^{3+} for the available, permanent-charged sites. Hydrous oxides of Fe and Mn play a principal role in the retention of metals in soils (*76*). Solubility of Fe and Mn oxides is pH-related. Below pH 6, the Fe and Mn oxides dissolve; this releases sorbed metal ions to solution (*51*). In nearly every experiment, plants have been found to absorb more Cd and Zn at lower pH (*6, 93, 95, 152*).

Cadmium and Zn do not form highly stable complexes with organic matter. But organic matter is more important than hydrous oxides of Fe and Mn in moderating the effect of excessive soil Zn and limiting Zn and Cd uptake by plants (*172*). Stevenson (*151*) reported that, compared to Cu and Pb, Cd formed less stable complexes with humic acids derived from the A_I horizon of Harpster silt loam, the surface layer of an Ohio peat, and highly oxidized lignite (leonardite) from North Dakota. In this respect, Cd behaves much like Zn in its ability to complex with fulvic and humic acids. The Cd-retaining effect of organic matter was attributed by Haghiri (*63*) to an increase in cation exchange capacity, rather than to its capacity to chelate the metal. Similar observations were made recently by Hinesly and associates (*73*); but in their study the inverse relationship between Cd availability to plants and cation exchange capacity was true only when Cd was supplied as a soluble salt and not when it was supplied as a constituent of municipal sewage sludge.

Several studies have indicated that the activity of Cd and Zn in most soils may not be controlled as much by precipitation phenomena as by adsorption and chelation reactions (*89, 153*). Andersson and Nilsson (*6*) studied the adsorption of Cd by organic matter and Fe and Mn hydrous oxides and found that only organic matter was effective in adsorbing Cd at lower pH levels.

Garcia-Miragaya and Page (*55*) observed that the amount of Cd adsorbed on montmorillonite surfaces depended upon the exchangeable cation present on the clay, decreasing in order of Na-montmorillonite > K-montmorillonite > Ca-montmorillonite > Al-montmorillonite. Exchangeable Al was particularly effective in reducing Cd sorption by high energy sites on montmorillonite. Egozy (*48*) suggested that absorption of Cd by Na-montmorillonite was due to two classes of sites: constant charge sites and constant potential (oxide-like edge site) sites. But studies by Inskeep and Baham (*75*) indicated that constant potential edge sites may be primarily responsible for Cd adsorption at low adsorption densities.

Acid soils have a lower ability to retain Cd than neutral soils. Cavallaro and McBride (*28*) found that in the presence of 0.01 M $CaCl_2$ adsorption of Cd was considerably lower, suggesting competition by Ca for the adsorption sites. Their results suggested an ion-exchange mechanism of adsorption in surface soils and precipitation in calcareous subsoils. McBride (*97*) found that the quantity of Cd adsorbed in the presence of $CaCO_3$ depended upon the particle size of the $CaCO_3$. High levels of Cd apparently caused nucleation of $CdCO_3$ on $CaCO_3$. Chemiadsorption of Cd on $CaCO_3$ was very rapid, while $CdCO_3$ precipitation at higher Cd concentrations was slow. Thus, chemiadsorption probably is important in inhibiting Cd mobility in calcareous soil materials.

Because of the generally low geochemical abundances of Cd and Zn and the neutral to alkaline pH of most soil and overburden materials in the western United States, Zn and Cd toxicities are rarely encountered in the West under natural conditions. The high soil pH and low annual precipitation in the West also restricts leaching of Cd and Zn into groundwater systems. Studies of the fate of heavy metals applied to soil in sewage sludge have shown that Zn moves very little below the level of incorporation in surface soil (*80*).

Plant factors. Cadmium and Zn concentrations in plants generally increase with the increased availability of these elements in the substrate. However, plant species, genotypes, and ecotypes differ in their ability to absorb and accumulate Cd and Zn (*36, 37, 61, 92, 101*). Plants also differ in their sensitivity to Cd and Zn concentrations in the substrate and in plant tissues that cause symptoms of deficiency (Zn) or toxicity (Cd and Zn). There is some evidence that suggests that absorption and translocation mechanisms may be genetically controlled, but independent of each other (*37, 72*). Research has been conducted to identify plants that are tolerant to high levels of Cd and Zn (*13, 22, 37*) and plants that exhibit low susceptibility to Zn deficiency (*2, 54, 58, 124*). Indeed, some plant breeders have attempted to select and breed plants that are low accumulators of Cd and Zn and those that can withstand Zn deficiency.

Studies by Mahler and associates (*93*) and Bingham (*13*) have indicated that plants, such as spinach, lettuce, curly cress, and soybeans, are sensitive to Cd toxicity, whereas tomatoes, squash, cabbage, and paddy rice are relatively tolerant. Tolerance to excessive Zn has been reviewed briefly by Brown and associates (*25*). A number of studies show that metal-tolerant plants exhibit greater ability to bind the metals on root cell walls. These plants can thereby exclude toxic levels of metals from reaching the cytoplasm (*113, 159, 160*).

Generally, leafy vegetables tend to accumulate the greatest amounts of Cd, although cabbage has been found to be an apparent excluder of Cd compared to lettuce (*47, 57*). Tobacco is also an accumulator of Cd, particularly when grown in acid soils. The edible portions of many vegetable crops (such as squash, tomatoes, radishes, and potatoes) and cereal grains accumulate relatively low levels of Cd, compared to the nonedible or leafy portions (*9, 47, 57, 92*). In a recent study of corn grown on sludge-amended mine spoil, Pietz and associates (*114*) found that Cd and Zn concentrations in the leaves were 36 and 4 times higher than those in the grain, respectively. In a study by Giordano and Mays (*37*), soybean grain contained considerably higher amounts of Cd than corn grain or cotton seed. Also, Zn concentrations in soybeans were higher in both the leaf and grain fractions, compared to the other two crops. Decker and associates (*40*) reported significantly higher amounts of Cd and Zn in swiss chard and lettuce compared to soybeans or oat grains. Oats tend to accumulate greater amounts of Cd than do soybeans (*37*).

Differences in Cd and Zn uptake in various cultivars of soybeans and lettuce and in the inbred lines of corn have been reported (*17, 57, 72*). One study showed that the two lowest accumulators of Zn among lettuce cultivars studied were also the lowest accumulators of Cd (*57*). Another study, however, showed no correlation between Cd and Zn accumulation for corn inbred lines (*72*). Chang and associates (*30*) found no significant differences in Cd or Zn accumulation in four cultivars of barley when grown without or with low applications of sludge. However, under high rates of sludge application, Cd and Zn concentrations in some cultivars were significantly higher than in the other cultivars.

Cadmium toxicity in plants occurs at much lower concentrations than Zn toxicity. Plant Cd concentrations of 9, 17, 24, 37, and 43 μg/g reduced yields by 25 percent in sudangrass, clover, alfalfa, fescue, and bermudagrass, respectively (*15*). On a sewage-sludge-amended spoil, Pietz and associates (*114*) found no visible symptoms of toxicity with Cd concentrations up to 28.7 μg/g in corn leaves and 0.78 μg/g in corn grain. Also, no toxic effects were observed with Zn concentrations up to 200 μg/g in corn leaves and 45 μg/g in corn grain. In Nebraska, field experiments with Zn application rates up to 358 kg/ha on acid soil and 1,390 kg/ha on alkaline soil showed no toxicity symptoms in corn (*32*). On coastal plain soils in Florida the addition of 200 to 300 μg/g of Zn to the root medium was necessary to induce toxic symptoms in pine trees (*165*).

In summary, not only do various plant species and cultivars exhibit differences in their ability to absorb and respond to Cd and Zn, but they also differ in the distribution and accumulation of these elements within different tissues and organs. Generally, edible portions of most agricultural crops tend to accumulate lower concentrations of Zn and Cd than nonedible portions. The main exceptions are leafy vegetables, such as lettuce (*131*). Thus, leafy vegetables and forages generally present a greater heavy-metal hazard to the food supply.

Concentrations of Cd and Zn in plants, like most other elements, may vary significantly from year to year under the same soil environmental conditions. Also, concentrations of these elements in plants undergo changes during the growing season, and concentrations increase with increasing soil temperatures. As plants age, elemental concentrations in their tissues usually decrease. This may be due to changes in uptake rate, dilution with growth, or redistribution. However, all elements do not follow this trend. For example, Munshower and Neuman (*108*) observed that when Ponderosa pine needles aged Zn concentrations decreased while Cd concentrations remained unchanged. But in the Ponderosa pine fasicular sheaths both Cd and Zn increased with age. These increases, however, appeared to be related to the scavenging of airborne particulates by the fasicular sheaths and not to actual uptake from soils.

Various ionic interactions during absorption and translocation also affect Cd and Zn accumulation in plants. Interactions of Cd with other metals, such as Zn, Cu, Se, and Mo, have been observed in soybeans and corn (*36*). Haghiri (*63*) reported that large applications of Zn reduced Cd concentrations in corn shoots. However, Sheaffer and associates (*141*) found that Cd concentrations in soybean shoots increased after Zn application. Short-term isotopic studies have shown that Zn can serve as a competitive inhibitor of Cd absorption (*27*). Increases in plant Cd concentrations following phosphate applications have been well documented, but these increases result more from Cd contamination in phosphate fertilizers than from physiological interactions. In some studies, phosphate applications on P-deficient soils reduced Cd content in corn leaves (*36*). Many researchers have shown that P reduces absorption and translocation of Zn in plants (*111, 122, 125*). Antagonistic effects of Cd on Zn uptake have also been documented (*94*). Foy and associates (*54*) suggested that Cd toxicity in plants may be due in part to the ability of Cd to replace Zn in certain metabolic processes.

Researchers have attempted to establish criteria or formulas to predict accurately the heavy metal toxicity in plants based on extractable concentrations in the soil. Such attempts have resulted from the need to develop loading rates of heavy metals in sewage sludge applied to land. The first attempts to limit sewage-sludge-applied metals to soil were based on the concept of "Zn equivalent" (*33*). This concept assumes that toxicities of Cu and Ni can be expressed in terms of some multiple of Zn, and that the toxicities of these three elements to plants are additive (*36*). Since the introduction of the "Zn-equivalent" concept, new information shows that the toxicity of these elements generally is not additive. The formula does not apply to calcareous or high pH soils, or to a broad spectrum of plant species.

Chumbley (*33*) recommended 250 $\mu g/g$ of Zn as the level of Zn equivalent that is toxic to plants. Valdares and associates (*162*) found that seven times that amount did not affect plants growing in calcareous soil.

Another suggested method to limit sewage sludge application rates was based on the ratio of Zn to Cd in the sludge. This method is based on the premise that the ratio of Zn to Cd averages 500 for parent rocks and 100 for soils, and that sludge applications based on a ratio of Zn to Cd would result in high enough Zn concentrations in soil to kill plants before Cd could accumulate to levels considered hazardous to animals and humans. But research has shown that many plants grown on calcareous soils generally tolerate high levels of Zn as well as Cd and that, at a given ratio of Zn to Cd, the Cd concentration in plants increases with increasing Cd applications. To date no criteria or formulas have been developed which would satisfactorily predict plant toxicities from added Cd or Zn.

Management and environmental factors. The risk of increasing Cd and Zn in soils and plants is greatly associated with disposal and agricultural use of sewage sludges and other wastes, use of phosphate fertilizers, and naturally occurring or artificially created acidic soil environments. Examples of the last condition include acidic coal mine spoils, mostly prevalent in the eastern United States, and metal ore production and processing, conducted in both the eastern and western United States. Consideration of these factors is not only useful for understanding bioavailability of Cd and Zn in general, but also with respect to their implications for the reclamation of mined lands and metal tailings. Because sewage sludges and phosphate fertilizers are often applied during reclamation of disturbed lands, this consideration becomes urgent.

The effects of sewage sludges on Cd and Zn accumulation in soils and plants have been investigated in numerous studies (*36, 37*). Sewage sludges contain a variety of organic and inorganic forms of Cd and Zn. The levels of these elements depend upon the nature of the sludge and the treatment processes of the influent. Experiments with single and annually repeated applications of sludges to soils have shown increases in the Cd and Zn content of a majority of crops. But Cd and Zn accumulations in plants from repeated annual sludge applications have been either less than or similar to accumulations from single applications of sludge supplying Cd and Zn in amounts equivalent to the sum of the annual applications (*37*). Plant uptake of Cd and Zn is generally higher if these elements are supplied to soil in the form of inorganic salts than when supplied through sewage sludge (*43*). Availability of Cd and Zn, whether supplied through sludge or as inorganic salts, is generally higher in noncalcareous soils. But in a calcareous soil Baxter[1]

[1]Unpublished data from J. C. Baxter, Metropolitan Denver Sewage Disposal District No. 1, Denver, Colorado.

found greater uptake of Cd and Zn in corn following sludge treatment, compared to equivalent applications of Cd and Zn salts. This was due to a temporary reduction in soil pH by the sludge; Cd and Zn salts had no effect on soil pH. Chang and Page (*29*) found increased Cd and Zn contents in a number of vegetable crops grown on a calcareous Domino soil with increasing rates of sludge applications. Marked increases in Cd and Zn concentrations in corn leaves have also been observed with repeated applications of sludge to calcareous mine spoils (*114*). In this study the cumulative effects of sludge applications on Cd and Zn levels in corn did not decline rapidly with time. This was attributed to the pH and mineral composition of the spoil. Normally, the residual effects of Cd or Zn applications persist for a few succeeding crops, and the duration of such effects depends upon application rates, species and cultivars used in rotation, and the properties of the soils involved (*37, 126*).

Cadmium concentrations in soils and plants increase with both short- and long-term applications of some phosphate fertilizers (*102, 105, 118, 175*). Such increases result from phosphate fertilizers that contain varying amounts of Cd and other heavy metals as contaminations from phosphate rock. The degree of heavy metal contamination in phosphate fertilizers varies with the source of phosphate rock and the P concentration in the fertilizer (*103, 175*). Phosphate ores in the western United States contain considerably greater concentrations of Cd than ores from the southeastern United States. Application of superphosphates for more than 20 years on some Australian soils caused a nearly five-fold increase (from 0.046 μg/g to 0.212 μg/g) in the Cd concentration of topsoil (*175, 176*). Greenhouse studies indicated that some crops grown on phosphate-fertilized soils accumulated significantly higher amounts of Cd than those grown on soil collected from unfertilized fields. A linear relationship between Cd concentration in peas, radishes, and lettuce and Cd concentration in phosphate fertilizer has been observed by Reuss and associates (*118*). In a short-term study on wheat using diammonium phosphate fertilizers containing 2, 74, and 153 μg/g of Cd, Mortvedt and associates (*102*) found that levels of Cd in the wheat grain increased only with the application of the highest Cd-containing phosphate fertilizer. They concluded that "the daily intake of Cd in wheat products is not significantly changed by the phosphate fertilization and crop production practices in general use in the United States at this time" (*102*).

Plants growing on acid mine spoils, metal tailings, and in the vicinity of ore processing or smelting complexes often contain very high levels of heavy metals, including Cd and Zn. Acidic coal mine spoils are common in the eastern United States. In the semiarid and arid West, coal mine spoils are generally neutral to alkaline and pose fewer heavy metal toxicity problems. But some studies have indicated that the net acid production in some western coal mine spoils, resulting from pyrite oxidation, may exceed the neutralizing capacity afforded by the levels of alkalinity in such materials (*20*). In these situations the concentration of heavy metals may be relatively high.

Concentrations of Cd and Zn in some grasses growing near Colstrip, Montana, ranged from 0.04 to 0.20 μg/g and 6 to 35.5 μg/g, respectively (*132*). Generally, Zn concentrations in grasses growing on 2-year-old spoil were greater than those growing on native range. Grasses growing on 6-year-old or 50-year-old spoils contained intermediate levels of Cd and Zn. Erdman and Gough (*50*) found that two out of three samples of hard red winter wheat grown on topsoiled spoils at the Big Sky Mine in Montana contained significantly higher concentrations of Zn and other metals, compared to 17 samples of the same kind of wheat grain from farm storage bins in eastern Montana and northwestern South Dakota. Cadmium levels, however, were lower in the grain grown on the mine sites. Severson and associates (*139*) made a detailed study of trace element contents of plants in the unglaciated portion of the northern Great Plains. Observed ranges and geometric mean concentrations of Cd, respectively, were 0.01 to 0.08 μg/g and 0.03 μg/g for western wheatgrass, 0.09 to 0.67 and 0.30 μg/g for silver sagebrush, and 0.02 to 0.43 μg/g and 0.07 μg/g for the total biomass. Ranges and means for Zn, respectively, were 5.7 to 34 μg/g and 15 μg/g for western wheatgrass, 19 to 64 μg/g and 34 μg/g for silver sagebrush, and 19 to 41 μg/g and 27 μg/g for the total biomass. Most of these values cannot be considered hazardous to plant or animal life.

In the western United States, acidic soil conditions occur mostly in association with U, Pb, Cu, Co, Fe, Cr, and Pt mining. The sulfide minerals of Cd (greenokite) and Zn (sphalerite), the most abundant geological sources of these two elements, are typically found at the metal mine sites. Hence, the availability of Cd and Zn in the metal ore producing and processing areas is almost invariably high. Heavy metal toxicities resulting from acidification of metal tailings have been observed at the Blackbird Mine in central Idaho (*52*), in mine spoils in Utah (*110*), and at Beartooth Plateau in southern Montana (*78*). None of these reports specifically identified Zn toxicity as the major problem. However, high Zn levels likely contributed to poor vegetation in these areas. At the Blackbird Mine, "hot spots" contained 10 to 20 μg/g of total Cd. Also, Cd and Zn contamination was found near a smelting operation: 29 μg/g in soil samples collected two km from the smelter, compared to the normal background levels of 0.2 to 0.6 μg/g. Cadmium measurements in plant and animal tissues in the polluted area confirmed elevated levels of Cd throughout the food chain. At Kellogg, Idaho, soil surrounding a Pb smelting complex was highly enriched by Cd (up to 140 μg/g) and Zn (up to 29,000 μg/g), among other elements (*116*). Some metal-tolerant grasses growing at this site contained as high as 4,400 μg/g of Cd and 12,000 μg/g of Zn. A population of *Holcus lanatus* growing on a metal-contaminated site showed distinct tolerance to Cd, Zn, and Pb, while a population of the same species growing on an uncontaminated site did not (*35*).

Anderson and Keith (*4*) determined Cd and Zn content in sagebrush (*Artemisia tridentata*) growing near the Jim Bridger power plant in Wyoming. They found that 95 percent of the Cd and Zn concentrations ranged from 0.98 to 8.0 μg/g and 120 to 660 μg/g, respectively, with average values of 2.8 μg/g for Cd and 280 μg/g for Zn. Munshower and Newman (*108*) sampled a number of species before and after on-line operation of two electric generating units at Colstrip, Montana. Zinc and Cd concentrations varied greatly among species, with poplar and willow exhibiting much higher Cd and Zn concentrations than several grass and forb species. Significant annual fluctuations in elemental concentrations occurred in both the preoperational and postoperational collection periods. Directional or distance variability in elemental content existed prior to plant operation and continued unchanged after the power plants were on line. Schwab and associates (133) found that Cd concentrations in Indian ricegrass (*Oxyzopsis hymenoides*), fourwing saltbush (*Atriplex canescens*), western wheatgrass (*Agropyron Smithii*), Sherman big bluegrass (*Poa ampla*), Utah sweetvetch (*Heydsarum boreale*), and winterfat (*Ceratoides lanata*) growing on soil covered with retorted shale were not hazardous to livestock.

Munshower and Newman (*108*) observed decreases in Zn and Cd concentrations in some wildlife species sampled in the postoperational period at two power generating plants near Colstrip, Montana. In another study goats and lambs were fed corn silage that contained 5.26 μg/g of Cd and 107.0 μg/g of Zn (*46*). Dry matter intake, daily milk production, and feed efficiency were not affected by 3 years of consumption of silage high in Cd. The Cd concentration in goat milk did not increase in spite of a daily dietary intake of as much as 5 mg of Cd in the corn silage (*45*).

Although Cd concentrations in primary herbivores may not be toxic, Cd could be concentrated in the food chain. Grasshoppers on western rangeland have been found to accumulate both Cd and Zn (*108*). Beyer and associates (*12*) reported that earthworms from a soil containing only 1.9 μg/g of Cd contained 101 μg/g of Cd. In a sewage-sludge-treated soil containing 8.2 μg/g of Cd, earthworms contained 118 μg/g of Cd. In this study, liming reduced Cd concentrations in earthworms slightly. Thus, while liming may prevent excessive accumulation of Cd in plants, elevated levels of Cd may still move into food chains as a result of predation of earthworms by other animals. Research in England has shown that as much as 10 percent of the dry matter intake by cattle is soil (*51*). Cattle grazing on contaminated soil may ingest excessive Cd or Zn, even when the trace metal content of the forage is low.

Measurements of cadmium and zinc in soil and overburden

Total Cd and Zn concentrations in rock and soil samples are generally determined using hydrofluoric acid and nitric acid as extractants. But correlations between total concentrations of these elements in soils and their uptake by plants are generally poor. Thus, analysis of soils for total Cd or Zn, though useful and often necessary in geochemical studies, does not provide a good indication of the extent of their availability to plants. Considerable research has been conducted to develop appropriate methods for extracting and determining the plant-available fractions of these elements in different types of soils. The extractants tested for assessing Cd and Zn availability in soils include dilute acids, neutral salts, and complexing and chelating agents.

Development of soil tests for assessing Zn availability and their calibration against plant growth response and Zn uptake have received greater attention than those for Cd. This is mainly because of the widespread deficiency of Zn encountered in many areas of the world. Cox and Kamprath (*38*) have reviewed many of these methods and discussed the soil factors that must be considered in interpreting the analytical results. The factors that must be quantified to interpret soil tests for Zn properly include pH, free lime, and the level of soil P or P fertilizer application rates.

A number of extractants have been used with varying degrees of success in predicting Zn deficiency in soils. These include a two-phase system of aqueous ammonium acetate (CH_3COONH_4) and dithizone in carbon tetrachloride (CCl_4) (*140, 167*); 0.lN HCl (*10*); a mixture of 0.05 N HCl and 0.025 N H_2SO_4 (*171*); EDTA-$(NH_4)_2CO_3$ (*157*); and neutral salts, such as NH_4NO_3, KCl, and $MgCl_2$ (*117*).

Recently, use of 0.005 M diethylenetriaminepentaacetic acid (DTPA) + 0.01 M $CaCl_2$ + 0.1 M triethanolamine (TEA), buffered at pH 7.3, has gained as much acceptance as a soil extractant for a number of micronutrients, including Zn and Cd (*90*). DTPA-extractable Zn has proved to be a good predictor of Zn uptake by plants in numerous studies. The critical concentration below which Zn deficiency in plants is known to occur ranges from 0.5 to 1 $\mu g/g$ (*24, 38, 66*). This method has also been used to assess Zn availability in soils treated with manure and sewage sludge (*14, 56, 142, 153, 170*). But in a study of 46 Ontario soils varying in the degree of metal contamination, there was better correlation between Zn uptake by swiss chard and CH_3COONH_4-extractable Zn compared to EDTA- or DTPA-extractable Zn (*67*).

Extractants used successfully to assess plant-available Cd in soils include HCl (*79, 84*), CH_3COONH_4 (*63, 77*), DTPA (*14*), and acetic acid (CH_3COOH) (*67*). Haq and associates (*67*) found CH_3COOH to be the best predictor of Cd availability for swiss chard when pH was used as the independent variable in the prediction equation. Mahler and associates (*94, 95*) used water saturation extracts to predict Cd availability to plants growing on acid and calcareous soils. These researchers found that Cd concentrations were higher in saturation extracts of some calcareous soils compared to acid soils, but Cd availability to plants was generally higher in acid soils.

Uptake of Cd by plants correlated equally well with the total Cd in saturation extracts, free ionic Cd, or the activity of Cd (*95*). However, the DTPA extraction method for assessing Cd availability has been quite successful in many studies (*14, 153*). The DTPA soil test has a sound theoretical basis for the extraction of Cd (*148*). A recently developed, modified DTPA procedure includes NH_4HCO_3 in the extracting solution (*146, 147*). This procedure provided good correlation with plant uptake of several elements, including Zn and Cd (*18, 19, 69, 144, 145*).

Very little work has been done to develop and calibrate soil tests for assessing trace metal availability to naturally growing plants on uncultivated soils or mine spoils. Severson and associates (*139*) found that DTPA, EDTA, and $(NH_4)_2C_2O_4$ extractants were of limited use in predicting the availability of several trace metals to native plants growing on uncultivated soils in the northern Great Plains. Gough and associates (*60*) urged caution in defining plant-available levels for trace metals in mined lands based on DTPA or EDTA extraction procedures. In spite of these difficulties the DTPA procedure still offers the greatest potential of any extractant to predict the availability of Cd or Zn on mined lands with reasonable accuracy.

In many western mined-land situations the testing of soil, overburden, or spoil materials for Cd and Zn may not be necessary because of the generally alkaline nature of the materials. Munshower (*107*) suggested that Cd and Zn analysis should not be routinely required on western mined lands unless there are geochemical factors that warrant such analysis; for example, where overburden materials contain significant amounts of pyritic minerals or where acid production from pyrite oxidation is a problem analysis for heavy metals, including Cd and Zn, should be routinely conducted.

Presently, only New Mexico's surface mining regulations require mining companies to test for both Cd and Zn in overburden materials or in materials to be used as a substitute for or supplement to soil (*11*). Colorado requires Zn analysis for overburden materials. Since the potential for acid formation does exist in some calcareous overburden materials in the West, it is desirable that Cd and Zn assays should be performed in conjunction with acid-potential (weathering) studies. Alternatively, extraction of the soil or overburden materials with dilute acid can be used to provide an indication of the Cd and Zn hazards that may unleash in the weathering process.

Measurement of Cd in soil solutions by atomic absorption spectroscopy has been found to give significantly lower values of Cd if the solution is not acidified (*143*). Differences between nonacidified and acidified soil solutions with respect to Cd determination are more pronounced when Cd concentrations in solution are low.

In sampling soil and overburden materials for Cd or Zn analysis normal precautions should be taken to prevent contamination of the samples. Use of galvanized buckets or cans can seriously contaminate samples for either Zn or Cd analyses. Grease used on drill pipe sections can contaminate deep cores obtained in overburden sampling (*44*). At least 90 percent of the commercially available drilling greases have a high potential to contaminate overburden samples with either Cu or Zn. Two drilling greases have been developed that will not contaminate overburden samples (*44*); both have excellent make-up and break-out torques performance.

Some conclusions and recommendations

Cadmium and Zn exhibit many similarities in their chemical reactivity, but the two elements differ greatly in their significance to biological systems. Whereas Zn is an essential nutrient element for plants and animals, Cd is potentially toxic to both and is not known to be essential to either. Concentrations at which Zn is toxic to plants or animals far exceed those of Cd. Zinc toxicity is relatively less common than its deficiency. But under some environmental and management conditions the risk of Cd and Zn toxicities is greatly enhanced.

Cadmium and Zn concentrations in the lithosphere average 0.18 μg/g and 80 μg/g, respectively. But the inherent concentrations of these elements vary widely in different soils, depending upon the nature of parent materials and their degree of weathering. Soils contaminated by metal mine tailings; or that receive sewage sludges, phosphate, and Zn fertilizers; or that are subject to industrial and automobile pollution contain generally high levels of Cd and Zn. Vegetation growing on such soils can increase the loading of these elements in food chains, creating a health risk for animals and humans.

A number of soil, plant, and management factors influence the bioavailability of Cd and Zn. The activities of these elements in the soil are controlled by adsorption and chelation reactions, and their solubility increases as the pH of the soil decreases. Other soil factors that affect the availability of these elements to plants include carbonates, organic matter, hydrous metal oxides, and cation exchange capacity. Availability of Cd and Zn to plants is generally less in alkaline and calcareous soils than in acidic soils. Applications of these elements to soils, either directly in the form of their salts or indirectly as contaminants in sewage sludges, fertilizers, or other sources, increases their concentrations in plants. These increases are substantially greater in noncalcareous soils than in calcareous soils.

Considerable variation exists among plant species, cultivars, and genotypes in their ability to absorb and translocate Cd and Zn. Moreover, accumulation of these elements is generally higher in the foliage and other vegetative tissues than in the fruit, grain, or tubers. Cd and Zn concentrations in plant tissues change with the age of the plant and may vary considerably from year to year under similar growing conditions. A number of antagonistic and synergistic interactions with other elements also affect their uptake by plants. Moreover, the addition of Zn to soil may alleviate Cd toxicity symptoms in plants, while Cd may reduce Zn uptake.

Plants as well as animals markedly differ in their susceptibility to Zn deficiency and in their tolerance to Cd and Zn toxicities. Accumulation of Cd and Zn in animal organs and tissue is influenced not only by the levels of these elements in their diet but also by digestion efficiency and a complex set of interactions with other dietary factors. Cadmium tends to accumulate in the kidneys of animals and humans and can cause acute and chronic illness.

In view of these considerations, there should be some concern about possible Cd and Zn problems in mined-land environments. Mining disturbance, resulting in the exposure of overburden materials to oxidizing condition, can trigger a number of chemical reactions. For example, oxidation of pyrites in the overburden materials results in acid mine drainage, which in turn increases the availability of heavy metals, including Cd and Zn, to toxic levels. However, acid-mine drainage with its attendant problem of heavy metal toxicities is more common in the eastern United States. Overburden materials in the West are generally neutral to alkaline and contain calcite in amounts often sufficient to neutralize any acid generated from pyrite oxidation. Consequently, the likelihood of Cd or Zn toxicity occurring in these environments is small.

At some abandoned and active mines in the West, however, acidic spoils do exist. Soil, spoil, or overburden that exhibits potential for acid generation should be analyzed for Cd and Zn. Soil or spoil materials with alkaline pH values do not warrant routine Cd and Zn analysis, unless there is reason to believe that such materials are low enough in available Zn to cause Zn deficiency problems in plants or in animals feeding upon such plants.

Soil-test methods devised to assess and predict the availability of Zn or Cd to plants are based mostly on calibration studies involving agricultural soils and cultivated plant species. There is very little information about the suitability of these testing methods for mine soils and spoils or to what extent they accurately predict the uptake of these elements by native rangeland or introduced pasture species. The DTPA-extraction procedure of Lindsay and Norvell (*90*) and the DTPA-NH_4HCO_3 procedure of Soltanpour and Schwab (*146*) have been used for mine soils and spoils with some success. But more research is needed on testing procedures and the response of naturally occurring plant species to Cd and Zn.

REFERENCES

1. Allaway, W. H. 1968. *Agronomic controls over the environmental cycling of trace metals.* Advances in Agronomy 20: 235-274.
2. Ambler, J. E., and J. C. Brown. 1969. *Cause of differential susceptability to zinc deficiency in two varieties of navy beans (*Phaseolus vulgaris *L.).* Agronomy Journal 61: 41-43.
3. Ambler, J. E., J. C. Brown, and H. G. Gauch. 1970. *Effect of zinc on translocation of iron in soybean plants.* Plant Physiology 46: 320-323.
4. Anderson, B. M., and J. R. Keith. 1976. *Soil and sagebrush chemistry near the Jim Bridger Powerplant.* In *Geochemical Survey of the Western Energy Re-*

gions. Open-file Report 76-729. U.S. Geological Survey, Washington, D.C. pp. 37-47.

5. Anderson, B. M., J. R. Keith, and J. J. Connor. 1975. *Antimony, arsenic, germanium, lithium, mercury, selenium, tin and zinc in soils of the Powder River Basin.* In *Geological Survey of the Western Coal Regions.* Open-file Report 75-436. U.S. Geological Survey, Washington, D.C. pp. 50-57.
6. Andersson, A., and K. O. Nilsson. 1974. *Influence of lime and soil pH on cadmium availability to plants.* Ambio 3: 198-200.
7. Antonovics, J., A. D. Bradshaw, and R. G. Turner. 1971. *Heavy metal tolerance in plants.* Advances in Ecological Research 7: 1-85.
8. Baes, C. F., Jr. 1973. *The properties of cadmium.* In W. Fulkerson and H. E. Goeller [editors] *Cadmium, the Dissipated Element.* ORNL/NSF/EP-21. Oak Ridge National Laboratory, Oak Ridge, Tennessee.
9. Baker, D. E., M. C. Amacher, and R. M. Leach. 1979. *Sewage sludge as a source of cadmium in soil-plant-animal systems.* Environmental Health Perspectives 28: 45-49.
10. Barrows, H. L., and M. Drosdoff. 1960. *A rapid polarographic method for determining extractable zinc in mineral soils.* Soil Science Society of America Proceedings 24: 169-171.
11. Berg, M. 1983. *State requirements for soil and overburden—an overview.* In M. Berg [editor] *Soil and Overburden Requirements for Successful Revegetation.* U.S. Office of Surface Mining, Denver, Colorado. pp. 5-17.
12. Beyer, W. N., R. L. Chaney, and B. M. Mulhern. 1982. *Heavy metal concentrations in earthworms from soil amended with sewage sludge.* Journal of Environmental Quality 11: 381-385.
13. Bingham, F. T. 1979. *Bioavailability of Cd to food crops in relation to heavy metal content of sludge-amended soil.* Environmental Health Perspectives 28: 39-43.
14. Bingham, F. T., A. L. Page, R. J. Mahler, and T. J. Ganje. 1975. *Growth and cadmium accumulation of plants grown on a soil treated with a cadmium-enriched sewage sludge.* Journal of Environmental Quality 4: 207-211.
15. Bingham, F. T., A. L. Page, R. J. Mahler, and T. J. Ganje. 1976. *Yield and Cd accumulation of forage species in relation to Cd content of sludge-amended soil.* Journal of Environmental Quality 5: 57-60.
16. Bittell, J. E., D. E. Koeppe, and R. J. Miller. 1974. *Sorption of heavy metal cations by corn mitochondria and the effects on electron and energy transfer reactions.* Physiologia Plantarum 30: 226-230.
17. Boggess, S. F., S. Willavize, and D. E. Koeppe. 1978. *Differental response of soybean varieties to soil cadmium.* Agronomy Journal 70: 756-760.
18. Boon, D. Y. 1984. *The ammonium bicarbonate—DTPA soil test (AB-DTPA) for determination of plant available Pb, Cd, Ni, and Mo in mine tailings and contaminated soils.* In Proceedings, Sixth High Altitude Revegetation Conference. Water Resources Research Institute, Colorado State University, Fort Collins, Colorado. pp. 128-147.
19. Boon, D. Y. 1985. *Lead, cadmium and zinc: Evaluation of contamination using the NH_4CO_3-DTPA soil test.* M.S. thesis. Colorado State University, Fort Collins, Colorado.
20. Boon, D. Y., and P. J. Smith. 1985. *Carbonaceous materials problems associated with reclamation.* In Proceedings, Second Annual Meeting of the American Society for Surface Mining and Reclamation, Denver, Colorado. pp. 248-252.
21. Bowen, H.J.M. 1966. *Trace elements in biochemistry.* Academic Press, New York, New York.
22. Bradshaw, A. D., T. S. McNeilly, and R.P.G. Gregory. 1965. *Industrializa-*

tion, evolution, and the development of heavy metal tolerance in plants. In G. T. Goodman, R. W. Edwards, and J. M. Lambert [editors] *Ecology and the Industrial Society.* British Ecological Society Symposium 5: 327-343. Blackwell Scientific Publishers, Oxford, England.

23. Brenchley, W. E. 1927. *Inorganic plant poisons and stimulants.* Cambridge University Press, London, England.
24. Brown, A. L., J. Quick, and J. L. Eddings. 1971. *A comparison of analytical methods for soil zinc.* Soil Science Society of America Proceedings 35: 105-107.
25. Brown, J. C., J. E. Ambler, R. L. Chaney, and C. D. Foy. 1972. *Differential responses of plant genotypes to micronutrients.* In J. J. Mortvedt, P. M. Giordano, and W. L. Lindsay [editors] *Micronutrients in Agriculture.* Soil Science Society of America, Madison, Wisconsin. pp. 389-418.
26. Carroll, R. E. 1966. *The relationship of cadmium in the air to cardiovascular disease death rates.* Journal American Medical Association 198: 267-269.
27. Cataldo, D. A., and R. E. Wildung. 1979. *Soil and plant factors influencing the accumulation of heavy metals by plants.* Environmental Health Perspectives 27: 149-159.
28. Cavallaro, N., and M. B. McBride. 1978. *Copper and cadmium adsorption characteristics of selected acid and calcareous soils.* Soil Science Society of America Journal 42: 550-556.
29. Chang, A. C., and A. L. Page. 1979. *Effects of repeated soil application of sludges on the Cd and Zn levels of affected vegetable crops.* Final report. Wastewater Solids Management Program, Los Angeles/Orange County Metropolitan Area, University of California, Riverside.
30. Chang, A. C., A. L. Page, J. E. Warneke, M. R. Resketo, and T. E. Jones. 1983. *Accumulation of cadmium and zinc in barley grown on sludge-treated soils: A long-term field study.* Journal of Environmental Quality 12: 391-397.
31. Chapman, H. D. 1966. *Zinc.* In H. D. Chapman [editor] *Diagnostic Criteria for Plants and Soils.* University of California, Berkeley. pp. 489-499.
32. Chesnin, L. 1967. *Corn, soybeans and other Great Plains crops. The micronutrient manual.* Farm Technology 23: 6.
33. Chumbley, G. G. 1971. *Permissible levels of toxic metals in sewage used on agricultural land.* Advisory Service of Agricultural Division, Ministry of Agriculture, London, England. 12 pp.
34. Compliance Program Evaluation. 1974. *Total diet studies (7320.08).* Bureau of Foods, Food and Drug Administration, Washington, D.C.
35. Coughtry, P. J., and M. H. Martin. 1978. *Tolerance of* Holeus lanatus *to lead, zinc, and cadmium in factorial combination.* New Phytologist 81: 147-154.
36. Council for Agricultural Sciences and Technology. 1976. *Application of sewage sludge to cropland: Appraisal of potential hazards of the heavy metals to plants and animals.* Report No. 64. Ames, Iowa.
37. Council for Agricultural Sciences and Technology. 1980. *Effects of sewage sludge on the cadmium and zinc content of crops.* Report No. 83. Ames, Iowa.
38. Cox, F. R., and E. J. Kamprath. 1972. *Micronutrient soil tests.* In J. J. Mortvedt, P. M. Giordano, and W. L. Lindsay [editors] *Micronutrients in Agriculture.* Soil Science Society of America, Madison, Wisconsin. pp. 289-317.
39. Dean, W. E. 1976. *Geochemistry of the Green River shale.* In *Geochemical Survey of the Western Energy Regions.* Open-file Report 76-729. U.S. Geological Survey, Washington, D.C. pp. 48-56.
40. Decker, A. M., C. H. Darrah, D. J. Wehner, E. Strickling, M. F. Rotschild, L. R. Gouin, C. B. Link, J. B. Shanks, J. T. Davison, R. C. Hammond, S. B. Mohanty, R. L. Chaney, J. J. Murry, D. L. Kern, and T. S. Rumsey. 1980. *Feasibility of using sewage sludge for plant and animal production.* Final

Report for 1978 and 1979. Suburban Sanitary Commission, Washington, D.C.
41. DeFilippis, L. F. 1979. *The effect of heavy metal compounds on the permeability of* Chlorella *cells.* Zeitschrift fuer Pflanzenphysiologie 92: 39-49.
42. DeFilippis, L. F., and C. K. Pallaghy. 1976. *The effect of sub-lethal concentrations of Hg and Zn on* Chlorella. *2. Photosynthesis and pigment composition.* Zeitschrift fuer Pflanzenphysiologie 78: 314-322.
43. Dijkshoorn, W., and J. E. Lampe. 1975. *Availability for rye grass of cadmium and zinc dressings of sewage sludge.* Netherlands Journal of Agricultural Science 23: 328-344.
44. Dollhopf, D. J., and J. D. Goering. 1983. *Noncontaminating drilling grease for overburden sampling.* Special Report SR-2. Montana Agricultural Experiment Station, Bozeman.
45. Dowdy, R. H., B. J. Bray, and R. D. Goodrich. 1983. *Trace metal and mineral composition of milk and blood from goats fed silage produced on sludge-amended soil.* Journal of Environmental Quality 12: 473-478.
46. Dowdy, R. H., B. J. Bray, R. D. Goodrich, G. C. Marten, D. E. Pamp, and W. E. Larson. 1983. *Performance of goats and lambs fed corn silage produced on sludge amended soil.* Journal of Environmental Quality 12: 467-472.
47. Dowdy, R. H., and W. E. Larson. 1975. *The availability of sludge-borne metals to various vegetable crops.* Journal of Environmental Quality 4: 278-282.
48. Egozy, Y. 1980. *Adsorption of cadmium and cobalt on montmorillonite as a function of solution composition.* Clays and Clay Minerals 28: 311-317.
49. Erdman, J. A. 1978. *Potential toxicologic problems associated with stripmining.* In *Geochemical Survey of the Western Energy Regions.* Open-file Report 78-1105. U.S. Geological Survey, Washington, D.C. pp. 117-125.
50. Erdman, J. A., and L. P. Gough. 1979. *Mineral composition of wheat from coal-mine spoils and farms in the Northern Great Plains, USA.* In M. K. Wali [editor] *Ecology and Coal Resources Development. Volume 2.* Pergamon Press, New York, New York. pp. 575-585.
51. Esser J., and N. ElBassam. 1981. *On the mobility of cadmium under aerobic soil conditions.* Environmental Pollution. Series A. Ecological and Biological. Applied Science Publishers, Ltd., Essex, England. pp. 15-31.
52. Farmer, E. E., B. Z. Richardson, and R. W. Brown. 1976. *Revegetation of acid mining wastes in central Idaho.* Research paper INT-178. U.S. Forest Service, Ogden, Utah.
53. Fleischer, M., A. F. Sarofim, D. W. Fassett, P. Hammond, H. T. Shacklette. I.C.T. Nisbet, and S. Epstein. 1974. *Environmental impacts of cadmium—A review by the panel on hazardous trace substances.* Environmental Health Perspectives 19: 253-323.
54. Foy, C. D., R. L. Chaney, and M. C. White. 1978. *The physiology of metal toxicity in plants.* Annual Review of Plant Physiology 29: 511-566.
55. Garcia-Miragaya, J., and A. L. Page. 1977. *Influence of exchangeble cation on the sorption of trace amounts of cadmium by montmorillonite.* Soil Science Society of America Journal 41: 718-721.
56. Gaynor, J. D., and R. L. Halstead 1976. *Chemical and plant extractability of metals and plant growth on soils amended with sludge.* Canadian Journal of Soil Science 56: 1-8.
57. Giordano, P. M., D. A. Mays, and A. D. Behel, Jr. 1979. *Soil temperature effects on uptake of cadmium and zinc by vegetables grown on sludge-amended soil.* Journal of Environmental Quality 8: 233-236.
58. Gladstones, J. S., and J. F. Loneragan. 1967. *Mineral elements in temperate crop and pasture plants. I. Zinc.* Australian Journal of Agricultural Research

18: 427-446.
59. Goldschmidt, V. M. 1954. *Geochemistry.* The Clarendon Press, Oxford, England.
60. Gough, L. P., J. M. McNeal, and R. C. Severson. 1980. *Predicting native plant copper, iron, manganese, and zinc levels using DTPA and EDTA soil extractants, Northern Great Plains.* Soil Science Society of America Journal 44: 1,030-1,036.
61. Gregory, R.P.G., and A. D. Bradshaw. 1965. *Heavy metal tolerance in populations of* Agrostis tenuis *Sibth, and other grasses.* New Phytologist 64: 131-143.
62. Gunn, S. A. 1967. *Metals in pathology, I. Bulletin.* Pathogens 8: 42-43.
63. Haghiri, F. 1974. *Plant uptake of cadmium as influenced by cation exchange capacity, organic matter, zinc and soil temperature.* Journal of Environmental Quality 3: 180-183.
64. Halvorson, G. A. 1984. *Zinc and phosphorus fertilizer experiments with corn on reclaimed and undisturbed land.* North Dakota Farm Research 41(6): 16-17.
65. Hampp, R., K. Beulich, and H. Ziegler. 1976. *Effects of Zn and Cd on photosynthetic CO_2 fixation and Hill activity of isolated spinach chloroplasts.* Zeitschrift fuer Pflazenphysiologie 77: 336-344.
66. Haq, A. U., and M. H. Miller. 1972. *Prediction of available soil Zn, Cu, and Mn using chemical extractants.* Agronomy Journal 64: 779-782.
67. Haq, A. U., T. E. Bates, and Y. K. Soon. 1980. *Comparison of extractants for plant-available zinc, cadmium, nickel, and copper in contaminated soils.* Soil Science Society of America Journal 44: 772-777.
68. Harter, R. D. 1983. *Effect of soil pH on adsorption of lead, copper, zinc, and nickel.* Soil Science Society of America Journal 47: 47-51.
69. Havlin, J. L., and P. N. Soltanpour. 1981. *Evaluation of the NH_4HCO_3-DTPA soil test for iron and zinc.* Soil Science Society of America Journal 45: 70-75.
70. Heil, R. D., and P. C. Deutsch. 1979. *Soil-plant correlation research in mined land reclamation: Characterization of materials overlying coal seams as plant growth media—EMRIA EMERY, Utah site.* Bureau of Land Management, U.S. Department of the Interior, Washington, D.C.
71. Henkin, R. I. 1971. *Newer aspects of copper and zinc metabolism.* In W. Mertz and E. W. Cornatzer [editors] *Newer Trace Elements in Nutrition.* Marcel Dekkar, New York, New York. pp. 255-312.
72. Hinesly, T. D., D. E. Alexander, E. L. Ziegler, and G. L. Barrett. 1978. *Zinc and Cd accumulation by corn inbreds grown on sludge amended soil.* Agronomy Journal 70: 425-428.
73. Hinesly, T. D., K. E. Redborg, E. L. Ziegler, and J. D. Alexander. 1982. *Effect of soil cation exchange capacity on the uptake of cadmium by corn.* Soil Science Society of America Journal 46: 490-497.
74. Huang, Chi-Ying, F. A. Bazzaz, and L. A. Vanderhoef. 1974. *The inhibition of soybean metabolism by cadmium and lead.* Plant Physiology 54: 122-124.
75. Inskeep, W. P., and J. Baham. 1983. *Adsorption of Cd(II) and Cu(II) by Na-Montmorillonite at low surface coverage.* Soil Science Society of America Journal 47: 660-665.
76. Jenne, E. A. 1968. *Controls on Mn, Fe, Co, Ni, Cu, and Zn concentrations in soils and water: The significant role of hydrous Mn and Fe oxides.* Advances in Chemistry Series 7: 337-387.
77. John, M. K., C. J. Van Laerhoven, and H. H. Chuah. 1972. *Factors affecting plant uptake and phytotoxicity of cadmium added to soils.* Environmental Science and Technology 6: 1,005-1,009.

78. Johnston, R. S., R. W. Brown, and J. Cravians. 1975. *Acid mine rehabilitation problems at high elevations.* In Proceedings, Watershed Management Symposium. American Society of Civil Engineers, New York, New York.
79. Jones, R. L., T. D. Hinesly, E. L. Ziegler, and J. J. Tyler. 1975. *Cadmium and zinc contents of corn leaf and grain produced by sludge-amended soil.* Journal of Environmental Quality 4: 509-514.
80. King, L. D., and H. D. Morris. 1972. *Land disposal of liquid sewage sludge: II. The effect on soil pH, manganese and zinc and growth and chemical composition of rye (*Secale cerale *L.).* Journal of Environmental Quality 1: 425-429.
81. Koeppe, D. E. 1977. *The uptake, distribution, and effect of cadmium and lead in plants.* Science of the Total Environment 7: 197-206.
82. Kubota, J., and W. H. Allaway. 1972. *Geographic distribution of trace element problems.* In J. J. Mortvedt, P. M. Giordano, and W. L. Lindsay [editors] *Micronutrients in Agriculture.* Soil Science Society of America, Madison, Wisconsin. pp. 525-554.
83. Kuo, S., and A. S. Baker. 1980. *Sorption of copper, zinc and cadmium by some acid soils.* Soil Science Society of America Journal 44: 969-974.
84. Lagerwerff, J. V. 1971. *Uptake of cadmium, lead and zinc by radish from soil and air.* Soil Science 111: 129-133.
85. Lagerwerff, J. V. 1972. *Lead, mercury, and cadmium as environmental contaminants.* In J. J. Mortvedt, P. M. Giordano, and W. L. Lindsay [editors] *Micronutrients in Agriculture.* Soil Science Society of America, Madison, Wisconsin. pp. 593-636.
86. Lee, C. R., and G. R. Craddock. 1969. *Factors affecting plant growth in high zinc medium: II. Influence of soil treatments on growth of soybeans on strongly acid soil containing zinc from peach sprays.* Agronomy Journal 61: 565-567.
87. Lee, K. C., B. A. Cunningham, G. M. Paulsen, G. H. Liang, and R. B. Moore. 1976. *Effects of cadmium on respiration rate and activities of several enzymes in soybean seedlings.* Physiologia Plantarum 36: 4-6.
88. Lewis, T. R. 1972. *Effects of air pollution on livestock and animal products.* In *Helena Valley, Montana Area Environmental Pollution Study.* Office of Air Program Publication AP-91. U.S. Environmental Protection Agency, Washington, D.C.
89. Lindsay, W. L. 1979. *Chemical equilibria in soils.* John Wiley and Sons, New York, New York.
90. Lindsay, W. L., and W. A. Norvell. 1978. *Development of a DTPA soil test for zinc, iron, mangnese, and copper.* Soil Science Society of America Journal 42: 421-428.
91. Lund, L. J., E. E. Betty, A. L. Page, and R. A. Elliott. 1981. *Occurrence of naturally high cadmium levels in soils and its accumulation by vegetation.* Journal of Environmental Quality 10: 551-556.
92. Maclean, A. J. 1976. *Cadmium in different plant species and its availability in soils as influenced by organic matter and additions of lime, P, Cd, and Zn.* Canadian Journal of Soil Science 56: 129-138.
93. Mahler, R. J., F. T. Bingham, and A. L. Page. 1978. *Cadmium-enriched sewage sludge application to acid and calcareous soils: Effect on yield and cadmium uptake by lettuce and chard.* Journal of Environmental Quality 7: 274-281.
94. Mahler, R. L., F. T. Bingham, A. L. Page, and J. A. Ryan. 1982. *Cadmium enriched sewage sludge application to acid and calcareous soils: Effect on soil and nutrition of lettuce, corn, tomato, and swiss chard.* Journal of Environmental Quality 11: 694-700.

95. Mahler, R. J., F. T. Bingham, G. Sposito, and A. L. Page. 1980. *Cadmium enriched sewage sludge application to acid and calcareous soils: Relationship between treatment, cadmium in saturation extracts, and cadmium uptake.* Journal of Environmental Quality 9: 359-364.
96. Majmundar, H. 1977. *Trace elements in stream sediments, Monterey County, California.* Open File No. 77-5 SF. Division of Mines and Geology, Department of Conservation, San Francisco, California.
97. McBride, M. B. 1980. *Chemisorption of Cd^{2+} on calcite surfaces.* Soil Science Society of America Journal 44: 26-28.
98. McBride, M. B., and J. J. Blasiak. 1979. *Zinc and copper solubility as a function of pH in an acid soil.* Soil Science Society of America Journal 43: 866-870.
99. Miles, L. J., and G. R. Parker. 1979. *The effect of soil-added cadmium on several plant species.* Journal of Environmental Quality 8: 229-232.
100. Miller, R. J., J. E. Bittell, and D. E. Koeppe. 1973. *The effect of cadmium on electron and energy transfer reactions in corn mitochondria.* Physiologia Plantarum 28: 166-171.
101. Millikan, C. R. 1961. *Plant varieties and species in relation to occurrence of deficiencies and excesses of certain nutrient elements.* Journal. Australian Institute of Agricultural Science 27: 220-233.
102. Mortvedt, J. J., D. A. Mays, and G. Osborn. 1981. *Uptake by wheat of cadmium and other heavy metal contaminants in phosphate fertilizers.* Journal of Environmental Quality 10: 193-197.
103. Mortvedt, J. J., and P. M. Giordano. 1977. *Crop uptake of heavy metal contaminants in fertilizers.* In R. A. Wildung and H. Drucker [editors] *Biological Implications of Heavy Metals in the Environment.* Conference Report 750929. Energy Research Development Agency, Oak Ridge, Tennessee.
104. Mortvedt, J. J., P. M. Giordano, and W. L. Lindsay, editors. 1972. *Micronutrients in agriculture.* Soil Science Society of America, Madison, Wisconsin.
105. Mulla, D. J., A. L. Page, and T. J. Ganje. 1980. *Cadmium accumulations and bioavailability in soils from long-term phosphorus fertilization.* Journal of Environmental Quality 9: 408-412.
106. Munshower, F. F. 1977. *Cadmium accumulation in plants and animals of polluted and nonpolluted grasslands.* Journal of Environmental Quality 6: 411-413.
107. Munshower, F. F. 1983. *Soil and overburden trace element concentrations.* In M. G. Berg [editor] *Soil and Overburden Requirements for Successful Revegetation.* U.S. Office of Surface Mining, Western Technical Center, Denver, Colorado. pp. 115-149.
108. Munshower, F. F., and D. R. Neuman. 1983. *The effects of stack emissions on the range resource in the vicinity of Colstrip, Montana. After five years of operation of power plant units I & II.* Special Report No. 1. Montana State University, Bozeman.
109. Munshower, F. F., E. J. DePuitt, and D. R. Neuman. 1978. *The effects of stack emissions on the range resource in the vicinity of Colstrip, Montana.* Research Report 126. Montana Agricultural Experimental Station, Bozeman.
110. Nielson, R. F., and H. B. Peterson. 1972. *Treatment of mine tailings to promote vegetation stabilization.* Bulletin 485. Agricultural Experiment Station, Utah State University, Logan.
111. Olsen, S. R. 1972. *Micronutrient interactions.* In J. J. Mortvedt, P. M. Giordano, and W. L. Lindsay [editors] *Micronutrients in Agriculture.* Soil Science Society of America, Madison, Wisconsin. pp. 243-264.
112. Page, A. L., F. T. Bingham, and C. Nelson. 1972. *Cadmium absorption and growth of various plant species as influenced by solution cadmium concentrations.* Journal of Environmental Quality 1: 288-291.

113. Peterson, P. J. 1969. *The distribution of zinc-65 in* Agrostis tenuis *Sibth. and* A. stolonifera *L. tissues.* Journal of Experimental Botany 20: 863-875.
114. Pietz, R. I., J. R. Peterson, T. D. Hinesly, E. L. Ziegler, K. E. Redborg, and C. Lue-Hing. 1983. *Sewage sludge application to calcareous strip-mine spoil: II. Effects on spoil and corn cadmium, copper, nickel and zinc.* Journal of Environmental Quality 12: 463-467.
115. Prasad, A. S., A. Miale, Jr., Z. Farid, H. H. Sandstead, and A. R. Schulert. 1963. *Zinc metabolism in patients with the syndrome of iron deficiency anemia, hepatosplenomegaly, dwarfism and hypogonadism.* Journal of Laboratory and Clinical Medicine 61: 537-549.
116. Ragiani, R. C., H. R. Ralston, and N. Roberts. 1977. *Environmental trace metal contamination in Kellogg, Idaho, near a lead smelting complex.* Environmental Science and Technology 11: 773-781.
117. Ravikovitch, S., M. Margolin, and J. Navrot. 1968. *Zinc availability in calcareous soils: I. Comparison of chemical extraction methods for estimation of plant "available" zinc.* Soil Science 105: 57-61.
118. Reuss, J. O., H. L. Dooley, and W. Griffis. 1978. *Uptake of cadmium from phosphate fertilizers by peas, radishes, and lettuce.* Journal of Environmental Quality 7: 128-133.
119. Ringrase, C. D., R. W. Klusman, and W. E. Dean. 1976. *Soil chemistry in the Piceance Creek Basin.* In *Geochemical Survey of the Western Energy Region.* Open-file Report 76-729. U.S. Geological Survey, Washington, D.C. pp. 101-111.
120. Robinson, W. O., H. W. Lakin, and L. E. Reichen. 1947. *The zinc content of plants on the Friedensville zinc slime ponds in relation to biogeochemical prospecting.* Economic Geology 42: 572-582.
121. Ryan, J. A., H. R. Pahren, and J. B. Lucas. 1982. *Controlling cadmium in the food chain. A review and rationale based on health effects.* Environmental Research 28: 251-302.
122. Safaya, N. M. 1976. *Phosphorus-zinc interaction in relation to absorption rates of phosphorus, zinc, copper, manganese, and iron in corn.* Soil Science Society of America Journal 40: 719-722.
123. Safaya, N. M. 1978. *Delineation of mineral stresses in mine spoils and screening plants for adaptibility.* In M. K. Wali [editor] *Ecology and Coal Resource Development, Volume 2.* Pergamon Press, New York, New York. pp. 830-839.
124. Safaya, N. M., and A. P. Gupta. 1979. *Differential susceptibility of corn cultivars to zinc deficiency.* Agronomy Journal 71: 132-136.
125. Safaya, N. M., and B. Singh. 1977. *Differential susceptibility of two varieties of cowpea (*Vigna unguiculata *(L.) Walp) to phosphorus—induced zinc deficiency.* Plant and Soil 48: 279-290.
126. Safaya, N. M., and N. Malakondaiah. 1981. *Responses of some crop species to applied and residual levels of zinc.* Journal of Plant Nutrition 3: 483-492.
127. Safaya, N. M., R. H. Bares, N. Malakondaiah, and S. A. Nicholson. 1982. *Ecological studies on the revegetation process of surface coal mined areas in North Dakota: 2. Physical and chemical properties of overburden and topsoil as affected by time and amendment.* Bureau of Mines, U.S. Department of the Interior, Washington, D.C.
128. Safaya N. M., U. C. Shukla, and S. S. Khanna. 1975. *Nutritional physiology of zinc and symptoms of its deficiency in some field crops.* Fertilizer News 20: 31-35.
129. Samarakoon, A. B., and W. E. Rauser. 1979. *Carbohydrate levels and photoassimilate export from leaves of* Phaseolus vulgaris *exposed to excess Cobalt, Nickel, and Zinc.* Plant Physiology 63: 1,165-1,169.

130. Santillan-Medrano, J., and J. J. Jurinak. 1975. *The chemistry of lead and cadmium in soil: Solid phase formation.* Soil Science Society of America Journal 39: 851-856.
131. Schauer, P. S., W. R. Wright, and J. Pelchat. 1980. *Sludge-borne heavy metal availability and uptake by vegetable crops under field conditions.* Journal of Environmental Quality 9: 69-73.
132. Schuman, G. E., F. F. Munshower, R. E. Ries, and E. G. Robbins. 1980. *Quality and heavy metal content of forage grown on reclaimed mined lands in the Northern Great Plains.* GPAC Number 98 Ag. Wyoming Agricultural Experimental Station RJ159, Laramie.
133. Schwab, A. P., W. L. Lindsay, and P. J. Smith. 1983. *Elemental contents of plants growing on soil-covered retorted shale.* Journal of Environmental Quality 12: 301-304.
134. Schwarz, K., and J. E. Spallholz. 1978. *Growth effects of small cadmium supplements in rats maintained under trace element-controlled conditions.* In Proceedings, First International Cadmium Conference. Metals Bulletin, Ltd. London, England. pp. 105-109.
135. Scott, M. L. 1972. *Trace elements in animal nutrition.* In J. J. Mortvedt, P. M. Giordano, and W. L. Lindsay [editors] *Micronutrients in Agriculture.* Soil Science Society of America, Madison, Wisconsin. pp. 555-591.
136. Severson, R. C. 1979. *Regional soil chemistry in the Bighorn and Wind River Basins, Wyoming and Montana.* Professional Paper 1134 B. U.S. Geological Survey, Washington, D.C.
137. Severson, R. C., J. M. McNeal, and L. P. Gough. 1978. *Total and extractable element composition of some Northern Great Plains soils.* In *Geochemical Survey of the Western Energy Regions.* Open-file report 78-1105. U.S. Geological Survey, Washington, D.C. pp. 87-117.
138. Severson, R. C., and L. P. Gough. 1981. *Chemical character of minespoils at one Alaskan and twelve western conterminous United States coal-stripmines.* Open File Report 81-243. U.S. Geological Survey, Washington, D.C.
139. Severson, R. C., L. P. Gough, and J. M. McNeal. 1977. *Availability of elements in soils to native plants, Northern Great Plains.* In *Geochemical Survey of the Western Energy Region.* Open-file Report 77-872. U.S. Geological Survey, Washington, D.C. pp. 98-143.
140. Shaw, E., and L. A. Dean. 1952. *The use of dithiozone as an extractant to estimate the zinc nutrient status of soils.* Soil Science 73: 341-347.
141. Sheaffer, C. C., A. M. Decker, R. L. Chaney, and L. W. Douglass. 1979. *Soil temperature and sewage sludge effects on metals in crop tissues and soils.* Journal of Environmental Quality 8: 455-459.
142. Silviera, D. J., and L. E. Sommers. 1977. *Extractability of copper, zinc, cadmium, and lead in soils incubated with sewage sludge.* Journal of Environmental Quality 6: 47-52.
143. Singh, S. S. 1977. *A note on the determination of cadmium in soil extracts by atomic absorption.* Communications in Soil Science and Plant Analysis 8: 177-181.
144. Soltanpour, P. N. 1983. *Use of NH_4HCO_3_DTPA soil test to determine deficiency or toxicity of some elements.* In M. G. Berg [editor] *Soil and Overburden Requirements for Successful Revegetation.* U.S. Office of Surface Mining, Western Technical Center, Denver, Colorado. pp. 89-95.
145. Soltanpour, P. N. 1985. *Use of ammonium bicarbonate DTPA soil test to evaluate elemental availability and toxicity.* Communications in Soil Science and Plant Analysis 16: 323-338.
146. Soltanpour, P. N., and A. P. Schwab. 1977. *A new soil test for simultaneous*

extraction of macro- and micronutrients in alkaline soils. Communications in Soil Science and Plant Analysis 8: 195-207.

147. Soltanpour, P. N., and S. M. Workman. 1979. *Modification of the NH_4HCO_3-DTPA soil test to omit carbon black.* Communications in Soil Science and Plant Analysis 10: 1,411-1,420.
148. Sommers, L. E., and W. L. Lindsay. 1979. *Effect of pH and redox on predicted heavy metal—chelate equilibria in soils.* Soil Science Society of America Journal 43: 39-47.
149. Special Studies Staff. 1975. *Environmental appraisal.* In J. R. Smith [editor] *Scientific and Technical Assessment Report on Cadmium.* EPA-600/6-75-003. U.S. Environmental Protection Agency, Washington, D.C.
150. Staker, E. V., and R. W. Cummings. 1941. *The influence of zinc on the productivity of certain New York peat soils.* Soil Science Society of America Proceedings 6: 207-214.
151. Stevenson, J. F. 1976. *Stability constants of Cu^{2+}, Pb^{2+}, and Cd^{2+} complexes with humic acids.* Soil Science Society of America Journal 40: 665-672.
152. Street, J. J., B. R. Sabey, and W. L. Lindsay. 1978. *Influence of pH, phosphorus, cadmium, sewage sludge, and incubation time on the solubility and plant uptake of cadmium.* Journal of Environmental Quality 7: 286-290.
153. Street, J. J., W. L. Lindsay, and B. R. Sabey. 1977. *Solubility and plant uptake of cadmium in soils amended with cadmium and sewage sludge.* Journal of Environmental Quality 6: 72-77.
154. Takkar, P. N., and T. Mann. 1978. *Toxic levels of soil and plant Zn for maize and wheat.* Plant and Soil 49: 667-669.
155. Thomas, C. D., editor. 1973. *Taber's cyclopedic medical dictionary.* F. A. Davis Co., Philadelphia, Pennsylvania.
156. Traynor, M. F., and B. D. Knezek. 1973. *Effects of nickel and cadmium contaminated soils on nutrient composition of corn plants.* In D. D. Hemphill [editor] *Trace Substances in Environmental Health.* University of Missouri, Columbia. pp. 83-87.
157. Trierweiler, J. F., and W. L. Lindsay. 1969. *EDTA-ammonium carbonate soil test for zinc.* Soil Science Society of America Proceedings 33: 49-54.
158. Turner, M. A. 1973. *Effect of cadmium treatment on cadmium and zinc uptake by selected vegetable species.* Journal of Environmental Quality 2: 118-119.
159. Turner, R. G. 1970. *The subcellular distribution of zinc and copper within the roots of* Agrostis tenuis *Sibth.* New Phytologist 69: 725-731.
160. Turner, R. G., and C. Marshall. 1972. *The accumulation of zinc by subcellular fractions of roots of* Agrostis tenius *Sibth., in relation to zinc tolerance.* New Phytologist 71: 671-676.
161. Underwood, E. J. 1971. *Trace elements in human and animal nutrition.* Academic Press, New York, New York.
162. Valdares, J.M.A.S., M. Gal., U. Mingelgrin, and A. L. Page. 1983. *Some heavy metals in soils treated with sewage sludge, their effect on yield, and their uptake by plants.* Journal of Environmental Quality 12: 49-57.
163. Vallee, B. L. 1976. *Zinc biochemistry: A perspective.* Trends in Biochemical Sciences (April): 88-91.
164. Vallee, B. L., and W.E.C. Wacker. 1970. *Metalloproteins.* In H. Neurath [editor] *The Proteins.* Volume 5. Academic Press, New York, New York.
165. Van Lear, D. H., and W. H. Smith. 1972. *Relationship between macro- and micro-nutrient nutrition of slash pine on three coastal plain soils.* Plant and Soil 36: 331-347.
166. Viets, F. G., Jr., L. C. Boawn, and C. L. Crawford. 1954. *Zinc contents and*

deficiency symptoms of 26 crops grown on a zinc-deficient soil. Soil Science 78: 305-316.

167. Viets, F. G., Jr., L. C. Boawn, C. L. Crawford, and C. E. Nelson. 1953. *Zinc deficiency in corn in central Washington.* Agronomy Journal 45: 559-565.
168. Vigue, G. T., I. L. Pepper, and D. F. Bezdicek. 1981. *The effect of cadmium on nodulation and $N_2(C_2H_2)$-fixation by dry beans.* Journal of Environmental Quality 10: 87-90.
169. Wallace, A., and W. L. Berry. 1978. *Trace elements in the environment effects and potential toxicity of those associated with coal.* In M. K. Wali [editor] *Ecology and Coal Resource Development.* Volume 1. Pergamon Press, New York, New York. pp. 95-114.
170. Wallingford, G. W., L. S. Murphy, W. L. Powers, and H. L. Manges. 1975. *Effects of beef-feedlot manure and lagoon water on iron, zinc, manganese, and copper content in corn and in DTPA soil extracts.* Soil Science Society of America Proceedings 39: 482-487.
171. Wear, J. I., and C. E. Evans. 1968. *Relationship of zinc uptake by corn and sorghum to soil zinc measured by three extractants.* Soil Science Society of America Proceedings 32: 543-546.
172. White, M. C., and R. L. Chaney. 1980. *Zinc, cadmium, and manganese uptake by soybean from two zinc- and cadmium-amended Coastal Plain soils.* Soil Science Society of America Journal 44: 308-313.
173. Whitney, D. A. 1980. *Micronutrient soil tests—zinc, iron, manganese, and copper.* In W. C. Dahnke [editor] *Recommended Chemical Soil Test Procedures for the North Central Region.* Bulletin No. 499 (Revised). North Dakota Agricultural Experiment Station, Fargo. pp. 18-21.
174. Wickliff, C., H. J. Evans, K. R. Carter, and S. A. Russell. 1980. *Cadmium effects on the nitrogen fixation system of red alder.* Journal of Environmental Quality 9: 180-184.
175. Williams, C. H., and D. J. David. 1973. *The effect of superphosphate on the cadmium content of soils and plants.* Australian Journal of Soil Research 11: 43-56.
176. Williams, C. H., and D. J. David. 1976. *The accumulation in soil of cadmium residues from phosphate fertilizers and their effect on the cadmium content of plants.* Soil Science 121: 86-93.
177. Wyn-Jones, R. G., and M. H. Sutcliff. 1972. *Some physiological aspects of heavy metal tolerance in* Argrostis tenuis. Welsh Soils Discussion Group Report No. 13: 1-15.
178. Yamamoto, T. 1975. *Coal mine spoil as a growing medium: AMAX Belle Ayr South Mine, Gillette, Wyoming.* In Proceedings, Third symposium on surface mining and reclamation, Volume 1. NCA/BCR Coal Conference and Expo II. National Coal Association, Washington, D.C.
179. Yamegata, N., and I. Shigematsu. 1970. *Cadmium pollution in perspective.* Institute of Public Health 19(1): 1-27. Tokyo, Japan.
180. Yerkes, R. F., and R. H. Campbell. 1980. *Geologic map of east-central Santa Monica Mountains, Los Angeles County, California.* Miscellaneous Investigations Section I-1146. U.S. Geological Survey, Washington, D.C.

Chapter 14

Mercury

Stephen E. Williams and Roger S. Funston

Mercury contamination of water, soils, or air is unlikely to occur in most mining operations. An obvious exception is the mining of Hg itself. While Hg is not a common parameter of concern in soils and overburden evaluation for mined land reclamation, it may be of interest in some circumstances. Much of western North America is in a mercuriferous zone that contains numerous Hg deposits and prospects (*12*). Hg can exist in local deposits or as a natural contaminant of ores to the extent that handling of such materials during mining could result in the release of some Hg forms into soils or water and some deposition in the food chain.

Natural concentrations of mercury

Naturally occurring mineral forms of Hg are mostly cinnabar (HgS), metacinnebar (also HgS), and livingstonite ($HgSb_4S_7$). Total Hg concentrations commonly reported for soils, sediments, and geologic deposits are less than 100 ng/g (10^{-9} g/g or ppb) (*49*). In sediments where geochemical processes have concentrated Hg, total Hg concentrations are generally in the range of 1 to 50 ng/g; they may be as high as 250 ng/g, however (*27*). Hg occurs in small amounts in most materials generated by mining, including coal and metal ores. Most Hg in coal is concentrated in that fraction with a heavier specific gravity (*8*). Removal of materials with a specific gravity greater than 1.60 provided concentration reductions of up to 64 percent. Tables 1, 2, and 3 summarize the reported concentrations of total Hg in geologic materials, soils, and mining products, respectively.

Most authorities only list total Hg concentrations. Normally, total Hg concentration has little relationship to extractable Hg, except that extractable (bioavailable) Hg is usually only a small fraction of the total. Proper selection of analytical methods and assessment of the specific environmental hazard represented by a particular Hg concentration depends upon an understanding of the concentration of extractable forms of Hg.

Analytical means of detecting mercury

Appropriate methods for Hg analysis require detection in the ng/g range. This is because of the low concentrations of naturally occurring Hg and the low critical limits set by regulatory agencies (Table 4). Care must be taken to prevent contamination of samples during sampling, storage in the laboratory, and analysis. Accurate determination of the total Hg concentration is complicated by serious losses that can occur during sample extraction and storage (*34*).

Sample collection and preparation procedures. Sample preparation for Hg analysis begins at the sampling site. Hg and some of its compounds are volatile. Air drying of soil samples for 2 days can result in the loss of as much as 40 percent of the Hg (*29*). The most volatile and the most toxic compound of Hg is methyl mercuric chloride (*42*). To minimize volatilization losses, samples should be stored at temperatures less than 4 °C, in glass containers with tight-fitting closures (*30, 38*). Sample storage time should be minimized. To minimize losses of Hg by evaporation, the drying of samples prior to analysis should be avoided. Results can be reported later on a corrected dry basis. If samples must be dried, temperatures less than 10 °C should be used (*43*). Grinding of dried soil samples to pass a 15-μm sieve will not cause further Hg losses (*45*). Mercury is ubitiquitous in many laborato-

Table 1. Total mercury content of representative geological materials (*27*).

Material	Total Hg (ng/g) Range	Mean
Stream, river, and lake sediments	10-700	73
Sandstone, arkose, conglomerate, etc.	10-300	55
Ocean sediments	10-2,000	100
Carbonaceous and bituminous shales	100-3,250	437
Limestone, dolomite, etc.	10-220	40
Gypsum, anhydrite	10-60	25
Halite, sylrite, etc.	20-200	30
Quarzites	10-100	53
Amphibolites	30-100	50
Hornfels	35-400	225
Schists	10-1,000	100
Gneisses	25-100	50
Marble, crystalline dolomite, etc.	10-100	50

Table 2. Total and extractable mercury from representative soils.

Site/Description	Hg (ng/g) Total	Hg (ng/g) Extractable $NaHCO_3$	Hg (ng/g) Extractable NH_4OAc	Hg (ng/g) Extractable DTPA	Hg (ng/g) Extractable HCl	Soil Properties: Soil pH	Soil Properties: CEC (meq/100 gm)	Soil Properties: Organic Carbon (%)	Soil Properties: Organic Matter (%)	Reference
Taken 20 km from Hg smelter, Spain	2,300	10	300			5.6	10.5			*32*
Taken 1 km from Hg smelter, Spain	97,000	960	6,000			5.6	10.5			*32*
Riverwash soil contaminated with Cinnabar, California	540,000	1		1	26	6.8	9.9	2.4		*21*
Manitoba soils A horizon	33									*36*
Manitoba soils C horizon	39									*36*
Ontario soils, unimproved land and pastures	80									*17*
Ontario soils, field crops	70									*17*
Ontario soils, organic	410									*17*
Ontario soils, sandy (crops)	60									*17*
Ontario soils, loam	90									*17*
Ontario soils, clay	80									*17*
Ontario soils, apples crop	390									*17*
British Columbia soils										
Northern B.C. (mineral)	80					6.3			5.4	*26*
Ashcroft Savona	66					7.7			3.8	*26*
Muck, Fraser Valley	109					4.6			54.0	*26*
Aridic Argiborolls/Haploborolls	22				1.1			1.6		*15*
Mesic Argiborolls/Haploborolls	23				1.2			3.3		*15*
Udic Argiborolls/Haploborolls	22				0.8			5.1		*15*
Boralfs (mineral soil)	24				3.0			0.5		*15*

ries and will contaminate unprotected samples exposed for any length of time (*8*).

Analytical procedures. Analysis for total Hg is accomplished by conversion of all Hg compounds to the Hg (II) oxidation state. Hg exists in three oxidation states: elemental Hg(O), mercurous Hg(I), and mercuric Hg(II). Conversion is accomplished by oxidation with $KMnO_4$, $K_2S_2O_8$, or $K_2Cr_2O_7$ and a mineral acid, such as HCl, HNO_3, $HClO_4$, or H_2SO_4. An aliquot of the digest is then reduced to Hg vapor by the cold water atomic absorption technique. Open-system or closed-system atomic absorption both provide reliable results. Concentrations in the range of 0.05 to 100 ng/g can be determined on most atomic absorption spectrophotometers. Complete analytical techniques are described elsewhere (*43*).

Several factors can complicate accurate determination of total Hg concentrations. Flameless atomic absorption spectrophotometry will not detect organic Hg. Samples with high chloride ion concentrations will interfere with atomic absorption testing. During sample oxidation, chloride ions are converted to Cl_2, which absorbs at precisely the same wavelength as Hg

Table 3. Total mercury contents of unprocessed fossil fuel, derived wastewaters, fossil fuels, and metal ores.

	Mean (ng/g)	Range (ng/g)	Remarks	References
Water used to transport coal in slurry form	0.77	0.12-7.60		*39*
Groundwater from in situ shale oil recovery	1.20	0.20-4.0		*24*
Retort water from 10-ton simulated retort	10			*24*
Black trona water (*13*)	0.39	0.20-3.0		*24*
Coal (Rosebud Seam)	200	80-610	Hanna Basin, WY	*39*
Coal		10-8,500		*11*
Coal (mercuriferous basins)		20-300,000	USSR	*11*
Coal	33,000		Richland Co., MT	*25*
Coal	510		Pierce Co., WA	*25*
Coal	18,600		Campbell Co., WY	*25*
Coal	220		Pitkin Co., CO	*25*
Coal	110		LeFlore Co., OK	*25*
Coal (composite)	60		Western USA	*8*
Coal	380		Harrison Co., TX	*25*
Peat		60-300		*11*
Bitumens, solid hydrocarbons, asphalts, etc.		2,000-900,000		*11*
Petroleum crudes in mercuriferous belts		1,900-21,000	California	*11*
Crude oils		20-2,000	Libya and USA	*11*
Copper, lead and zinc ores general	40,000	4,400-200,000	Turkey	*28*
Lead-zinc, sulfide ore	6,300	300-40,000		*28*
Oxidized lead ore	130,000	300-1,000,000		*28*

Table 4. Critical levels for mercury.

Comments	Critical Level	Reference
Total soil	50 ng/g	5
Soil solution	1 ng/g	5
Drinking water for livestock	10 ng/g	5
For the protection of human health from the toxic properties of mercury ingested through water and contamination aquatic organisms, the ambient water criterion is determined to be	144 ng/liter*	*46*
For the protection of human health from the toxic properties of mercury ingested through contaminated aquatic organisms alone, the ambient water criterion is determined to be	146 ng/liter*	*46*

*The procedures described in the "Guidelines for Deriving Numerical National Water Quality Criteria for the Protection of Aquatic Organisms and Their Uses" indicate that, except where a locally important species is very sensitive, freshwater aquatic organisms and their uses should not be affected unacceptably if the 4-day average concentration of mercury does not exceed 0.012 μg/liter more than once in a 3-year period; the edible portion of the consumed species should be analyzed to determine whether the concentration of methylmercury exceeds the FDA action limit. (*47*).

(*43*). Use of inductively coupled plasma for analysis can circumvent some of these problems. The following extractants are commonly used to extract Hg from soils: DTPA solutions (*21, 33*), $NaHCO_3$ solutions (*21, 32*), HCl (*21, 22*), and NH_4OAc solutions (*32*).

DTPA extraction was used to measure plant-available Hg (*33*). $NaHCO_3$ extraction was used to measure Hg associated with soil organic matter, and NH_4OAc was used to measure cation-exchangeable Hg (*32*). HCl extraction includes Hg absorbed to clays, in addition to the other types of Hg specified above (*22*).

Plant-available Hg is the most direct measure of Hg availability in assessing environmental hazard. Soils suspected of having plant-available Hg should be planted with target plants and the Hg content in the aboveground portion compared to control plants.

The bioavailability of mercury

Solubility of mercury in soil solution. The solubility and form of Hg in the soil solution is influenced by numerous interrelated physical, chemical, and biological factors. Total Hg concentration in soil is a poor predictor of bioavailable Hg. The most important factors are the chemical form of Hg entering the soil, microbial activity, pH, reactive surface area and type and density of charges on colloids (cation exchange capacity), and concentration of anions that form stable complexes with Hg.

The chemical form of Hg that can be added to soil has some bearing on the fate of this toxicant (*7*). Certainly, organic compounds of Hg, such as ethyl, methyl, dimethyl, and phenyl, as well as mercuric chlorides are generally quite soluble and even volatile. Application of these to soils will result in Hg in the soil solution. However, the chemical, physical, and microbiological properties of the soil will ultimately determine the form the applied Hg will take. This will in turn influence the bioavailability of the Hg.

Methylated compounds of Hg deserve special attention. They are very toxic (*50*) and likely were the cause of the Minamata Catatrophe in the 1950s (*44*). In that catastrophe numerous Japanese were poisoned by eating fish containing methyl Hg, which formed from Hg discharged by industry into Minamata Bay.

Conversion of elemental or ionic forms of Hg to metylated forms occurs via biotic and abiotic methanisms (*4*). Abiotic methylation apparently depends upon the presence of the biological product methylcobalamine, a vitamin B_{12} analogue. Biological methylation is catalyzed by several enzyme systems: methionine synthetase, acetate synthetase, and methane synthetase (*4*). Numerous aerobic and anaerobic microorganisms are noted for their Hg methylation abilities. These include the fungus *Neurospora crassa* (*31*) and the bacteria *Pseudomonas fluorescens, Mycobacterium phlei, Bacillus megaterium, Escherichia coli* (*48*), *Enterobacter* (Aerobacter) *aerogenes* (*19*), and *Chlostridium cochlearium* (*51*), to name but a few.

Biological methylation of Hg can be viewed as a mechanism whereby Hg is removed from the site occupied by the microorganisms; this is because methylated Hg is quite volatile (*6, 31*). This characteristic of methyl Hg may result in its removal from soil or water systems where it is generated. However, methyl Hg has the insidious characteristic of being highly available to other biological systems; for example, it has the capacity to traverse the blood-brain barrier in mammals and concentrate in the central nervous system (*6*). Soil, overburden, and water contaminated with metal or ionic Hg thus have the capacity to be very toxic substances. Conversion of Hg to methylated forms occurs rapidly in soils because both biotic and abiotic methylation occurs simultaneously (*40, 41*). But it should be noted that methyl Hg is also demethylated via biological mechanisms. There is little information as to why this occurs (*4*).

Although the biological conversion of Hg and its ions to methylated Hg deserves emphasis, interconversions of Hg(O) to Hg(I) and Hg(II) and both returning to Hg(O) are all accomplished by bacteria and some plants (*6*).

Hg(II) follows the generalization for many other metals: enhanced solubility at low pH but reduced solubility at high pH (*5*). Under aerobic conditions ($E_h > 1.2$ volts), high pH, and low chloride concentrations, Hg exists primarily as $Hg(OH)_2$ (Figure 1). In many alkaline soils it is likely to exist also as the carbonate salt ($HgCO_3$). Both of these salts are essentially insoluble in water. At low soil pH and high chloride concentration, Hg exists primarily as $HgCl_2$, which is quite soluble in water (Figure 1). At low pH and low chloride concentration, Hg(II) exists largely as the free Hg(II) ion.

Under anaerobic conditions (E_h < O 1.2 volts) the chemistry of Hg is more complex. At 1.2 volts > E_h > 0.4 volts, Hg exists primarily as Hg(II): $HgCl_2$ below pH of 7.0 and $Hg(OH)_2$ above pH of 7.0. At E_h < 0.4 volts, Hg exists primarily as the metal Hg(o) or as HgS (*1*).

The ability of soils to sorb Hg ions relates to the reactive surface area of soils as well as organic matter content, clay mineralogy, and cation exchange capacity. Sorption maximums correlate as follows: surface area > organic matter content > cation exchange capacity (*37*).

Among clays, Hg is most strongly sorbed to montmorillonite and least strongly sorbed to kaolinite (*3*). But organic matter in soil generally will sorb more Hg over a wider pH range than any of the clays (*2*). Strong metal organic complexes are formed between Hg and organic matter components (*9*).

Downward and lateral movement of Hg in soils appears to be restricted, even in light-textured soils with a low organic matter content (*23*). In one test researchers worked with a soil that had been treated with an organomer-

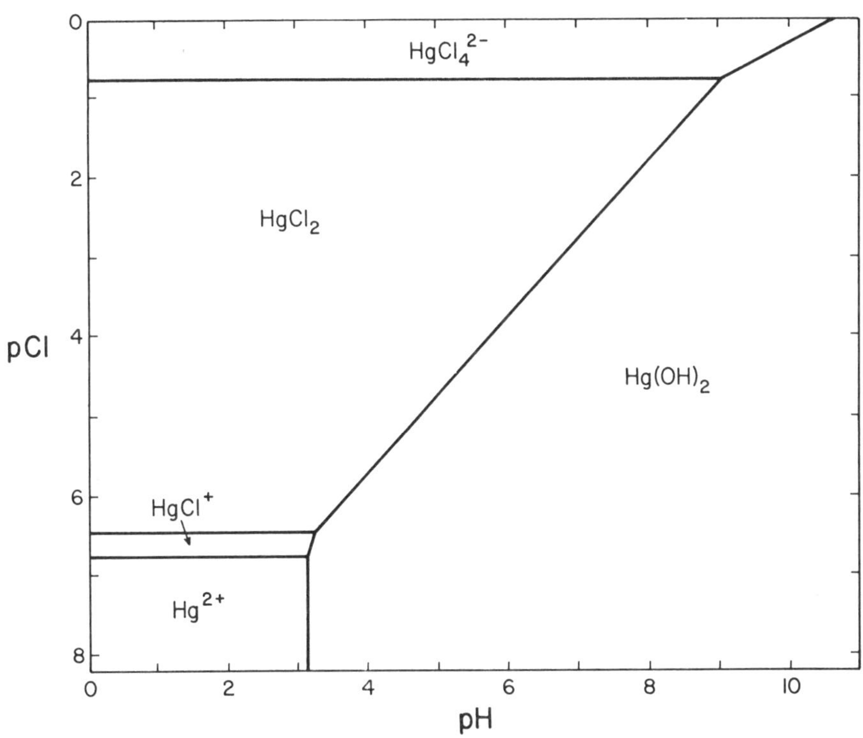

Figure 1. Speciation of mercury as a function of chloride and hydrogen ion concentrations (2).

curial fungicide annually for 15 years (*16*). The soil had a pH of 6.4, an organic matter content of 3.6 percent, and a cation exchange capacity of 12.1 meq/100 g. The surface 5 cm had a high total Hg concentration (453,000 ng/g). The concentration declined with depth (220,000 ng/g at 25 to 30 cm), and Hg showed little lateral movement off site.

Sulfide and hydroxide anions stabilize bivalent Hg as highly insoluble complexes. Hydrosulfides or sulfide compounds form even at low E_h. The mercuric ion forms $Hg(OH)_2$ as the predominant species at pH values greater than or equal to 4 for aqueous solutions containing less than 10^{-5} M chloride (*37*).

$HgCl_2$ and $Hg(OH)_2$ are notable among Hg compounds. $HgCl_2$ is the predominant mercuric compound at chloride concentrations above 1.1×10^{-3} ppm. However, $HgCl_2$ readily dissociates into constituent ions and, therefore, is among the most soluble of Hg salts (6.92 g/ml in cold water, 48.10 g/ml in hot water). $Hg(OH)_2$ does not readily disassociate into constituent ions and, therefore, has a low solubility product (10^{-25}). However, $Hg(OH)_2$ has the unusual property of high intrinsic solubility. Undissociated $Hg(OH)_2$ is soluble to the extent that 107 μg/g Hg can exist in solution (*18*). This theoretical treatment of $Hg(OH)_2$ solubility is not based on field observation. The model predicts that, under conditions of high chloride and/or high pH in a soil with low cation exchange capacity, low organic matter, and low carbonate or sulfate content, considerable Hg(II) might be in solution available for plant uptake. It seems unlikely that conditions would often exist resulting in this intrinsic solubility.

No research specifically addresses the issues of backfill quality and postmining groundwater quality issues. The factors affecting solubility presumably would be those that function also in soil.

Plant uptake of mercury. Numerous research documents attest that Hg compounds, including methyl Hg, can be taken up by plants (*10, 14, 20, 34*). But all of the researchers involved report Hg uptake by plants only from aqueous solutions. A salient point here is that increased Hg in soil will result in increased plant uptake of Hg (*2*). The mechanism controlling the uptake of chemicals other than nutrients in plants is under debate. Observations demonstrate that Hg uptake in plants depends upon site and soil conditions other than total Hg concentrations, as discussed above.

What Hg is available to plants (often equated to soluble or extractable Hg) may be quite different than the total Hg concentration. For example, one riverwash soil reportedly had very high total Hg concentrations but little soluble and plant-available Hg (*21*). Conversely, soil collected 1 km from a Hg smelter near Alamde'n, Spain, reportedly had high plant-available Hg, but the total soil Hg concentration was only 18 percent that in the previously mentioned riverwash soil (Table 2). Turfgrass (*Agrostis caniva*) grown on a site treated with soluble organomercurial fungicides contained only 1,650 ng/g Hg, versus 453,000 ng/g total Hg in the upper 5 cm of soil (*16*).

Management implications

Five conclusions can be drawn from the foregoing discussion:

1. The probability of any given mining operation having a potentially hazardous Hg problem because of high Hg levels in geological materials or in soils is small. This is true even in mercuriferous regions. It is not true, however, in the case of Hg mines.
2. Analytical methods for Hg must detect Hg in the ppb range (ng/g).
3. Total Hg in a substrate is not a reasonable indicator of extractable Hg (presumably bioavailable Hg). Extractable Hg is usually a very small portion of total Hg.
4. Bioavailability of Hg is determined largely by substrate pH. In alkaline soils Hg is often precipitated and insoluble. But there is some theoretical basis to suspect that considerable Hg may be bioavailable at high pH.
5. Though Hg problems are scarce in mining operations, where problems exist they can be devastating. Generally, a low-intensity sampling scheme should be undertaken at all mines. If Hg contamination is suspected, plants should be grown in suspect substrates and analyzed for total Hg.

REFERENCES

1. Andersson, A. 1970. *On the geochemical behavior of mercury* (English summary). Grundforbattring 23(5): 31-40.
2. Andersson, A. 1979. *Mercury in soils.* In J. O. Nriagu [editor] *The Biogeochemistry of Mercury in the Environment.* Elsevier/North-Holland Biomedical Press, Amsterdam, The Netherlands.
3. Aomine, S., and K. Inove. 1967. *Retention of mercury by soils II. Adsorption of phenylmercuric acetate by soil colloids.* Soil Science and Plant Nutrition 13: 195-200.
4. Bisogni, J. J. 1979. *Kinetics of metylmercury formation and decomposition in aquatic environments.* In J. O. Nriagu [editor] *The Biogeochemistry of Mercury in the Environment.* Elsevier/North-Holland Biomedical Press, Amsterdam, The Netherlands.
5. Bohn, H. L., B. L. McNeal, and G. A. O'Connor. 1985. *Soil chemistry.* John Wiley and Sons, New York, New York.
6. Carty, A. J., and S. F. Malone. 1979. *The chemistry of mercury in biological systems.* In J. O. Nriagu [editor] *The Biogeochemistry of Mercury in the Environment.* Elsevier/North-Holland Biomedical Press, Amsterdam, The Netherlands.
7. Cataldo, D. A., and R. E. Wildung. 1978. *Soil and plant factors influencing the accumulation of heavy metals by plants.* Environmental Health Perspectives 27: 149-159.
8. Cavallero, J. A., G. A. Gibbon, and A. W. Devrbrock. 1978. *A washability and analytical evaluation of potential pollution from trace elements in coal.* EPA 600/7-78-038. U.S. Environmental Protection Agency, Washington, D.C 75 pp.
9. Cheam, V., and D. S. Gamble. 1974. *Metal-fulvic acid chelation equilibrium in*

aqueous $NaNO_3$ solution. Canadian Journal of Soil Science 54: 413-417.

10. Chigbo, F. E., R. W. Smith, and F. L. Shore. 1982. *Uptake of arsenic, cadmium, lead and mercury from polluted waters by the water hyacinth* Eichornia crassipes. Environmental Pollution 27: 31-36.
11. D'Itri, F M. 1972. *The environmental mercury problem.* CRC Press, Cleveland, Ohio.
12. D'Itri, P. A., and F. M. D'Itri. 1977. *Mercury contamination: A human tragedy.* John Wiley and Sons, New York, New York.
13. Dana, G. F., and J. W. Smith. 1972. *Oil yields and stratigraphy of the Green River formations tipton member at bureau of mines sites near Green River, Wyoming.* Report of Investigation 7681. U.S. Bureau of Mines, Washington, D.C.
14. Davis, R. D., P.H.T. Beckett, and E. Wollan. 1978. *Critical levels of twenty potentially toxic elements in young spring barley.* Plant and Soil 49: 395-408.
15. Dudas, M. J., and S. Powluk. 1976. *The nature of mercury in Chernozemic and Luvisolic soils in Alberta.* Canadian Journal of Soil Science 56: 413-423.
16. Estes, G. O., W. E. Knoop, and F. D. Houghton. 1973. *Soil-plant response to surface applied mercury.* Journal of Environmental Quality 2: 451-452.
17. Frank, R., K. Ishida, and P. Suda. 1976. *Metals in agricultural soils of Ontario.* Canadian Journal of Soil Science 56: 181-196.
18. Hahne, H.C.H., and W. Kroontje. 1973. *Significance of pH and chloride concentration on behavior of heavy metal pollutants: mercury (II), cadmium (II), zinc (II) and lead (II).* Journal of Environmental Quality 2: 444-450.
19. Hamdy, M. K., and O. R. Noyes. 1975. *Formation of methylmercury by bacteria.* Applied Microbiology 30: 424-432.
20. Haney, A., and R. L. Lipsey. 1973. *Accumulation and effects of methyl mercury hydroxide in a terrestrial food chain under laboratory conditions.* Environmental Pollution 5: 305-316.
21. Harsh, J. B., and H. E. Doner. 1981. *Characterization of mercury in a riverwash soil.* Journal of Environmental Quality 10: 333-337.
22. Hogg, T. J., J. R. Bettany, and J.W.B. Stewart. 1978. *The uptake of ^{203}Hg-Labeled mercury by bromegrass from irrigated, undisturbed soil columns.* Journal of Environmental Quality 7: 445-450.
23. Hogg, T. J., J.W.B. Stewart, and J. R. Bettany. 1978. *Influence of the chemical form of mercury on its adsorption and ability to leach through soils.* Journal of Environmental Quality 7: 440-444.
24. Jackson, L. P., R. E. Poulson, T. J. Spedding, T. E. Phillips, and H. B. Jensen. 1975. *Characteristics and possible roles of various waters significant to in situ oil shale processing.* Quarterly, Colorado School of Mines 70(4): 105-134.
25. Joensuu, O. I. 1973. *Fossil fuels as a source of mercury pollution.* In Mercury Poisoning: I MSS Information Corporation, New York, New York. pp. 102-104.
26. John, M. K., G. J. Van Laerhoven, V. E. Osborne, and I. Cotic. 1975. *Mercury in soils of British Columbia, a mercuriferous region.* Water, Air, and Soil Pollution 5: 213-220.
27. Jonasson, F. R., and R. W. Boyle. 1971. *Geochemistry of mercury.* In *Mercury in Man's Environment.* Geological Survey of Canada, Department of Energy, Mines and Resources, Ottawa, Ontario. pp. 5-21.
28. Klein, D. H. 1971. *Sources and present status of the mercury problem.* Presented at the Mercury in the Western Environment Conference, Portland, Oregon, February 24-26. (*Reviewed in reference 11*).
29. Koksoy, M., P.D.M. Bradshaw, and T. S. Tooms. 1967. *Notes on the determination of mercury in geological samples.* Institute of Mining and Metallurgy Bulletin 726: B121-124.
30. Landa, E. R. 1978. *Soil water content and temperature as factors in the volatile*

loss of applied mercury (II) from soils. Soil Science 126: 44-48.

31. Landner, L. 1971. *Biochemical model for the biological methylation of mercury suggested from methylation studies in vivo with* Neurospora crassa. Nature 230: 452-454.
32. Lindberg, S. E., D. R. Jackson, J. W. Huckabee, S. A. Jansen, M. J. Levin, and J. R. Lund. 1979. *Atmospheric emission and plant uptake of mercury form agricultural soils near the Almade'n mercury mine.* Journal of Environmental Quality 8: 572-578.
33. Lindsay, W. L., and W. A. Norvell. 1978. *Development of a DTPA soil test for zinc, iron, manganese and copper.* Soil Science Society of America Journal 42: 421-428.
34. Lipsey, R. L. 1975. *Accumulation and physiological effects of methyl mercury hydroxide on maize seedlings.* Environmental Pollution 8: 149-155.
35. Lo, J. M., and C. M. Wal. 1975. *Mercury loss from water during storage: Mechanisms and prevention.* Analytical Chemistry 47: 1,869-1,870.
36. Mills, J. G., and M. A. Zwarich. 1975. *Heavy metal content of agricultural soils in Manitoba.* Canadian Journal of Soil Science 55: 295-300.
37. Moore, J. W., and S. Ramamoorthy. 1984. *Heavy metals in natural waters: Applied monitoring and impact analysis.* Springer-Verley, New York, New York.
38. Newton, D. W., and R. Ellis, Jr. 1974. *Loss of mercury (II) from solutions.* Journal of Environmental Quality 3: 20-23.
39. Peavy, H. S. 1981. *Water pollution potential of coal-slurry pipelines.* EPA-600/S7-81-082. U.S. Environmental Protection Agency, Washington, D.C.
40. Rogers, R. D. 1976. *Methylation of mercury in agricultural soils.* Journal of Environmental Quality 5: 454-458.
41. Rogers, R. D. 1977. *Abiological methylation of mercury in soil.* Journal of Environmental Quality 6: 463-467.
42. Siegel, B. Z., and S. M. Siegel. 1979. *Biological indicators of atmospheric mercury.* In J. O. Nriagu [editor] *The Biogeochemistry of Mercury in the Environment.* Elsevier/North-Holland Biomedical Press, Amsterdam, The Netherlands.
43. Stewart, J.W.B., and J. R. Bettany. 1982. *Mercury.* In A. L. Page, R. H. Miller, and D. R. Keeney [editors] *Methods of Soil Analysis* (part 2). American Society of Agronomy, Madison, Wisconsin. pp. 367-384.
44. Takizawa, Y. 1979. *Epidemiology of mercury poisoning.* In J. O. Nriagu [editor] *The Biogeochemistry of Mercury in the Environment.* Elsevier/North-Holland Biomedical Press, Amsterdam, The Netherlands. pp. 325-326.
45. Ure, A. M., and C. A. Shand. 1974. *The determination of mercury in soils and related materials by cold vapour atomic absorption spectrometry.* Analytica Chemica Acta 72: 63-77.
46. U.S. Environmental Protection Agency. 1980. *Ambient water quality criteria for mercury.* EPA 440/5-80-058. Washington, D.C.
47. U.S. Environmental Protection Agency. 1985. *Water quality criteria; availability of documents.* Federal Register 50: 30,784-30,796.
48. Vonk, J. W., and A. Kaars Sijpesteijn. 1973. *Studies on the methylation of mercuric chloride by pure cultures of bacteria and fungi.* Antoine van Leeuwenhoek 39: 505-513.
49. Warren, H. V., R. E. Delavault, and J. Barasko. 1966. *Some observations on the geochemistry of mercury as applied to prospecting.* Economical Geology 61: 1,010-1,028.
50. Wood, J. 1972. *A progress report on mercury.* Environment 14: 33-39.
51. Yamada, M., and K. Tonomura. 1972. *Formation of methylmercury compounds from inorganic mercury by* Clostridium cochlearium. Journal of Fermentation Technology 50: 151-166.

Index

A

B

C

K

L

M

P

R

S

X

Y

Z